森林·环境与管理

林木生理与生态水文
Forest Physiology and Ecohydrology

陈存根 编著

科学出版社
北京

内 容 简 介

深入揭示环境变化过程中林木生理及其生态水文效应,对于改善森林群落结构、增强森林水文调节能力及提升综合生态防护功能均有十分重要的意义。本书主要采用了野外定位观测、盆栽控制实验、数量模型模拟、实地勘察和文献综述等多种技术手段和方法,分析了典型林木的光合生理指标及其对复杂环境因子胁迫的响应,土壤呼吸动态与环境因子的关系,降雨特征及小气候对林冠层降雨再分配的作用,林木植被类型和降雨格局对地表水文过程的影响等。

本书可供生态学、农林科学、地理、环境等相关领域的科研院所及高等院校师生参考。

图书在版编目(CIP)数据

林木生理与生态水文 / 陈存根编著. —北京:科学出版社,2018.9
(森林·环境与管理)
ISBN 978-7-03-057849-5

Ⅰ. ①林… Ⅱ. ①陈… Ⅲ. ①森林–生理生化特性 ②森林生态系统 Ⅳ. ①S718.48 ②S718.57

中国版本图书馆 CIP 数据核字(2018)第 129477 号

责任编辑:李轶冰 / 责任校对:彭 涛
责任印制:徐晓晨 / 封面设计:无极书装

科学出版社 出版
北京东黄城根北街 16 号
邮政编码:100717
http://www.sciencep.com

北京厚诚则铭印刷科技有限公司 印刷
科学出版社发行 各地新华书店经销

*

2018 年 9 月第 一 版　开本:787×1092　1/16
2019 年 7 月第二次印刷　印张:19 1/2
字数:460 000
定价:238.00 元
(如有印装质量问题,我社负责调换)

作者简介

陈存根，男，汉族，1952年5月生，陕西省周至县人，1970年6月参加工作，1985年3月加入中国共产党。先后师从西北林学院张仰渠教授和维也纳农业大学 Hannes Mayer 教授学习，获森林生态学专业理学硕士学位（1982年8月）和森林培育学专业农学博士学位（1987年8月）。西北农林科技大学教授、博士生导师和西北大学兼职教授。先后在陕西省周至县永红林场、陕西省林业研究所、原西北林学院、杨凌农业高新技术产业示范区管委会、原陕西省委教育工作委员会、原陕西省人事厅（陕西省委组织部、陕西省机构编制委员会）、原国家人事部、重庆市委组织部、重庆市人民代表大会常务委员会、中央和国家机关工作委员会等单位工作。曾任原国家林业部科学技术委员会委员、原国家林业部重点开放性实验室——黄土高原林木培育实验室首届学术委员会委员、原国家林业局科学技术委员会委员、中国林学会第二届继续教育工作委员会委员、中国森林生态专业委员会常务理事、普通高等林业院校教学指导委员会委员、陕西省林业学会副理事长、陕西省生态学会常务理事、《林业科学》编委和《西北植物学报》常务编委等职务。著有《中国森林植被学、立地学和培育学特征分析及阿尔卑斯山山地森林培育方法在中国森林经营中的应用》（德文）、《中国针叶林》（德文）和《中国黄土高原植物野外调查指南》（英文）等论著，编写了《城市森林生态学》《林学概论》等高等教育教材，主持了多项重大科研课题和国际合作项目，在国内外科技刊物上发表了大量学术文章。曾获陕西省教学优秀成果奖一等奖（1999年）、陕西省科学技术进步奖二等奖（1999年）、中国林学会劲松奖、陕西省有突出贡献的留学回国人员（1995年）和国家林业局优秀局管干部（1998年）等表彰。1999年下半年，离开高校，但仍不忘初心，始终坚持对我国森林生态系统保护、森林生产力提高、森林固碳和退化生态系统修复重建等方面的研究，先后指导培养硕士研究生、博士研究生38名。

留学奥地利维也纳农业大学

与博士导师Prof. Dr. Hannes Mayer（右一）及博士学位考核答辩小组教授合影留念

奥地利维也纳农业大学博士学位授予仪式

1987年8月，获得奥地利维也纳农业大学博士学位

获得博士学位，奥地利维也纳农业大学教授表示祝贺

获得博士学位，奥地利维也纳留学生和华人表示祝贺

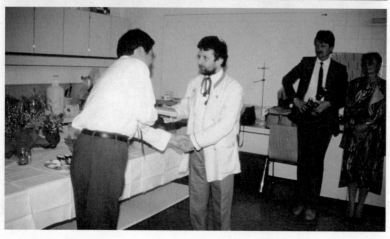

获得博士学位，维也纳农业大学森林培育教研室聚会表示祝贺

留学期间参加同学家庭聚会

留学期间在同学家里过圣诞节

与奥地利维也纳农业大学的同学们合影

与西北农学院(西北农业大学、西北农林科技大学前身)的老师们合影

与西北林学院学科带头人合影

撰写学术论文

在母校西北林学院和导师张仰渠教授亲切交谈

参加课题组学术研讨活动

1987年7月20日～8月1日，参加德国西柏林第十四届国际植物学大会

访问奥地利葛蒙顿林业中心

与德国慕尼黑大学Fisher教授共同主持中德黄土高原水土流失治理项目第一次工作会议

就主持的中德科技合作项目接受电视台采访

奥地利国家电视台播放采访画面

接受国内电视台采访

参加林木病虫害防治课题成果鉴定会议

参加纪念于右任先生诞辰120周年海峡两岸学术研讨会议

参加科技创新报效祖国动员暨先进表彰会议

2002年7月6日，陪同第十一届全国政协副主席（陕西省副省长）陈宗兴（左三）先生考察秦岭火地塘生态定位研究站

2003年6月27日，陪同陕西省省长贾治邦（左四）先生考察秦岭火地塘生态定位研究站

在西北林学院会见到访的外国专家

考察国外森林气象研究站

考察奥地利国家苗圃

对陕西杨凌农业高新区科技示范园温室栽培进行技术指导

林冠空隙降水监测

青杆林树干茎流监测

森林碳通量监测

指导大学生野外实习

指导国外大学生野外实习

介绍国外现代温室栽培技术

森林小气候自动观测设备及技术

林分环境因子梯度观测

沙棘茎流速率观测

研究生进行榆属树种采种和土壤采样

研究生采集树木实验样品

实验室样品测试分析

沙地滴灌造林实验

榆科不同种类种子活性测定

研究生布设树干茎流收集管

研究生进行不同榆树品种苗木繁育试验

研究生野外采集运输土壤样品

侧柏树干茎流观测

梯田油松林小气候观测

黄土高原小梯田营造小叶杨治理水土流失

彩图A

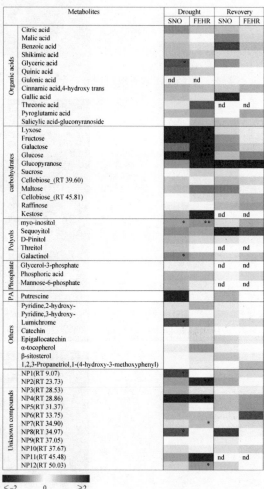

彩图B

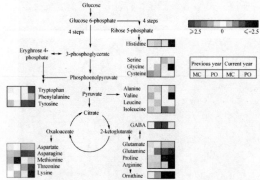

彩图C

序　一

陈存根教授送来《森林·环境与管理》书稿请我作序，我初阅书稿后又惊又喜。惊的是我知道陈教授已从政多年，竟然不忘科研初心，专业研究与培养学生没有间断过；喜的是，自己当年看好的青年才俊，一生结出了硕果累累，使我欣慰。

我和陈存根教授是 1990 年在四川成都国际林业研究组织联盟举办的国际亚高山森林经营研讨会上认识的。当时，我是这次在中国举办的国际会议的主持人。他提交的论文正符合大会主题，脉络清晰、观点独到。在野外考察活动中，他对川西的林木和草本很熟悉，能说出拉丁学名。他曾留学奥地利，因此能用流利的英语和德语与外宾交流。他的表现，使我在后来主持国家自然科学基金第一个林学的重大项目"中国森林生态系统结构与功能规律研究"时，毅然把他的团队纳入骨干研究力量。

他师从西北林学院我的好友张仰渠先生，公派到欧洲奥地利学习并获得博士学位，是当时生态学研究领域青年中的佼佼者。当他作为西北林学院森林生态学科带头人，谋划学科发展时，我给予了支持帮助，数次参加过他指导的博士生的毕业答辩。1999 年，得知他被组织安排到杨凌农业高新技术产业示范区管委会工作时，觉得很可惜，认为他将离开会有所建树的科研事业了。

令人宽慰的是，学校为他保留了从事科研和培养研究生的机制，所以后来总能在学术期刊上看到他的署名文章。他后来调到北京工作，后又到重庆等领导岗位，我们都见过几次面，逢年过节，他都给我问候。

他经常送来他指导的博士研究生的毕业论文让我审阅，这些论文涉及面很宽。从秦岭和黄土高原的植被到青藏高原草地植被；从宏观到微观，涉及景观生态学、生态系统生态学、群落学、种群学、个体等各个层面，甚至还涉及森林动物研究；他还有国外来华的留学生。这么多年，他之所以持之以恒地坚持生态学研究，是因为他割舍不下对专业的这份感情和挚爱！

他的书稿就像他的人生阅历，内容丰富、饱满精彩，且有不少独到之处。如通过剖析秦岭主要用材树种生产力特征，为培育大径材、优质材林分，提高森林生产功能、生态功能提供了技术指引；通过分析高山、亚高山森林植被群落学特征，为天然林保护、国家级自然保护区建设和国家森林公园管理提供了科学佐证；通过研究黄土高原植被演替与水土流失关系，为区域水土流失治理和植被生态恢复提供了科技支撑，等等。他的研究工作，学以致用、研以实用，研究成果能直接指导实际生产，产生经济效益、社会效益和生态效益。

他的书稿即将出版，正逢中央大力推进生态文明建设之际。习近平总书记指出，"绿水青山就是金山银山""绿色发展是生态文明建设的必然要求""人类发展活动必须尊重自然、顺应自然、保护自然""要加深对自然规律的认识，自觉以对规律的认识指导行动""广大科技工作者要把论文写在祖国的大地上，把科技成果应用在实现现代化的伟大事业中"。党的十九大报告更是对加快生态文明体制改革，建设美丽中国和促进科技成果转化，建设创新型国家提出了明确要求。当前，我国经济发展的基本特征就是从高速增长阶段转向高质量增长阶段，我国生态建设在新时代也面临提质增效的重大考验。我想，陈存根教授的《森林·环境与管理》丛书出版正当其时，完全符合中央的大政方针和重大部署，所以予以推荐，希望广大科研人员、管理人员、生产人员和读者能从中有所启迪和收益。

是为序。

中国科学院院士
中国林业科学研究院研究员
2018 年春于北京

序　二

对于陈存根先生，我是很早就结识了的。当年国家林业局直属的6所林业高等院校分别是北京林业大学、东北林业大学、南京林业大学、中南林学院、西南林学院和西北林学院，我负责北京林业大学的工作，陈存根先生负责西北林学院的工作，我们经常一起开会研讨林业高等教育发展问题。后来，陈存根先生走上从政的道路，先后在杨凌农业高新技术产业示范区管委会、原陕西省委教育工作委员会、原陕西省人事厅、原国家人事部、重庆市委组织部、重庆市人民代表大会常务委员会、中央和国家机关工作委员会等不同的岗位上工作，但我们之间的学术交流和专业探讨从未间断过，所以，也算是多年的挚友了。这次他送来《森林·环境与管理》书稿让我作序，我很高兴，乐意为之，就自己多年来对陈存根先生在创事业、干工作、做研究等方面的了解和感受略谈一二。

我对陈存根先生的第一个印象就是他创事业敢想敢干、思路广、劲头足。西北林学院是当时六所林业高等院校中建校较晚的一个，地处西北边远农村，基础设施、师资配备、科研力量等方面都相对薄弱。陈存根先生主持西北林学院工作后，呕心沥血，心无旁骛，积极争取上级部门的鼎力支持，广泛借鉴兄弟院校的先进经验，大力推动学校的改革发展。他曾多次与我深入探讨林业高等院校的学科设置及未来发展问题，在我和陈存根先生的共同努力下，北京林业大学和西北林学院开展了多方面共建与合作，极大地促进了两个学校的交流和发展。在陈存根先生的不懈努力下，西北林学院的教师队伍、学科设置、学生培养、办学条件等方面都上了一个大台阶，学校承担的国家重大科技研究项目不断增多，国际交流与合作日益广泛，整个学校面貌焕然一新，事业发展日新月异。

我对陈存根先生的第二个印象就是他干工作爱岗敬业、有激情、懂方法。这点还要从中国杨凌农业高新科技成果博览会"走向全国，迈出国门"说起。2000年初，陈存根先生已经到杨凌农业高新技术产业示范区管委会工作了，他因举办博览会的事宜来北京协调。他对我讲他要向国务院有关部委汇报，要把这个博览会办成国内一流的农业高新科技博览会，办成一个有国际影响力的盛会。当时我感到很惊讶，在我的印象中，这个所谓的博览会原来也就是农村小镇上每年一次主要只有陕西地市参加的冬季农业物资交流会，这要花多大的气力才能达到这个目标啊！但 2000 年 11 月博览会的盛况大家都看到了，不仅有十多个国家部委主办和参与支持，同时世界银行、联合国开发计划署、联合国粮食及农业组织、联合国教育、科学及文化组织和欧盟等多个国际机构参与协办，并成功举办了首届国际农业高新科技论坛，杨凌农高会不仅走出了陕西，走向了全国，而且迈出了国门，真正成了中国农业高新科技领域的奥林匹克博览盛会。杨凌——这个名不见经传的小镇，一举成为国家实施西部大开发战略、国家农业高新技术开发的龙头和国家级的农业产业示范区。这些成绩的取得，我认为饱含着陈存根先生不懈的努力和辛勤的付出！

  我对陈存根先生的第三个印象就是他做研究精益求精、标准高、重实用。我应邀参加过陈存根先生指导的博士研究生的学位论文审阅和答辩工作，感受到了他严谨缜密的科研态度和求真务实的学术精神。陈存根先生带领的科研团队，对植被研究延伸到了相关土壤、水文、气候及历史人文变迁的分析，对动物研究拓展到了春夏秋冬、白天黑夜、取食繁衍等方方面面的影响，可以说研究工作非常综合、系统和全面。长期以来，他们坚持与一线生产单位合作，面向生产实际需要开展研究，使科研内容非常切合实际，研究成果真正有助于解决生产问题。近年来国家在秦巴山区实施的天然林保护工程、近自然林经营、大径材林培育，高山、亚高山脆弱森林植被带保护，黄土高原水土流失治理与植被恢复重建，以及国家级自然保护区管理和国家森林公园建设等重大决策中，都有他们科研成果的贡献。

  我对陈存根先生的第四个印象就是他的科研命题与时俱进、前瞻强、创新好。1987年他留学归来就积极倡导改造人工纯林为混交林、次生林近自然经营等先进理念，并在秦岭林区率先试验推广，这一理念与后来世界环境与发展大会提出推进森林可持续经营不谋而合。他研究林木异速生长规律有独到的方法，我记得当年学界遇到难以准确测定针叶面积问题，陈存根先生发明了仅测定针叶长度和体积两个参数即可准确快捷计算针叶面积的方法，使得这一难题迎刃而解，我们曾就这一问题一块儿进行过深入探讨。他在森林生产力研究方面也很有见地，开发积累了许多测算森林生物量、碳储量的技术方法，建造了系列测算和预测模型，提出了林业数表建设系统思路，这些工作为全面系统开展我国主要森林碳储量测算打下了基础，也为应对全球气候变化、推进国际碳排放谈判、签署《京都议定书》、参与制定巴黎路线图、争取更大经济发展空间、建设人类命运共同体做出了积极贡献。当前，中央大力推进生态文明建设，推动经济发展转型和提质增效。习近平总书记强调，实现中华民族伟大复兴，必须依靠自力更生、自主创新，科学研究要从"跟跑者"向"并行者""领跑者"转变。我想，陈存根先生在科研方面奋斗的成效，真正体现了习总书记的要求，实现了科研探索从学习引进、消化吸收到创新超越的升华。

我仔细研读了送来的书稿，我感到这个书稿是陈存根先生积极向上、永不疲倦、忘我奉献、一以贯之精神的一个缩影。《森林·环境与管理》丛书内容丰富饱满，四个分册各有侧重。《森林固碳与生态演替》分册侧重于森林固碳、森林群落特征、森林生物量和生产力方面的研究，《林木生理与生态水文》分册侧重于植被光合生理、森林水文分配效应等方面的研究，《森林资源与生境保护》分册侧重于森林内各类生物质、鸟类及栖息地保护方面的研究，《森林经营与生态修复》分册侧重于近自然林经营、森林生态修复、可持续经营与综合管理等方面的研究。各分册中大量翔实的测定数据、严谨缜密的分析方法、科学客观的研究结论，对当今的生产、管理、决策以及科研非常有价值，许多研究成果处于国内领先或国际先进水平。各分册内容互为依托，有机联系，共同形成一部理论性、技术性、应用性很强的研究专著。陈存根先生系列著作的出版，既丰富了我国森林生态系统保护的理论与实践，也必将在我国生态文明建设中发挥应有的作用，推荐给各位同仁、学者、广大科技工作者和管理人员，希望有所裨益。

有幸先读，是为序。

中国工程院院士
北京林业大学原校长
2018 年春于北京

自　　序

  时光如梭，犹如白驹过隙，转眼间从参加工作到现在已经四十七个春秋。这些年，我曾在基层企业、教育科研、产业开发、人事党建等不同的部门单位工作。回首这近半个世纪的历程，尽管工作岗位多有变动，但无论在哪里，自己也算是朝乾夕惕，恪尽职守，努力工作，勤勉奉献，从未有丝毫懈怠，以求为党、国家和人民的事业做出自己应有的贡献。特别是对保护我国森林生态系统和提高森林生产力的研究和努力，对改善祖国生态环境和建设美丽家园的憧憬与追求，一直没有改变过。即使不在高校和科研院所工作后，仍然坚持指导博士研究生开展森林生态学研究。令人欣慰的是，这些年的努力，不经意间顺应了时代发展的潮流方向，秉持了习近平总书记"绿水青山就是金山银山"的科学理念，契合了十八大以来党中央关于建设生态文明的战略部署，响应了十九大提出的推动人与自然和谐发展的伟大号召。因此，我觉得有必要对这些年的研究工作进行梳理和总结，以为各位同仁做进一步研究提供基础素材，为以习近平同志为核心的党中央带领全国人民建设生态文明尽绵薄之力。

  参加工作伊始，我就与林业及生态建设结下了不解之缘。1970 年，我在陕西省周至县永红林场参加工作，亲身体验了林业工作的艰辛，目睹了林区群众的艰难，感受到了国家经济建设对木材的巨大需求，以及森林粗放经营、过度采伐所引起的水土流失、地质灾害、环境恶化、生产力降低等诸多环境问题。如何既能从林地上源源不断地生产出优质木材，充分满足国家经济建设对木材的需求和人民群众对提高物质生活水平的需要，同时又不破坏林区生态环境，持续提高林地生产力，做到青山绿水、永续利用，让我陷入了深思。

  1972 年，我被推荐上大学，带着这个思索，走进了西北农学院林学系，开始求学生涯。1982 年，在西北林学院张仰渠先生的指导下，我以华山松林乔木层生物产量测定为对象，研究秦岭中山地带森林生态系统的生产规律和生产力，获理学硕士学位。1985 年，我被国家公派留学，带着国内研究的成果和遇到的问题，踏进了欧洲著名的学术殿堂——奥地利维也纳农业大学。期间，我一边刻苦学习欧洲先进的森林生态学理论、森林培育技术和森林管理政策，一边潜心研究我国森林培育、森林生态的现实状况、存在的主要问题以及未来发展对策，撰写了《中国森林植被学、立地学和培育学特征分析及阿尔卑斯山山地森林培育方法在中国森林经营中的应用》博士论文，获得农学博士学位。随后，我的博士论文由奥地利科协出版社出版，引起了国际同行的高度关注，德国《森林保护》和瑞士《林业期刊》分别用德文和法文给予了详细推介，并予以很高的评价。世界著名生态学家 Heinrich Walter 再版其经典著作《地球生态学》中，以 6 页篇幅详细引用了我的研究成果，在国际相关学术领域产生了积极影响。

  欧洲先进的森林经营管理理念、科学的森林培育方法和优美的森林生态环境，增强了我立志改变我国落后森林培育方式、提高林区群众生活水平和改善森林生态环境的梦想和追求。1987 年底，我分别婉言谢绝了 Hannes Mayer 教授让我留校的挽留和冯宗炜院士希望我到中国科学院生态环境研究中心工作的邀请，毅然回到了我的母校——西北林学院，这所地处西北落后贫穷农村的高校。作为学校森林生态学带头人之一，在此后的 30 多年间，我和我的学生们以秦巴山脉森林和黄土高原植被为主要对象，系统地研究了其生态学特征、群落学特征和生产力，及其生态、经济和社会功能，取得了许多成果，形成了以秦巴山地和黄土高原植被为主要对象的系统研究方法，丰富了森林生态学和森林可持续经营的基础理论，提出了以森林生态学为指导的保护方法，完善了秦巴山地森林经营利用和黄土高原植被恢复优化的科学范式。

  在研究领域上，以森林生态学研究为基础，不断拓展深化。一是聚焦森林生态学基础研究，深入探索森林群落学特征、森林演替规律及其与生态环境的关系，如深入地研究了太白红杉林、巴山冷杉林、锐齿栎林等的群落学特征，分析了不同群落类型生态种组、生态位特点，及与环境因子的关系。二是在整个森林生态系统内，研究不断向微观和宏观两个方面拓展。微观方面探索物种竞争、协作、繁衍、生息及与生态环境的关系，包括物种的内在因素、基因特征等相互作用和影响，如分析了莺科 11 属 37 种鸟类的 *cyt b* 全基因序列和 *COI* 部分基因序列，构建了 ML 和 Bayesian 系统发育树。宏观方面拓展到森林生态系统学和森林景观生态学，如大尺度研究了黄土高原次生植被、青藏高原草地生态系统植被的动态变化。三是研究探索森林植被与生态环境之间相互作用的关系，如对山地森林、城市森林、黄土高原植被等不同植被类型的固肥保土、涵养水源、净化水质、降尘减排、固碳释氧、防止污染、森林游憩、森林康养等多种生态、社会功能进行了分析。四是研究森林生态学理论在森林经营管理中的应用，如研究提出了我国林业数表的建设思路，探讨了我国林业生物质能源林培育与发展的对策，研究了我国东北林区森林可持续经营问题，以及黄土高原植被恢复重建的工艺技术，为林业生态建设的决策和管理提供科学依据。

在技术路线和研究方法上，注重引进先进理论、先进技术和先进设备，并不断消化、吸收、创新和应用。一是引进欧洲近自然林经营理论，结合我国林情建立多指标评价体系，分析了天然林和人工林生物量积累的差异性，以及不同林分的健康水平和可持续性，提出了以自然修复为主、辅以人工适度干预的生态恢复策略，为当前森林生态系统修复重建提供了方法路径。二是为提高林木生物量测定精度，对生物量常规调查方法进一步优化，采取分层切割和抽样全挖实体测定技术，以反映林木干、枝、叶、果、根系异速生长分化特征。针对欧洲普遍采用的针叶林叶面积测定技术中存在的面积测定繁难、精度不高的问题，我们创新发明了只需测定针叶长度和体积两个参数即可准确快捷计算针叶面积的可靠方法。三是重视引进应用新技术，如引入了土壤花粉图谱分析技术，研究地质历史时期森林植被发展演替；引入高光谱技术、植物光合测定技术，测定植物叶绿素含量、光合速率、胞间CO_2浓度、气孔导度等生理生态指标，分析其与生态环境的关系，深入研究树种光合作用特征和生长环境适应性，为树种选择提供科学依据。四是引入遥感、地理信息系统等信息技术进行动态建模，创新分析技术和方法，使对高寒草地生态系统植被动态变化研究由平面空间上升到立体空间，更加生动地揭示了大尺度范围植被的动态演化特征。

在科研立项上，坚持问题导向，瞄准关键技术，注重结合生产，实行联合协作，积极争取多方支持。一是按照国家科研项目申报指南积极申请科研课题，研究工作先后得到了国家科学技术部、国家林业局、德国联邦科研部、奥地利联邦科研部、陕西省林业厅、陕西省科学技术厅等单位的大力支持，在此深表感谢。二是研究工作与生产实践紧密结合，主动和陕西省森林资源管理局、陕西太白山国家级自然保护区、陕西省宁东林业局、黄龙桥山森林公园、延安市林业工作站、榆林市林业局、火地塘实验林场等一线生产单位合作，面向生产实际需要，使我们的研究工作和成果应用真正解决生产问题。三是加强国际交流合作，先后和德国慕尼黑大学、奥地利维也纳农业大学围绕秦岭山地森林可持续经营和黄土高原沟壑区植被演替规律及水土流失综合治理等进行科技合作，先后有 7 名欧洲籍留学生来华和我的研究生一起开展研究工作。

多年的辛勤耕耘和不懈努力结出了丰硕成果，我们先后在国内外科技刊物上发表或出版学术论文（著）千余篇（部），《中国针叶林》（德文，1999）、《中国黄土高原植物野外调查指南》（英文，2007）等论著相继出版，国际科技合作和学术交流渠道更加通畅。研究成果大量应用于生产实践，解决了生产中许多急需解决的难题，产生了很好的经济效益、社会效益和生态效益。例如，对华山松林、锐齿栎林等主要用材树种生产力的深入研究，为培育大径材、优质材林分，提高森林经济功能、生态功能提供了坚实的技术支撑。对秦岭主要植被类型群落学特征、生态功能和经营技术的研究，为国家在秦巴山脉实施天然林保护工程，发挥其涵养水源功能提供了强有力的理论支撑。对高山、亚高山森林植被的研究，为天然林保护、国家级自然保护区管理和国家森林公园建设提供了充分的科学佐证。对黄土高原植被演替与水土流失关系的研究，为区域水土流失治理和植被生态恢复提供了科学理论和生产技术支撑，等等，这里就不一一枚举。卓有成效的国际学术交流合作也促进了中国、奥地利两国之间友好关系的发展，2001年奥地利总统克莱斯蒂尔先生访华时，我作为特邀嘉宾参加了有关活动。

抚摸着每一份研究成果，当年自己和学生们一起开展野外调查的场景历历在目。当时没有便捷的交通工具，也没有先进的导航仪器，更没有防范不测的野外装备，我们爬陡坡、淌急流、翻山越岭、肩扛背背，将仪器设备、锅碗瓢勺以及帐篷干粮等必需物资运入秦巴山脉深处。搭帐篷、起炉灶，风餐露宿，一待就是数月，进行野外调查。为调查林分全貌和真实状况，手持简易罗盘穿梭密林深处，常常"远眺一小沟，抵近是悬崖"，不慎跌摔一跤，缓好久才爬起来，拄根树枝继续前行。打植被样方，挖土壤剖面，做树干解析，全是手工作业，又脏又累，但绝不草率马虎，始终精细极致。为测定植被生物量和碳储量，手持简陋笨拙的农用工具，伐树、刨根、分类、称重、取样，挥汗如雨，却也顾不得衣服挂破扯烂和手掌上磨出血泡的疼痛。为监测森林水文，顶着大雨疾行抢时间，赶赴森林深处测量林分径流。为观测森林野生动物，悄然进入人迹罕至处，连续数日守望观察。这些野外调查长年累月、夜以继日，每次都是为了充分利用宝贵外出时间，天未亮就做准备工作，晨光熹微已到达现场，漫天繁星才收工返回。头发湿了，上衣湿了，裤子湿了，鞋子湿了，也辨不清挂在额头的是汗水、雾水，还是雨水、露水。渴了，捧一掬山泉，饿了，啃一口馒头，晚上回到营地时，已饥肠辘辘、疲惫不堪，还要坚持整理完一天所采集的全部数据和样本。伴随这些的，是蚊群的围攻、蚂蟥的叮附、野蜂的突袭、毒蛇的威胁，以及与野猪、黑熊、羚牛等凶猛野生动物的不期遭遇。但是，当获取了第一手宝贵的数据，所有的紧张与忙碌、艰辛与疲惫、疼痛与危险，都化作内心深处丝丝的甜蜜、欣慰和喜乐。个中酸甜苦辣，也唯有亲历者方能体会。

这次是对以往研究的主要成果进行汇编，虽然有些文章发表时间较早，但依然不失学术价值，文中大量翔实的测定数据、严谨缜密的分析方法、科学客观的研究结论，对当今的生产、管理、决策以及教学科研仍有参考和借鉴价值，许多研究成果依然处于领先水平。所以，将文章整理编辑成册，方便有关学者、研究人员、管理者、生产者查阅，这既是对我们研究工作的一个阶段性总结，同时，多少能够发挥这些研究成果的作用，造福国家和人民，也是我长久以来的心愿。

本丛书以《森林·环境与管理》命名，共收录论文106篇，总字数150万字。按研究内容和核心主题的侧重点不同，我们将其编辑为四个分册。第一分册为《森林固碳与生态演替》，共收录论文23篇，主要侧重森林固碳、群落特征刻画以及生物量积累和生产力评价方面的研究；第二分册为《林木生理与生态水文》，共收录论文20篇，主要侧重植被光合生理和森林水文分配效应等方面的系统研究成果；第三分册为《森林资源与生境保护》，共收录论文34篇，主要侧重介绍森林内各类生物质能源和鸟类栖息地及其保护的相关研究成果；第四分册为《森林经营与生态修复》，共收录论文29篇，主要介绍与近自然林规划设计、生态修复策略、森林可持续经营与综合管理等有关的研究成果。四部分册有机联系，互为依托，共同形成一部系统性和针对性较强、能够服务森林生态系统经营管理的专业丛书。

本丛书的出版发行得到了科学出版社的大力支持，以及中国林业科学研究院专项资金"陕西主要森林类型空间分布及其生态效益评价"（CAFYBB2017MB039）的资助，同时得到该院惠刚盈研究员的大力支持和热情帮助。本丛书的编辑中，我的研究生龚立群、彭鸿等37位学生给予了大力协助，白卫国、卫伟不辞劳苦，做了大量琐碎具体工作。正是学生们的通力协作，本丛书最终得以成功出版，在此一并予以衷心感谢。但限于时间仓促，错讹之处在所难免，恳请各位同仁不吝赐教、批评指正。

陈存根
2017年底于北京

前　言

　　林木是森林生态系统的最小单元和重要组分，其赖以生存的水分、光照、土壤等环境因子的变化，会导致林木从生理生态和外在形态等方面逐步形成复杂适应对策，以确保最大程度降低环境变化对其自身的不利影响。在适应环境变化的过程中，各类林木也发挥着降雨拦蓄、林冠截流、水源涵养和水土保持等多重服务功能。深入揭示环境变化过程中林木生理及其生态水文效应，对于改善森林群落结构、增强森林水文调节能力及提升综合生态防护功能均有十分重要的意义。

　　鉴于林木生理和生态水文的重要研究价值，本册以此为关键切入点，系统收录了团队已发表的 20 篇相关论文。相关研究主要采用了野外定位观测、盆栽控制实验、数量模型模拟，实地勘察和文献综述集成等多种技术手段和方法，开展了以下几个方面的研究工作：①典型林木的光合生理指标及其对复杂环境因子胁迫的响应，②土壤呼吸动态与环境因子的关系，③降雨特征及小气候对林冠层降雨再分配的作用，④林木植被类型和降雨格局对地表水文过程的影响等。

　　通过对林木植被生理活动的观测研究，发现其自身生理特点对光照、水分、土壤类型、重金属污染等外在环境因素有重要的响应和适应。其中，重点探讨了典型林木叶片衰老期间影响光合生理的限制因子；模拟了白榆、小叶杨和旱榆等重要植被类型对干旱胁迫的敏感性；测定了不同水分处理对林木幼苗叶绿素含量的影响；分析了典型植被类型在土壤重金属污染治理中的重要作用；揭示了天然降水格局与植被类型对水土流失过程的综合影响，明确了穿透雨和树干茎流的延滞性以及地表产流动态对极端降雨和坡面植被格局的响应规律。

　　本书的出版得益于所有署名作者的辛苦努力，在此对他们表示感谢。本书有望为从事森林生态、生态水文、植物生理和林业科技等领域的工作人员提供参考，期望能为我国森林管理和区域可持续发展提供依据。限于时间和作者水平，疏漏乃至错讹之处在所难免，敬请读者朋友批评赐教。

陈存根
2018 年 1 月于北京

目 录

秦岭林区锐齿栎林木个体生长分析 ··· 1

水分胁迫对 6 种苗木光合生理特性的影响 ·· 10

黄土高原四种人工植物群落土壤呼吸季节变化及其影响因子研究 ································ 20

不同土壤类型白榆种子贮藏物质动态变化的研究 ·· 30

基于高光谱参数的枫杨叶绿素含量估算模型优化 ·· 38

重金属铅、镉在白榆中分布规律和累积特性研究 ·· 46

A coastal and an interior Douglas fir provenance exhibit different metabolic strategies to deal with drought stress ·· 55

Elevated temperature differently affects foliar nitrogen partitioning in seedlings of diverse Douglas fir provenances ·· 82

Effects of copper-carbon core-shell nanoparticles on the physiology of bald cypress (*Taxodium distichum*) seedlings ··· 104

降雨特征及小气候对秦岭油松林降雨再分配的影响 ·· 117

秦岭天然次生油松林冠层降雨再分配特征及延滞效应 ·· 128

半干旱黄土丘陵沟壑区降水特征值和下垫面因子影响下的水土流失规律 ·················· 139

黄土丘陵沟壑区极端降雨事件及其对径流泥沙的影响 ·· 149

黄土小流域水沙输移过程对土地利用/覆被变化的响应 ··· 157

Application of Gash analytical model and parameterized Fan model to estimate canopy interception of a Chinese red pine forest ··· 164

Land use change: trends, drivers and consequences on surface hydrological processes ········ 182

Effects of surficial condition and rainfall intensity on runoff in a loess hilly area, China ···· 193

Effects of rainfall change on water erosion processes in terrestrial ecosystems: a review ···· 217

The effect of land uses and rainfall regimes on runoff and soil erosion in the semi-arid loess hilly area, China ··· 232

Responses of water erosion to rainfall extremes and vegetation types in a loess semiarid hilly area, NW China ··· 252

目 录

秦岭林区锐齿栎林木个体生长分析*

陈存根　龚立群　彭　鸿

摘要

应用植物生长分析方法，通过破坏性取样，对锐齿栎林木的光合器官、非光合器官的生长过程以及物质生产效率进行了研究。结果表明，光合器官的增长过程与非同化器官的生长过程并不同步，锐齿栎在展叶的前20d内，主要靠消耗树体先年贮存的营养物质维持生长；展叶20d以后，生长则依赖当年的光合产物。

关键词：锐齿栎；生长分析；logistic 模型

锐齿栎（*Quercus aliena* var. *acuteserrata*）林是秦岭林区主要的森林类型，在涵养水源保持水土和木材生产方面具有非常重要的作用。为了探讨锐齿栎林的生长规律，提高经营水平和林分生产力，应用植物生长分析方法研究了锐齿栎林木个体生长过程，以便为天然锐齿栎林的经营管理提供科学依据。

1 研究区的自然概况

秦岭林区天然锐齿栎林主要分布于海拔1400～1800m的中山地带。分布区年均气温8～13℃，≥10℃的积温2500～4000℃，年降水量700～1200mm，相对湿度65%～70%。土壤为火成岩、变质岩以及石灰岩母质上发育的中到弱酸性山地棕壤，土层厚度50～70cm。热量充足，气候适宜，土壤肥沃，生境具有较高的生产力。

2 研究方法

在锐齿栎林分布区划分了7个调查区。在每个调查区按林分起源、林分类型、海拔、坡向、坡度、土壤以及人为干扰程度设置标准地。7个调查区共设标准地33块，标准地的面积为0.04～0.15hm²。

对标准地的林木进行检尺，确定林分平均标准木和径级标准木。共砍伐250株标准木，采用"分层切割法"[1]测定地上部分的现存量，并进行树干解析。根系采用全挖

---

* 原载于：西北林学院学报，1994，9（1）：1-7。

法测定。干、枝、根按照粗度分级取样，在现场用塑料袋密封，带回室内称量鲜重后于105℃下烘至恒重，通过测定样品的含水率计算林木各器官的总干重。

在主要分布区的宝鸡市辛家山林场，根据研究要求，设置了3块固定标准地。在标准地内选取了林分平均标准木，固定其胸高位置，在整个生长季节每10d一次用测微尺准确测定其胸径生长。通过破坏性取样，定期测定叶面积和叶重量的增长。叶样每次在标准木上从不同层次和方位采集50片，袋装密封带回室内，测定样叶的长、宽、面积和鲜重，然后置于105℃下烘至恒重后测定干重。

对伐倒木的资料用计算机进行统计分析处理，建立干、皮、枝、根、叶的生物量最佳估计方程。利用生物量估计方程和在固定标准地测定的直径资料，研究锐齿栎个体林木年生物量的动态变化。

对上述资料应用生长分析方法[2]进行研究分析。

3 结果及分析

3.1 光合器官的年发育过程

光合器官的生长，是林木干物质生产的基础，直接决定现实生产力。对锐齿栎单叶面积（S_l）和单叶干重（W_l）生长的测定表明，其过程均呈S形，但各有自己的特点。展叶的最初一周，叶重量增长比叶面积快；嗣后，叶面积增长则快于叶重量。整个过程表现为叶面积增长迅速，较早达到稳定；叶干重增加缓慢，持续时间较长。当两者的相对增长百分率达到50%时，叶重增长需要25d时间，而叶面积仅需16.5d。以相对增长百分率99.95%为叶子生长发育的稳定态，达到这一百分率叶面积约需55d，叶重则需90d，说明叶面积和叶重生长并不同步（图1）。

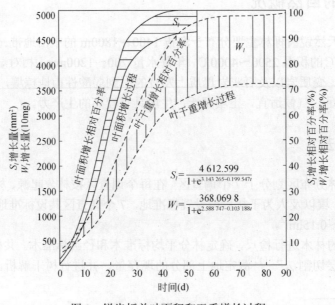

图1 锐齿栎单叶面积和干重增长过程

比较单叶的绝对增长速率可知，叶面积在展叶 16.5d 时绝对增长速率达到最大，为 219mm²/d，此时的叶面积恰好为稳定时叶面积的一半，即 2306mm²；叶重绝对增长速率在展叶 25d 时达到最大，为 9.6mg/d，此时叶重也为稳定时叶重的一半，即 184mg（图 2）。可见，无论是叶面积还是叶重，当其增长达到稳定量的一半时，绝对增长速率都达到最大值。由于叶面积比叶重提前 8.5d 达到其稳定量的一半，所以与叶重绝对增长速率的变化特点相比，叶面积绝对增速率不仅上升快，而且下降迅速，生长持续时间也短（图 2）。比较相对增长速率可知，它们也具有相似的特点。

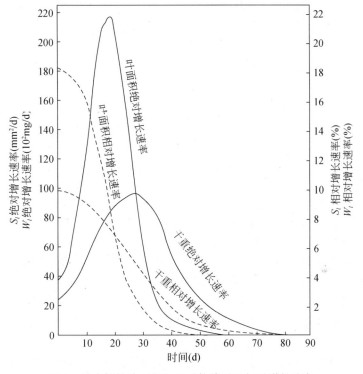

图 2　锐齿栎单叶面积和干重的绝对及相对增长速率

叶面积和叶重的生长，是相互独立而又有联系的过程，其变化虽然都呈 S 形曲线，但由于不是同步增长，因而造成叶面积干重比（W/S，单位面积所生产的净干物质）的变化曲线不完全呈 S 形（图 3）。在展叶的前 20d 内，随着叶面积迅速增加，叶面积干重比急骤下降；大约到第 20d 时，叶面积干重比降至最低值，仅 0.025mg/mm²，此时叶面积和叶重的相对增长速率也趋于相等。随后，叶面积干重比随时间而增加。展叶 90d 后，叶重生长趋于稳定，此时叶面积干重比为 0.082mg/mm²，此值为展叶 20d 时的 3.28 倍；展叶 55d 后，叶面积生长达到稳定，此时叶面积干重比为 0.076mg/mm²，占叶面积生长稳定时叶面积干重比的 93%。

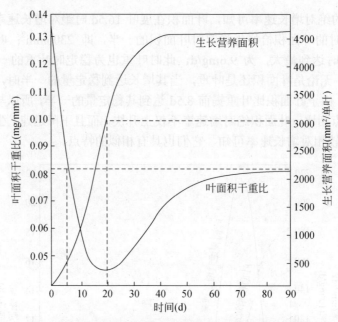

图3　锐齿栎叶面积干重比与生长营养面积的变化过程

叶面积干重比变化的另一特点是，展叶 90d 后，叶重增加趋于停止，此时的叶面积干重比较展叶最初一周时为低。在展叶第 7d 时，不仅叶面积和叶重的相对增长百分率与生长停止时的相等，而且这两个时期的叶面积干重比也相等，其原因有待进一步研究。

叶面积干重比随时间而呈阶段性的变化表明，叶子的生长在展叶后的前 20d 和展叶 20d 以后所需的营养物质的来源不同。展叶的前 20d，主要依赖树体先年贮存的营养物质来满足其叶组织分化、叶面积和重量的增加；展叶 20d 后，叶面积已达生长停止时叶面积的 65%，叶面积干重比也降至最低值。可以这样认为，从此时以后，植物叶子以及树体的生存和生长，不再依赖先年贮存的营养物质，而靠自身的光合生产。因之此时的叶面积可看作维持当年植物体生存必需的最小叶面积，称为生存营养面积；而把叶生长停止时树体上现存的叶面积与生存营养面积之差称为生长营养面积。所以展叶 20d 后，树体的生长则依赖于生长营养面积的光合作用。

图 3 也说明了叶面积干重比与生长营养面积的关系，在展叶的最初 7d 内，叶芽中贮存的营养物质主要用于叶组织的分化，叶面积生长非常缓慢，其值还小于生存营养面积，叶面积干重比呈急促下降趋势；7d 以后，叶面积生长加快，其值与生存营养面积的差距逐步缩小，光合产物增加，则进一步促进了叶组织的分化和叶面积的扩张，由于可利用当年的光合产物，从而对先年贮存物质的消耗速率下降，叶面积干重比的下降趋势也逐步减弱。展叶 20d 时，叶子已具有维持自身所需的生存营养面积，约 $3028.1mm^2$/单叶。由于不再需要树体内先年贮存的营养物质，其叶面积干重比也达到了最小值。展叶 20d 后，随着叶子的生长，生长营养面积不断增加，其光合产物除维持自身的呼吸消耗外，还有剩余和积累，从而使叶面积干重比不断增加，直到叶重稳定不再增长时，叶面积干重比也趋于稳定，并且其值与展叶第 7d 时的叶面积干重比相等。可见，展叶最

初一周内叶面积干重比其所以一直大于其后整个生长季的叶面积干重比，是由于这一时期叶面积生长缓慢，而体内先年贮存的营养物质又大量向叶子输送的结果。

3.2 非光合器官的生长过程

锐齿栎非光合器官现实生产力的年变化用 Logistic 方程可以很好地描述（图4）。与光合器官比较，非光合器官生长缓慢且延续的时间长。测定结果表明，非光合器官的干重增长几乎持续整个生长季节。

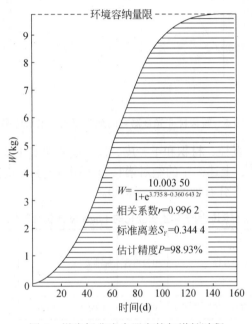

图4 锐齿栎非光合器官的年增长过程

与光合器官的绝对生长速率相比，非光合器官的干重绝对生长速率变化更为缓慢（图5）。由前面的分析可知，展叶 25d 时，单叶重绝对增长速率达到最大值，即全树总叶重的绝对增长速率也达到最大值，为 34.9g/d；此时非光合器官的干重绝对生长速率为 59.0g/d，是前者的 1.68 倍。展叶 60d 后，非光合器官的绝对增长速率达到最大值，为 154g/d；而此时全树叶干重绝对增长速率为 3.64g/d，前者是后者的 93.9 倍。展叶 90d 后，叶重绝对增长速率趋于零，但非光合器官的增长速率仍保持在 70g/d 左右。

与光合器官的相对生长速率比较，非光合器官的干重相对生长速率则呈与绝对生长速率相反的变化情况，即相对生长速率显著小于光合器官的相对生长速率。这是因为非光合器官的绝对重量远远大于光合器官的缘故。在锐齿栎的干物质现实生产量中，叶现实生产量仅占总现实生产量的 20%。可见，大的植物体具有大的绝对增长速率和小的相对增长速率[2]。

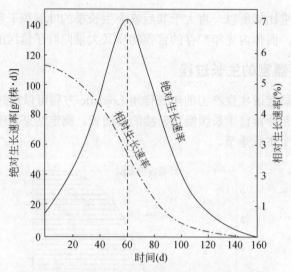

图 5 锐齿栎非光合器官的绝对和相对生长速率

分析锐齿栎树高、直径、材积和干、皮、枝、根的现实生长量表明，其年生长过程也均呈 Logistic 变化[3]。上述因子的绝对增长速率稍有差异，而相对增长速率基本保持相同的量值，这是一个十分有趣的现象（表1）。

表 1 锐齿栎主要测树因子年生长变化的 Logistic 方程及其生长变化的主要特征值

估计分量	Logistic 方程	相关系数	标准高差	估计精度（%）	最大绝对增长率（Δ/d）	1/2 相对增长速率 [Δ/(Δ·d)]	生长趋于停止的天数
直径（cm）	$\Delta D = 7.343\,03/(1+e^{3.443\,56-0.062\,10t})$	0.996 1	0.255 6	98.92	0.114	0.031	144
树高（cm）	$\Delta H = 3.417\,76/(1+e^{3.450\,40-0.062\,01t})$	0.996 6	0.111 2	99.01	0.053	0.031	141
断面积（cm²）	$\Delta S = 13.634\,03/(1+e^{3.670\,59-0.001\,64t})$	0.996 1	0.474 8	99.15	0.21	0.030 8	146
材积（dm³）	$\Delta V = 9.877\,29/(1+e^{3.682\,03-0.061\,94t})$	0.996 2	0.342 3	98.92	0.152 9	0.031	145
干重（kg）	$\Delta W_a = 1.987\,37/(1+e^{3.678\,66-0.061\,73t})$	0.996 1	0.174 3	98.91	0.076 9	0.030 8	146
皮重（kg）	$\Delta W_{Ba} = 1.190\,52/(1+e^{3.636\,67-0.041\,67t})$	0.996 6	0.038 9	98.99	0.018 4	0.030 8	145
枝重（kg）	$\Delta W_B = 1.447\,36/(1+e^{3.655\,28-0.041\,45t})$	0.996 4	0.048 7	96.69	0.022 2	0.030 7	146
根重（kg）	$\Delta W_R = 2.378\,31/(1+e^{3.695\,37-0.414\,0t})$	0.996 2	0.082 6	98.92	0.036 6	0.030 8	146

注：样本数均为 10。

3.3 锐齿栎的物质生产效率

研究表明，锐齿栎的现实生产量随时间的变化符合 Logistic 方程。其现实叶面积（S_L）、叶干重（W_L）和总现实生产量（W）的增长过程可表示为

$$S_L = 16.490\,04/(1+e^{3.143\,27-0.189\,55t})$$

$$W_L = 1.315\,80/(1+e^{2.598\,75-0.103\,49t})$$

$$W = 11.31408 / (1 + e^{3.15107 - 0.05782t})$$

式中，t 为时间。

净生产力的大小取决于植物的光合产量和呼吸消耗。叶重和植物体总干重之比（W_L/W，叶重比 LWR）在某种意义上反映了光合物质和呼吸物质的比率[2]。它的倒数（W/W_L）可理解为单位叶重量所生产的净干物质，即以叶重表示的物质生产效率。从图6可以看出，在展叶25d内，物质生产效率下降，而叶重比呈现上升趋势。这一时间恰好与叶干重绝对生长速率达到最大值的时间是一致的（比较图2）。这说明在这一时间之前，锐齿栎主要靠消耗体内先年贮存的物质来建造光合器官，由于叶组织幼嫩，同化能力差而呼吸旺盛，使得物质生产效率下降。展叶25d之后，即叶重绝对生长速率达到最大值后，光合物质有了积累，使得物质生产效率增加，叶重比下降。之后，随着叶子的进一步生长发育，光合能力不断增强，物质生产效率也随之提高。直到叶子停止发育时，物质生产效率才逐步趋于稳定。由此可知，锐齿栎完全发育成熟的叶子，才具有最大的物质生产效率。

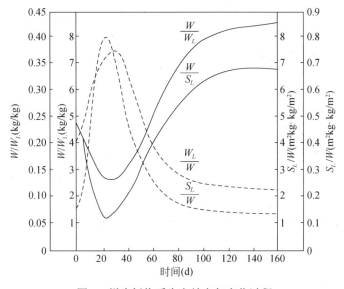

图6 锐齿栎物质生产效率年变化过程

与叶重比相同，叶面比（S_L/W）及其倒数（W/S_L，叶面积干重比）也可用来表示物质生产效率[2]。从图6可知，叶面比随时间变化的趋势具有与叶重比相似的规律。所不同的是叶面比最大值出现的时间早于叶重比最大值，大约在展叶20d后，这是因为叶面积绝对增长速率最大值出现的时间早于叶重绝对增长速率最大值的缘故。

与物质生产效率相比，同化作用的平均速率，即单位叶面积某一时刻净同化作用的速率（简称单位叶速率），能更好地反映出各个时刻物质生产的变化情况（图7）[2]。单位叶速率与叶面积和叶重增长密切相关。展叶初期，锐齿栎消耗先年贮存的物质维持光合器官生长，幼嫩的叶组织光合作用微弱，单位叶速率呈下降趋势。展叶25d后，当叶面积达到稳定态叶面积的80%以上，叶重达到了稳定态叶重的50%时，全树叶重的绝对增长速率也达到了最大值34.9g/d，此时单位叶速率停止下降，其值为6.09g/(m²·d)。这标志着主要靠消耗体内物质维持生长阶段的结束，锐齿栎开始了依赖当年叶子的光合产物进行干物质生产。展叶55天时，叶面积已经稳定，叶重也达到了稳定态叶重的95%，此时叶速率达到最大值10g/(m²·d)。展叶150d以后，叶子经充分发育并趋于衰老，光合能力也逐渐丧失，单位叶速率也降到最低限度，仅为0.16g/(m²·d)。

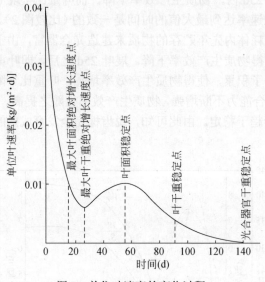

图7 单位叶速率的变化过程

4 结论

1）研究锐齿栎年生长过程表明，叶面积、叶重和非同化器官的生长遵从Logistic方程。

2）叶面积和叶重增长过程不同步。叶面积先期扩展迅速，其最大相对速率是叶重的2倍。展叶16.5d时，单位叶面积相对增长百分率即达到50%；55d后，即可达到99.95%，几乎完全趋于稳定。而单叶叶重前期增加缓慢，相对增长速率要达到上述值分别需要25d和90d。绝对增长速率叶面积在展叶16.5d时最大，约219mm²/d，此时的叶面积为生长停止时叶面积的一半，即2306mm²；叶重绝对增长速率在展叶25d时最大，约9.6mg/d，此时的叶重也恰为生长停止时叶重的一半，为184mg。

3）锐齿栎在展叶后的前25d内，主要靠消耗体内先年贮存的物质维持生长，此阶段叶面积叶重比呈迅速下降趋势，光合器官的干物质生产效率也随之下降，达到最低值。此后，叶子的光合产物有了盈余，叶面积干重比也随之增加，展叶90d后趋于稳定，此时叶子的干物质生产效率也达到最大值。叶面积比率随时间的变化也表现出与叶面积叶重比相似的规律，只是在时间上有所提前。

4）单位叶速率在展叶25d时停止下降，其值为6g/(m²·d)；展叶55d时，叶面积趋于稳定，叶重也达到停止生长时叶总重的95%，此时叶速率达到最大值，约10g/(m^2·d)；展叶150d时，叶子成熟并趋于衰老，单位叶速率仅为0.16g/(m^2·d)，标志着一年生长的结束。

致谢　此文是在张仰渠教授指导下完成的，特此致谢。

参 考 文 献

[1] 陈存根. 秦岭林区华山松林生物量的测定分析. 杨陵：西北林学院硕士学位论文，1982.
[2] R. 亨特. 植物生长分析. 陈宪辉译. 北京：科学出版社，1980：19-34.
[3] S.B. 查普曼，等. 植物生态学的方法. 阳含熙等译. 北京：科学出版社，1980：74-128.

水分胁迫对6种苗木光合生理特性的影响*

王　强　陈存根　钱红格　彭晓邦

> **摘要**
>
> 土壤控水条件下，测定了2年生小叶杨、沙地白榆、旱榆、河南白榆、旱柳和紫穗槐等6个树种苗期叶片的气体交换参数、叶绿素相对含量和叶片水势的变化值，并对其经行比较分析。结果表明：供测的6种苗木随水分胁迫程度的加剧叶片净光合速率、蒸腾速率、胞间CO_2浓度、气孔导度和瞬时水分利用效率降低的幅度并不相同，6种苗木上述5个指标的降幅排序依次为河南白榆>小叶杨>旱柳>紫穗槐>旱榆>沙地白榆，说明了不同树种在水分胁迫条件下的光合能力及水分利用能力差别较大。6种苗木的叶绿素相对含量随水分胁迫程度的加剧变幅也不相同，河南白榆在严重的水分胁迫下叶绿素相对含量降低显著，而沙地白榆即使在极端严重的干旱胁迫下叶绿素相对含量的降幅也较小，干旱胁迫对不同树种光合机构的损伤程度有显著差异。6种苗木叶片水势的测定结果表明河南白榆和小叶杨对干旱胁迫较敏感，最先受到干旱胁迫的伤害，而沙地白榆和旱榆具有很强的忍耐脱水的性能，对水分胁迫反应迟缓，叶片遭受干旱伤害的程度较轻。
>
> **关键词**：水分胁迫；光合特性；土壤含水量；叶片水势

　　植物的光合作用与其生存环境密切相关，研究植物光合生理特性是揭示植物对不同生存环境适应性机制的有效途径。近年来，随着植物生理生态测试技术的发展，便携式光合测定系统的出现使快捷有效的测定植物的气体交换过程、诊断植物光合机构的运转状况成为可能[1]。从而为我们更加准确迅速地了解植物在不同生境下的生理生态特性提供了科学有效的手段。水分亏缺是限制植物生长的重要因素之一[2]，尤其在干旱半干旱的西北地区，水分一直是我们进行植被恢复的主要限制因子。而植物在干旱胁迫条件下的光合生产力是鉴定植物耐旱能力的主要指标之一[3,4]。许多研究表明，光合作用对叶片水分亏缺非常敏感，轻度的干旱胁迫就会使植物的光合速率下降，使植物的生长发育受到明显的抑制[5,6]。因此，对植物在水分胁迫下光合生理特性的研究，对西北干旱半

---

* 原载于：水土保持通报，2009，29（2）：144-149.

干旱地区的植被恢复和生态建设具有十分重要的意义。本研究采用盆栽试验,以干旱和半干旱地区的主要造林树种小叶杨(*Populus simonii* Carr.)、沙地白榆(*Ulmus pumila* var. *sabulosa*)、旱榆(*Ulmus glaucescens* Franch)、河南白榆(*Ulmus pumila* L.)、旱柳(*Salix matsudana* Koidz)、紫穗槐(*Amorpha fruticosa* L.)为研究对象,探讨水分胁迫对6种树种光合生理特性的影响,评价各树种在干旱胁迫条件下的适应能力,进一步了解植物适应干旱的生理生态机制,为西北干旱区抗旱造林技术和树种的选择提供理论依据。

1 材料和方法

1.1 试验区概况

试验地设在神木县大堡当镇沙地苗圃园内(38°38′36″N,109°58′16″E)。该区为风沙草滩区,属半干旱大陆性季风气候。年平均气温8.9℃,无霜期175d。年平均日照时数为2875h,年平均降水量360mm,其中65%的降雨量集中在7~8月份。年均蒸发量为1380mm。试区土壤为沙壤。

1.2 试验材料

试验材料为小叶杨、沙地白榆、旱榆、河南白榆、旱柳和紫穗槐两年生苗木,均取自榆林沙地苗圃基地。将苗木移植在口径40cm,高60cm的塑料桶内,桶内用毛乌素沙地阿罗太沙区沙壤土装满。桶内土壤性质见表1。将桶底均匀剔出10个直径为1cm的圆孔,然后铺上三层细纱布,以方便桶内外气体交换,防止桶底水分过多。将苗木置于苗圃防雨棚内进行培育,经正常水分培育后,选择个体生物量基本一致的苗木进行控水处理。

表1 桶内土壤性质

采土地点	坐标	土壤类型	粒径(mm)	土壤容重(g/cm^3)	风干含水量(%)	最大持水量(%)	最大吸湿系数(%)
毛乌素沙区	38°38′N 109°58′E	沙壤	0.05~0.01	1.44	2.11	18.5	2.34

1.3 试验设计

本研究共6个处理,分别为处理Ⅰ(对照组)全灌水,使土壤含水量达到土壤田间持水量的90%±3.0%;处理Ⅱ土壤含水量为田间持水量的75%±3.0%;处理Ⅲ土壤含水量为田间持水量的60%±3.0%;处理Ⅳ土壤含水量为田间持水量的45%±3.0%;处理Ⅴ土壤含水量为田间持水量的30%±3.0%;处理Ⅵ为持续干旱,不灌水。

在测试开始前3d,对所有供试幼苗连续每天灌水,使桶内土壤含水量保持在较高的范围内。测定开始后,每3d灌水一次,灌水时间为测定前两天傍晚,每次灌水前用TDR(HH_2 moisture meter 美国,Spectrum技术公司)测定桶内土壤水分含量,以保证桶内土壤水分达到实验要求。经过水分胁迫处理后,各处理土壤水分相对含量(TDR测定值)和土壤水分绝对含量方差分析见表2。

表 2 不同处理的土壤水分含量

处理方式	处理 I	处理 II	处理 III	处理 IV	处理 V	处理 VI
土壤水分含量（TDR 值：%）	28.73A	22.24B	17.14C	12.56D	7.86E	4.54F
土壤水分含量（绝对含量：%）	16.78A	13.56B	10.38C	8.15D	5.65E	4.06F
持水量（%）	90	75	60	45	30	
胁迫程度	无胁迫	轻微胁迫	轻度胁迫	中度胁迫	严重胁迫	极端严重胁迫

注：同一列数值后的不同大写字母代表各处理有显著差异（$P<0.01$）。

1.4 测定指标和方法

1.4.1 光合生理参数的测定

选择晴朗无云的天气，利用 Li-6400 便携式光合测定系统（美国，LI-COR 公司）对各处理的苗木进行观测。测试时选取长势相近的健康植株，选取充分伸展、无病虫害的健康叶，每株选 3 片叶子，分别挂牌标记，并用铅笔轻画一条线以保证叶室每次夹在同一位置，测定时保持叶片自然生长角度不变。测定参数主要包括：叶片净光合速率[P_n，μmol/(m²·s)]、蒸腾速率[T_r，mmol/(m²·s)]、胞间 CO_2 浓度（C_i，μmol/mol）、气孔导度[g_s，mmol/(m²·s)]及环境因子光合有效辐射[PAR，μmol/(m²·s)]、大气相对湿度（R_H，%）、大气温度（T_a，℃）、叶温（T_{Leaf}，℃）等。测定时间从 8：00 至 11：00，每个指标测 3 个重复，每个重复记录 3 个数据，最后取平均值。

1.4.2 叶绿素含量的测定

在测定光合作用的同时利用 CM1000 chlorophyⅡ meter（美国，Spectrum 技术公司）测定挂牌标记叶片的叶绿素相对含量，每次 3 个重复，每个重复记录 3 个数据，最后取平均值。

1.4.3 叶水势和土壤湿度的测定

植物叶片水势利用 WP4 露点水势仪在早晨 6：00～8：00 进行测定，测定前仪器预 30min。用 6440FS TDR-100 soil moisture probe（美国，Spectrum 技术公司）测定桶内土壤湿度，每次重复 3 次，最后取平均值。求出不同控水条件下各桶土壤相对湿度（SWC）。

1.5 数据处理

所有数据均通过 Microsoft Office Excel 进行整理，方差分析、相关性分析以及其他统计分析处理均采用 SAS 和 SPSS 软件，图表处理采用 Excel 软件。

2 结果与分析

2.1 水分胁迫对各苗木光合生理的影响

6 种测试苗木在不同处理下叶片净光合速率 P_n（图 1）、蒸腾速率 T_r（图 2）、胞间

CO_2浓度 C_i（图3）、气孔导度 g_s（图4）以及瞬时水分利用效率 WUE（图5）。

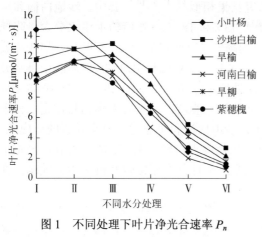

图1 不同处理下叶片净光合速率 P_n

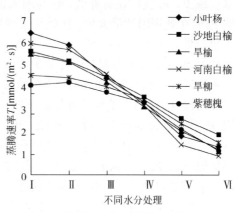

图2 不同处理叶片的蒸腾速率 T_r

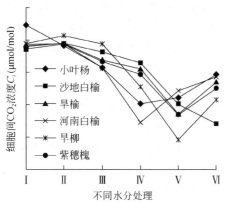

图3 6种苗木不同处理胞间 CO_2 浓度 C_i

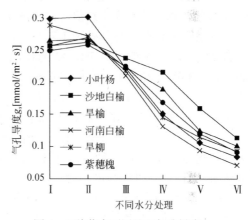

图4 6种苗木不同处理气孔导度 g_s

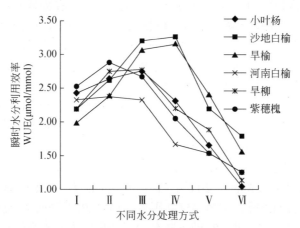

图5 6种苗木不同处理瞬时水分利用效率 WUE

6种苗木在不同控水处理条件下，叶片净光合速率（P_n）变化并不相同（图1），除河南白榆外，其他5种苗木随土壤含水量减少，净光合速率值均先增大，然后随土壤含水量减少，P_n值均显著降低，这与前人的研究结果相似[7, 8, 9, 10]；其中，沙地白榆和旱榆 P_n 值在处理Ⅲ达最大值，比处理Ⅰ分别增大13%和18.6%，然后随土壤水分含量降低 P_n 值降低，在处理Ⅵ P_n 值最小，比处理Ⅰ分别降低73.5%和79.9%；小叶杨，旱柳和紫穗槐 P_n 值在处理Ⅱ值最大，比处理Ⅰ分别增大8.1%，20.0%和18.1%，然后随土壤水分含量降低，P_n 值急剧降低，处理Ⅵ P_n 值最小，比处理Ⅰ分别降低了91.7%，83.7%和86.3%；河南白榆 P_n 值随土壤水分降低而下降，在处理Ⅰ P_n 值最大，然后下降趋势加剧，处理Ⅵ P_n 最小，比处理Ⅰ下降93.8%。由此看出，沙地白榆和旱榆对干旱有较强的适应性，这不仅表现在这两种苗木在一定的干旱条件下仍然能保持较高的 P_n 值，而且在严重干旱胁迫下其 P_n 值下降相对较小；河南白榆与其他5种苗木相比较其抗旱能力较弱，在严重干旱胁迫处理下，其 P_n 值下降最大。

从图2看以得出，随着干旱胁迫程度的加重，各苗木的蒸腾速率（T_r）的总体变化趋势为缓慢减小，但是，紫穗槐 T_r 值在处理二达到最大值，以后依次减小；河南白榆各处理 T_r 值分别与其处理Ⅰ T_r 值对比下降了4.9%、23.4%、46.7%、76.9%、85.4%，沙地白榆各处理 T_r 值分别与其处理Ⅰ T_r 值对比下降了8.3%、21.1%、37.5%、54.8%、67.0%，旱榆分别下降了6.7%、23.5%、42.9%、62.5%、74.7%，小叶杨分别下降了8.8%、31.7%、49.6%、73.0%、80.8%，旱柳分别下降了3.2%、12.7%、25.2%、48.6%、68.1%，紫穗槐各处理 T_r 值分别与其处理Ⅰ T_r 值对比的变化值依次为-2.8%、9.0%、19.8%、49.6%、72.8%。由此得出，在干旱胁迫下6种苗木 T_r 的变化幅度大小依次为：河南白榆>小叶杨>旱榆>紫穗槐>旱柳>沙地白榆；而蒸腾速率的大小以及下降幅度反映了树种控制蒸腾失水的能力[11]，所以六种苗木叶片控制蒸腾失水的能力大小依次为：沙地白榆>旱柳>紫穗槐>旱榆>小叶杨>河南白榆。

6种苗木不同处理条件下胞间CO_2浓度（C_i）的变化比较复杂（图3），沙地白榆随土壤水分的降低 C_i 值先小幅度增大然后持续降低，在处理Ⅵ C_i 值最小，与其处理Ⅰ比较降低了35.4%；河南白榆随土壤水分的降低 C_i 值先增大，在处理Ⅱ达到最大值，比其处理Ⅰ增加了1.5%，然后随土壤水分含量降低 C_i 值急剧降低，在处理Ⅳ达到最小值，比其处理Ⅰ降低了35.4%，最后随着土壤水分含量的降低，C_i 值反而回升，在处理Ⅵ只比其处理Ⅰ降低了14.5%；旱榆、旱柳、紫穗槐三种苗木随干旱胁迫程度的加重其 C_i 值变化相似，均在处理Ⅱ达到最大值，分别比其处理Ⅰ增加了2.7%、4.1%、2.1%，然后随土壤水分的减少，C_i 值急剧降低，在处理Ⅴ值最小，分别比其处理Ⅰ降低了30.4%、42.9%、30.2%，然后在极端严重的干旱胁迫下，C_i 值反而增大，处理Ⅵ的 C_i 值比其处理Ⅰ的 C_i 值仅分别降低了15.5%、24.8%、18.3%；小叶杨随干旱胁迫的加剧，其 C_i 值显著降低，在处理Ⅳ C_i 值最小，比处理Ⅰ降低了32.6%，随干旱胁迫程度的进一步加剧，C_i 反而增大，处理Ⅵ的 C_i 值比处理Ⅰ仅降低了20.6%。根据Farquhar和Sharkey提出的观点[12, 13]，P_n 下降伴随着 T_r 和 C_i 的下降，主要是气孔因素，如果 P_n 下降而 C_i 反而上升，则说明 P_n 的下降以非气孔因素为主，即 P_n 的下降是由于叶肉细胞的光合活性的下降造成的。由此得知试测的6种苗木在干旱胁迫的初期 P_n 的下降主要是气孔因素，随着胁迫程度的加重，除沙地白榆外其他5种苗木的非气孔因素逐渐成为主导因素，在严重干旱胁迫下这 5 种苗木其叶片光合机构可能遭到了破坏,使非气孔因素成为主要决定因素;而沙地白榆在处理Ⅵ其 P_n、T_r、C_i 值均为最小值,说明即使在极端缺水情况下,沙地白榆仍然能够通过气孔的开合维持正常的生命活动,其抗旱性最强。

不同控水处理条件对 6 种苗木气孔导度 g_s 的影响显著(图 2),除河南白榆外,小叶杨、沙地白榆、旱榆、旱柳、紫穗槐这 5 种苗木 g_s 值在处理Ⅱ都比其处理Ⅰ略有升高,然后随水分胁迫程度的加重,g_s 值迅速降低,在处理Ⅵ g_s 值都最小;但这 5 种苗木的降低幅度各不相同,上述 5 中苗木处理Ⅵ的 g_s 值比其处理Ⅰ的 g_s 值分别降低了 71.0%、56.3%、61.5%、64.0%、62.9%;河南白榆随着干旱胁迫程度的加剧,g_s 值急剧减小,处理Ⅵ的 g_s 值比处理Ⅰ的 g_s 值降低了 75.2%。由上可知,试测的 6 种苗木中河南白榆的降幅最大,高达 75.2%,而沙地白榆的降幅最小为 56.3%,沙地白榆较高的 g_s 保证了外部 CO_2 的供应,从而维持其相对较高的 P_n 值,这与綦伟等[14]的研究结果相同。

从图 5 可以看出,不同控水条件处理对 6 种苗木瞬时水分利用效率 WUE 的影响不同,沙地白榆和旱榆在处理Ⅳ WUE 达到最大值,比其处理Ⅰ的 WUE 值分别增大 45.5% 和 59.1%,然后迅速下降,处理Ⅵ的 WUE 值比其处理Ⅰ分别降低了 19.8% 和 20.7%;河南白榆,紫穗槐在处理Ⅱ WUE 值最大,比处理Ⅰ分别增大了 2.6% 和 14.8%,然后急剧下降,处理Ⅵ的 WUE 值比处理Ⅰ的值降幅分别达到 57.1%,50.0%;小叶杨,旱柳的 WUE 值在处理Ⅲ值最大,比处理Ⅰ分别增加了 15.2 和 25.5%,然后下降,在极端严重干旱胁迫下,WUE 值比处理Ⅰ降低了 56.7% 和 48.6%。由上可知,适度的干旱促进了 6 种苗木的 WUE 的提高,说明试测的 6 种苗木都有一定的抗旱性,但在严重干旱胁迫下,沙地白榆和旱榆的降幅较小,WUE 相对较高,而河南白榆降幅最大达 57.1%[15]。

2.2 水分胁迫对各苗木叶绿素相对含量的影响

水分胁迫条件下植物体叶绿素含量的变化,指示植物对水分胁迫的敏感性,并直接影响光合产量[16],在不同胁迫处理下,6 种苗木的叶绿素相对含量的变化也不相同,其中河南白榆和小叶杨的 chl 值变化相同,其处理Ⅱ的 chl 值和处理Ⅰ的 chl 值相比较都略有增大,但是二者差异不显著。随水分胁迫的加剧,chl 值都迅速降低,各处理间,以及和处理Ⅰ比较均有显著差异。旱柳和紫穗槐的 chl 值变化相同,其处理Ⅱ和处理Ⅲ的 chl 值与其处理Ⅰ相比较均有增加,但增幅不显著,随水分胁迫的加剧,chl 值都迅速降低,各处理间,以及和处理Ⅰ比较均有显著差异。旱榆在处理Ⅱ时的 chl 值比其处理Ⅰ值略有增大,但差异不显著,在处理Ⅲ时 chl 值达最大值,随水分胁迫的加剧,处理Ⅴ和处理Ⅵ,以及和处理Ⅰ比较均有显著差异。沙地白榆的叶绿素相对含量值在处理Ⅳ时最大,和处理Ⅰ、处理Ⅱ差异显著,和处理Ⅲ差异不大,随干旱胁迫程度的加剧,chl 值迅速减少,处理Ⅴ、处理Ⅵ、处理Ⅳ,以及前三个处理之间的 chl 值均有较大差异。6 种苗木相比较,变幅最大的是河南白榆,在严重干旱胁迫和极端严重干旱胁迫下其 chl 值分别是处理Ⅰ的 86.1% 和 83.1%,而沙地白榆其降幅较小在极端严重的干旱胁迫下其 chl 值是处理Ⅰ的 93.1%,说明沙地白榆即使在极端严重的干旱胁迫下仍然有较强的能力避免光合机构受损[17](表 3)。

表3 6种苗木叶片不同处理下叶绿素相对含量比较

树种\处理	处理Ⅰ	处理Ⅱ	处理Ⅲ	处理Ⅳ	处理Ⅴ	处理Ⅵ
小叶杨	476A	474BA	479A	467B	452C	419D
沙地白榆	393BC	392BC	399BA	401A	388C	366D
旱榆	396BA	399BA	404A	392B	371C	341D
河南白榆	402A	404A	400A	387B	346C	334D
旱柳	378A	379A	383A	377A	359B	331C
紫穗槐	363A	364A	367A	363A	348B	309C

注：同一列数值后的相同大写字母代表同一测定指标在 0.01 水平上不显著。

2.3 水分胁迫对各苗木叶水势的影响

植物水势反映了植物的水分状况和植物从土壤吸收水分的能力，是土壤-植物-大气连续体（soil plant atmosphere continuum，SPAC）中水分运转驱动力的重要组成部分和重要环节[18]。树种叶水势对干旱胁迫的不同反应，体现了各树种不同的耐旱机理和适应机制[19]。随干旱胁迫程度的加重，6 种试测苗木叶片水势变化趋势相同（图6～图11），

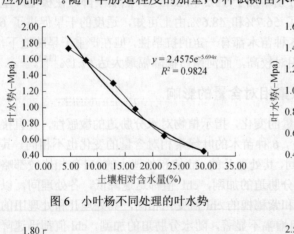

图6 小叶杨不同处理的叶水势

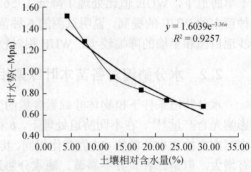

图7 沙地白榆不同处理的叶水势

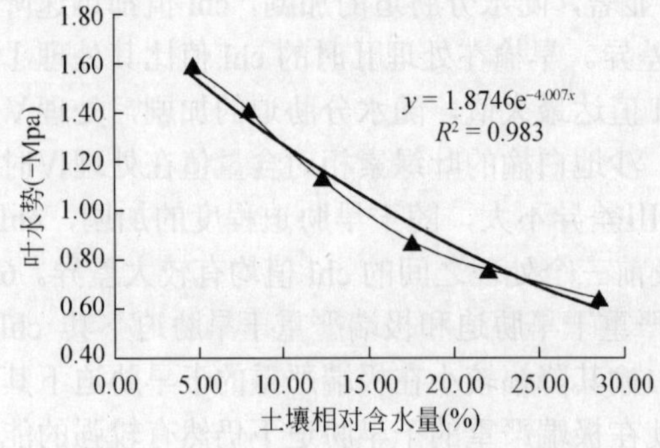

图8 旱榆不同处理的叶水势

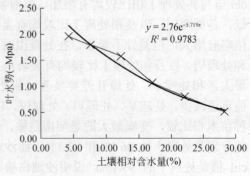

图9 河南白榆不同处理的叶水势

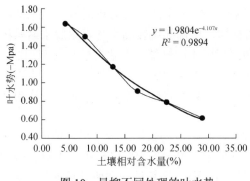
图10 旱柳不同处理的叶水势

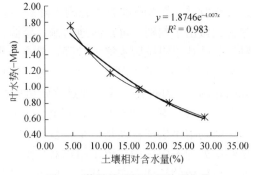
图11 紫穗槐不同处理的叶水势

极端严重的干旱胁迫下其叶片水势比处理Ⅰ的水势递减的幅度由大到小依次为河南白榆>小叶杨>旱柳>紫穗槐>旱榆>沙地白榆；其中河南白榆降幅达 1.44Mpa，而沙地白榆仅降低 0.84Mpa。说明河南白榆对外界水分胁迫较敏感，最先受到干旱胁迫的伤害，而沙地白榆具有很强的忍耐脱水的性能，对水分胁迫反应迟缓，叶片遭受干旱伤害的程度较轻，这与光合分析的结果相同。

3 讨论

研究表明水分胁迫对试测的 6 种苗木的光合作用影响极大，除沙地白榆外其余 5 种苗木随水分胁迫程度的增大，光合作用从气孔因素为主要限制因子阶段到非气孔因素为主要限制因子阶段，关义新等人的研究也证明了这一点[20]；各苗木在不同处理下光合作用的主要限制因子并不相同，河南白榆和小叶杨在处理Ⅴ时非气孔因素已经成为光合作用的主要限制因子，表现在随土壤含水量的下降 P_n、T_r、g_s、WUE 均在下降，而 C_i 的含量却在上升；而旱榆、旱柳、紫穗槐在处理Ⅵ时非气孔因素才成为光合作用的主要限制因子；而沙地白榆即使在处理Ⅵ时其 P_n、T_r、g_s、WUE 与 C_i 变化趋势一致，可见即使在极端严重的干旱胁迫下，沙地白榆光合作用的主导因子依然是气孔性限制因子。

叶绿素是植物体进行光合作用的色素，干旱胁迫条件下植物体叶绿素含量的变化，不仅能反映植物在逆境胁迫下同化物质的能力，而且可以指示出植物对水分胁迫的敏感性[21]。目前就水分胁迫对叶绿素含量的影响途径和机理还有待进一步研究，但夏尚光等人研究认为榆树在轻度水分胁迫下叶绿素含量会有所增加[22]，这与本实验结果一致，其中沙地白榆在中度的干旱胁迫下其叶绿素相对含量依然比处理Ⅰ高，从而保证了对光能的充分利用，通过提高转化率来保证碳同化，增强体内的代谢活动，这种生理反应，是树种抗旱能力的表现[23]。沙地白榆即使在极严重的干旱胁迫下叶绿素相对含量的降低幅度也较小，这可能是沙地白榆对干旱环境的长期适应，抗旱锻炼的结果；而河南白榆随干旱胁迫的加剧叶绿素相对含量的降低幅度最大，这可能与其种源地的环境有关。

树种叶水势对干旱胁迫的不同反应，体现了各树种不同的耐旱机理和适应机制[24]，6 个苗木的叶片水势都随干旱胁迫程度的加剧而降低，对供试苗木的叶水势与土壤含水率进行了双曲线拟合，结果表明，在水分胁迫初期，苗木叶水势变化平缓，随着干旱的进一步发展，苗木叶水势呈急剧下降趋势。由图6～图11中的曲线方程对6个苗木的耐旱方式做出了判别，可知沙地白榆、旱榆、紫穗槐、旱柳都属较高水势忍耐脱水耐旱树种，其叶片组织抗脱水能力较强，具有很强的忍耐脱水的性能，对水分胁迫反应迟缓，叶片最晚遭受干旱的伤害，河南白榆、小叶杨属较低水势耐脱水耐旱树种，这与李吉跃和张建国[19]的研究一致。通过比较可以得出，6种苗木的抗旱能力由大到小依次为沙地白榆>旱榆>紫穗槐>旱柳>小叶杨>河南白榆。

植物耐旱生理特性是一个很复杂的生理过程，它是植物长期适应干旱环境的结果。植物所处的不同环境导致了植物对环境适应机理的差异性[24]。以上6种苗木对我国西北干旱环境的长期适应，都表现出较强的抗旱能力，但由于其种间和所处环境的差异，抗旱能力的高低又有差别，因而对这些树种抗旱生理习性的研究对西北地区的植被恢复意义重大。

参 考 文 献

[1] 许大全. 光合作用效率. 上海：上海科学技术出版社，2002.

[2] 杨建伟，韩蕊莲，魏宇昆，等. 不同土壤水分状况对杨树、沙棘水分关系及生长的影响. 西北植物学报，2002，22（3）：579-586.

[3] 李吉跃. 植物耐旱性及其机理. 北京林业大学学报，1991，13（3）：92-100.

[4] 李忠武，蔡强国，唐政洪，等. 作物生产力模型及其应用研究. 应用生态学报，2002，13（9）：1174-1178.

[5] Harrison R D, Daniell J W, Cheshire J R. Net photosynthesis and conductance of peach seedlings and cutting in responses to changes in soil water potential. Journal of the American Society for Horticultural Science, 1989, 114: 986-990.

[6] Griffiths H, Parry M A J. Plant responses to water stress. Annals of Botany, 2002, 89: 801-803.

[7] 蒋高明，何维明. 毛乌素沙地若干植物光合作用、蒸腾作用和水分利用效率种间及生境间差异. 植物学报，1999，41（10）：1114-1124.

[8] 肖春旺，周广胜. 不同浇水量对毛乌素沙地沙柳幼苗气体交换过程及其光化学效率的影响. 植物生态学报，2001，25（4）：444-450.

[9] 李阳，齐曼·尤努斯，祝燕. 水分胁迫对大果沙枣光和特性及生物量分配的影响. 西北植物学报，2006，26（12）：2493-2499.

[10] 柯世省，金则新. 干旱胁迫和复水对夏蜡梅幼苗光合生理特性的影响. 植物营养与肥料学报 2007，13（6）：1166-1172.

[11] 李吉跃，周平，招礼军. 干旱胁迫对苗木蒸腾耗水的影响. 生态学报，2002，22（9）：1380-1385.

[12] Farquhar G D, Sharkey Y D. Stomatal conductance and photosynthesis. Annual Review of Plant Physiology and Plant Molecular Biology, 1982, 33: 314-336.

[13] 罗青红，李志军，伍维模，等. 胡杨、灰叶胡杨光合及叶绿素荧光特性的比较研究. 西北植物学报，2006，26（5）：983-988.

[14] 綦伟，谭浩，翟衡. 干旱胁迫对不同葡萄砧木光合特性和荧光参数的影响. 应用生态学报，2006，17（5）：835-838.

[15] 王旭军，吴际友，廖得志，等. 乐东拟单性木兰光合蒸腾及水分利用效率研究. 中国农学通报，

2008，24（10）：175-178.

[16] Reuveni J，Gale J，ZeroniM. Differentiating day from nigh t effects of high ambient CO_2 on the gas exchange and growth of *Xanthium strumarium* L. exposed to salinity stress. Annals of Botany，1997，79：191-196.

[17] 张永江，侯名语，李存东. 叶绿素荧光分析技术及在作物胁迫生理研究中的应用. 保定：中国作物生理第十次学术研讨会，2007.

[18] 康绍忠. 土壤-植物-大气连续体水分传输理论及其应用. 北京：水利电力出版社，1994：56-66.

[19] 李吉跃，张建国. 北京主要造林树种耐旱机理及其分类模型的研究（Ⅰ）——苗木叶水势与土壤含水量的关系及分类. 北京林业大学学报，1993，15（3）：1-11.

[20] 关义新，戴俊英，林艳. 水分胁迫下植物叶片光合的气孔和非气孔限制. 植物生理学通讯，1995，31（4）：293-297.

[21] 姜卫兵，高光林，俞开锦，等. 水分胁迫对果树光合作用及同化代谢的影响研究进展. 果树学报，2002，19（6）：416-420.

[22] 夏尚光，张金池，梁淑英. 水分胁迫下3种榆树幼苗生理变化与抗旱性的关系. 南京林业大学学报（自然科学版），2008，32（3）：131-134.

[23] 王金锡，许金铎. 长江上游高山高原林区迹地生态与营林更新技术. 北京：林业出版社，1995.

[24] 李吉跃. 太行山区主要造林树种抗旱特性的研究. 北京：北京林业大学博士学位论文，1990.

黄土高原四种人工植物群落土壤呼吸季节变化及其影响因子研究*

李红生 刘广全 王鸿喆 李文华 陈存根

摘要 以黄土高原侧柏、柠条、沙棘和油松人工植物群落为研究对象,研究了土壤呼吸日动态和季节变化及其与环境因子之间的关系,结果表明:4 种植物群落土壤呼吸速率具有典型的日变化和季节变化模式,4 个群落中,以侧柏 6 月土壤呼吸日变幅最大,沙棘 6 月土壤呼吸日变幅最小,大部分群落不同月份最大土壤呼吸与最小土壤呼吸倍数为 1.1~1.6。4 个群落中,以柠条土壤呼吸季节变幅最大,侧柏最小,最大土壤呼吸与最小土壤呼吸的倍数为 1.5~2.2 倍。同一植物类型土壤呼吸具有明显的季节变化特征,其具体变化趋势因植物类型而异。4 个植物群落土壤呼吸速率与土壤和大气温度以及土壤含水量的关系在不同季节表现为不同的关系,其中侧柏和柠条土壤呼吸与土壤温度均呈乘幂关系,与大气温度为指数关系。沙棘土壤呼吸与土壤温度和大气温度均为乘幂关系,油松土壤呼吸与土壤温度和大气温度均为线性关系。这表明土壤呼吸与土壤和大气温度之间的关系以及关系的紧密程度因植被类型而异。综合分析表明,同一气候区相同环境因子对不同植物群落土壤呼吸的影响作用大小不同,且因其自身具有明显的季节变化,从而导致对土壤呼吸的调控作用也具有明显的季节变异模式。

关键词:土壤呼吸;土壤温度;土壤水分;人工植物;黄土高原

人类活动和全球变化一定程度上改变了地球植被分布和全球碳循环[1]。土壤呼吸是全球碳循环的重要组成部分,也是陆生植物固定的 CO_2 返回大气的主要途径[2],大气中近 10% 的 C 由土壤产生,其微小变化就有可能对全球碳平衡产生重要的影响[3,4]。因此,研究不同植被类型土壤呼吸速率及其时空波动特征,阐明其影响因子和调控机制是目前生态学研究的重点内容之一[3,5]。

土壤呼吸主要由气候条件决定[6],但同一气候区域土壤呼吸常因植被状况的不同

---

* 原载于:生态学报,2008,28(9):4099-4106.

而存在差异[3, 7-9]。植被类型的差异一定程度上反映了土壤温度、水分等环境要素在时空上的分异，而这些都是影响土壤呼吸变化范围和季节动态的重要因子[8]。由于植被类型的差异，土壤微气候环境条件也显著不同，因此植被和土壤微气候环境的相互作用使得土壤呼吸的时空变异模式也因植被类型不同而存在差异。由于土壤呼吸不仅对环境因子变化敏感，而且具有高度的时空变异性[10]，因此探明同一气候区不同植被类型土壤呼吸的变化范围、季节动态及其影响因子，可为区域土壤碳估计和不同时间尺度植被作为碳源或汇的作用的准确预测提供一定的参考依据。

黄土高原在我国占据重要的地理位置，随着国家退耕还林（草）政策的实施，区域生态环境正在发生显著变化，为了探明该区域不同植被类型土壤呼吸的时间变异模式以及不同植被类型土壤呼吸的差异及产生这种差异的原因，本研究选择了黄土高原较为典型的 4 种人工植物群落为研究对象，从 2005 年 5 月到 9 月选择典型晴朗天气对土壤呼吸及其相关的环境因子进行日变化测定，主要是分析土壤呼吸及其相关环境因子（土壤水分，土壤温度和大气温度）的时间变异模式以及不同植被类型土壤呼吸之间的差异，阐明不同植被类型土壤呼吸与温度和土壤水分之间的关系，开展这一研究工作对于了解其他相似植被类型的土壤呼吸可提供一定的参考依据，也可为准确估测黄土高原地区土壤碳收支提供一定的数据积累。

1 研究区自然条件和研究方法

1.1 研究区自然概况

研究区位于陕西省吴起县境内，地处 107°38′57″E～108°32′49″E，36°33′33″N～37°24′27″N。该区属黄土高原典型梁状丘陵沟壑区，海拔为 1233～1809m。该区属温带大陆性季风气候，春季干旱多风，夏季旱涝相间，秋季温凉湿润，冬季寒冷干燥，年平均气温 7.8℃，极端最高气温 37.1℃，极端最低气温-25.1℃。年平均降雨量 478mm，7～9 月降雨量占年降雨量的 62%，年平均无霜期 146d。雨热同季，常有大风、暴雨、冰雹等灾害天气发生。土壤为淡灰绵土，质地为砂质壤土，石灰反应强烈，碳酸钙含量 13%左右，土壤 pH 8.5 左右。沙棘（*Hippophae reamnoides* L.）、柠条（*Caragana korshinskii* Kom.）、侧柏 [*Platycladus orientalis* (L.) Franco] 和油松（*Pinus tabulaefomis* Carr.）群落均为 1993 年人工栽植的实生苗，整地方式为水平阶，株行距为 2m×50cm，位于山地东南坡（半阳坡），地理位置为东经 108°10′50″和北纬 36°55′26″，海拔 1283.6m，群落基本特征如表 1 所示。

表 1 4 种人工植物群落基本概况

植物群落	坡向	坡度（°）	坡位	胸径或地径（cm）	树高（m）	主要伴生植物
侧柏 *P.orientalis*	东南 ES	12	中下部	1.87	1.9	山杏（*Prunus armeniaca*）、油松（*P. tabulaefomis*）、萎陵菜（*Potentilla chinensis* Ser）、胡枝子（*Lespedeza bicolor* Turcz）

续表

植物群落	坡向	坡度（°）	坡位	胸径或地径（cm）	树高(m)	主要伴生植物
柠条 C.korshinsk	东南 ES	12	中下部	2.82	3.8	紫菀（Aster tataricus Linn. f.）、冰草[Agropyron cristatum（L.）Gaertn]、铁杆蒿（Artemisia sacrorum）、茵陈蒿（Artemisia capillaris Thunb.）、长芒草（Stipa bungeana Trin）
沙棘 H.reamnoides	东南 ES	12	中下部	1.34	1.9	山杏（Prunus armeniaca）、山桃（P. davidian）、冰草[Agropyron cristatum（L.）Gaertn]、百里香（Thymus mongolicu）、针茅（Stipa breviflora）、胡枝子（L. bicolor Turcz）
油松 P.tabulaeformis	东南 ES	12	中下部	4.22	3.96	小叶杨（Populus simonii）、冷蒿（A. frigida Comm.）、长芒草（S. bungeana Trin）、甘草（Glycyrrhiza uralensis）

1.2 研究方法

在每一个人工群落内选择典型样地（5m×5m），采用美国 LI-cor 公司生产的 LI-6400 便携式光合作用测量系统和 LI6400-09 土壤呼吸室采集和储存数据。在样地内随机放置 5 个土壤隔离圈（0.008m^2），为了减小安放土壤隔离圈对土壤呼吸速率的影响，在土壤呼吸速率测定的前一周将土壤隔离圈埋入土壤约 2.0cm，并且在每次测定前 1 天，将测定点土壤隔离圈内的地表植被自土壤表层彻底剪除，但尽量不破坏土壤，以减少土壤扰动及根系损伤对测量结果的影响。从 2005 年 6 月到 9 月每月选择天气状况比较稳定日期进行土壤呼吸测量，每 2 h 测定一次，每个测点记录 5 个观测数据，每一月测定 3 天。在测定土壤呼吸速率的同时，利用便携式光合作用测量系统的温度探针同步测定地表土壤温度。大气温度由光合仪器自动测定记录。在测定土壤呼吸的同时，在各样地内随机打三个土钻用小铝盒取土样，测定 0～100cm 土层土壤含水量，每 20cm 测定一次，3 次重复。土壤含水量采用烘干法测定。

1.3 数据分析

采用 SPSS 统计分析软件包（SPSS 12.0 for Windows，Chicago，USA）对数据进行相关分析、回归分析和 One-Way ANOVA 方差分析，并用 LSD 法进行多重比较。

2 结果分析

2.1 不同植被类型土壤呼吸日变化动态

本研究中，不同群落类型土壤呼吸速率的日变化曲线存在差异，日变化幅度大小因群落类型而异（图1）。4 个群落中，以侧柏 6 月土壤呼吸日变幅最大 [0.83～2.05μmol/(m^2·s)]，以沙棘 6 月土壤呼吸日变幅最小 [1.24～1.39μmol/(m^2·s)]。大部分群落不同月份最大土壤呼吸与最小土壤呼吸倍数为 1.1～1.6，土壤呼吸日变化较为平缓，表明不同植物群落土壤呼吸日变化对同一气候区不同季节相似环境条件下差异不大。

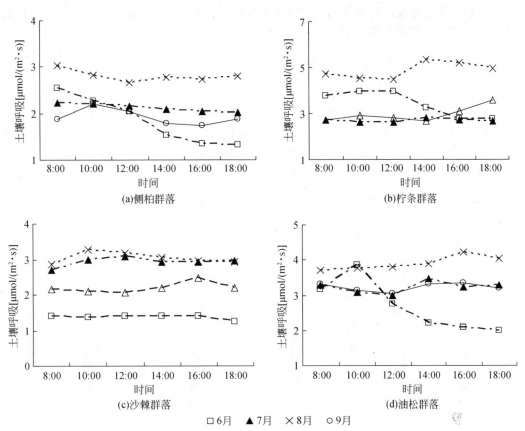

图1 不同植被类型土壤呼吸不同月份日变化动态

4个植物群落土壤温度和大气温度具有较为明显的日变化规律，基本表现为单峰曲线。大气温度一般在12:00出现最大值，土壤温度最高值出现的时间为14:00以后，滞后于大气温度。4个群落6～9月土壤呼吸日变化与气温和大气温度的变化趋势不太一致，表明可能存在其他因子调控土壤呼吸的日变化。

2.2 不同植被类型土壤呼吸以及环境因子的季节变化动态

2.2.1 同一植被类型不同月份土壤呼吸以及环境因子的变化动态

4个群落不同月份之间的土壤呼吸具有显著差异（$P<0.05$），变化范围各不同相同。侧柏土壤呼吸的季节变幅为1.34～2.31μmol/(m^2·s)，平均值为1.67μmol/(m^2·s)，柠条土壤呼吸季节变化幅度为2.69～4.68μmol/(m^2·s)，平均值3.48μmol/(m^2·s)，沙棘土壤呼吸变化幅度为1.36～3.03μmol/(m^2·s)，平均值为μmol/(m^2·s)，油松土壤呼吸季节变化幅度为2.67～3.90μmol/(m^2·s)，平均值为3.26μmol/(m^2·s)。其中，以柠条土壤呼吸季节变幅最大，侧柏的最小，最大土壤呼吸与最小土壤呼吸的倍数为1.5～2.2倍。

不同植物群落土壤温度一般小于同期大气温度，且存在显著差异（$P<0.05$），表明不同植被类型土壤温度和大气温度的季节差异较大，即使是同一植物群落，不同月份的土壤温度和大气温度也存在显著差异。

2.2.2 不同植被类型土壤呼吸以及环境因子的变化动态比较

侧柏、柠条、沙棘和油松群落6～9月土壤呼吸具有显著差异（$P<0.05$）（图2）。6月份柠条土壤呼吸速率显著大于侧柏、沙棘和油松，侧柏和沙棘土壤呼吸速率无显著差异，但均显著低于于油松，变化规律为柠条>油松>沙棘>侧柏。7月侧柏土壤呼吸显著低于柠条、沙棘和油松，后三种植物群落土壤呼吸无显著差异，变化规律为油松>柠条>沙棘>侧柏。8月和9月不同植物群落两两之间土壤呼吸速率均存在显著差异，土壤呼吸速率变化趋势分别为柠条>油松>沙棘>侧柏和油松>柠条>沙棘>侧柏。6月和8月、7月和9月土壤呼吸速率变化趋势一致，分别为柠条>油松>沙棘>侧柏和油松>柠条>沙棘>侧柏。表明不同植物群落土壤呼吸不同月份之间的差异以及变化趋势不一致。

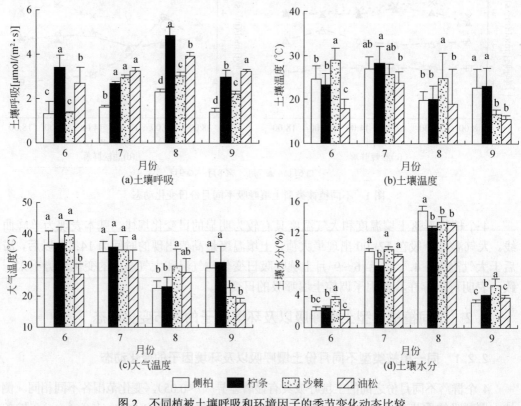

图2 不同植被土壤呼吸和环境因子的季节变化动态比较

注：图中a、b、c、d表示0.05水平上的差异显著性。

4种植物群落除了7月大气温度无显著差异外（$P>0.05$），其余月份大气温度、土壤温度和土壤水分在不同植被类型间均存在显著差异（$P<0.05$），表明植被类型对自身环境条件影响较大，并且具有季节变化。

2.3 不同植被类型土壤呼吸与环境因子之间的相关性分析

2.3.1 不同植物群落土壤呼吸速率与土壤温度和大气温度的相关性

4种植物群落中,侧柏、柠条、沙棘土壤呼吸速率对土壤温度和大气温度的响应较为一致,均随土壤温度和大气温度的增加而减少。油松群落土壤呼吸速率随土壤温度和大气温度的增加而增加,但关系不明显,可以用不同回归方程反映它们之间的关系(图3)。其中,侧柏群落土壤呼吸与土壤温度和大气温度,柠条土壤呼吸和大气温度的相关性都达到极显著水平($P<0.01$)。大气温度和土壤温度变化规律表现出较好的一致性,二者为正相关关系($R^2=0.81$)。

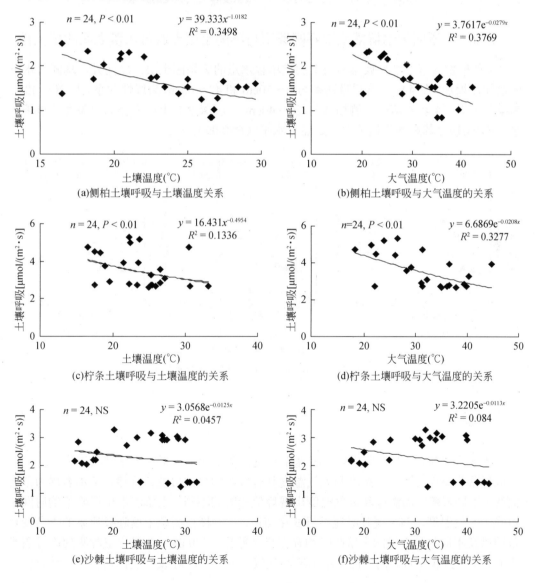

(a)侧柏土壤呼吸与土壤温度关系
(b)侧柏土壤呼吸与大气温度的关系
(c)柠条土壤呼吸与土壤温度的关系
(d)柠条土壤呼吸与大气温度的关系
(e)沙棘土壤呼吸与土壤温度的关系
(f)沙棘土壤呼吸与大气温度的关系

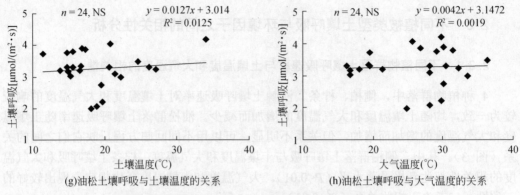

(g) 油松土壤呼吸与土壤温度的关系　　　　　　(h) 油松土壤呼吸与大气温度的关系

图3　不同植物土壤呼吸与土壤温度和大气温度的关系

2.3.2　不同植物群落土壤呼吸速率与不同土层土壤含水量之间的相关性

4种植物群落土壤呼吸与不同土层土壤含水量的关系因土层深度而异，均随含水量的增加而增加（图4）。土壤呼吸速率与0~20cm和40~60cm为线性关系，与20~40cm为乘幂关系。4种群落中，侧柏林与20~40cm土层深度的土壤水分的相关性达显著水平（$P<0.01$），其余群落相关性未到显著水平（$P>0.05$）。

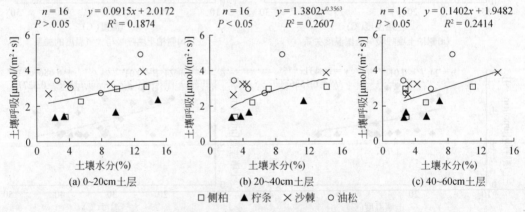

(a) 0~20cm土层　　　　　　(b) 20~40cm土层　　　　　　(c) 40~60cm土层

□ 侧柏　▲ 柠条　× 沙棘　○ 油松

图4　不同植物土壤呼吸与不同土层土壤含水量的关系

3　讨论

本研究表明，4种群落土壤呼吸速率日变化范围和最大最小土壤呼吸速率出现的时间因群落类型而异，土壤温度和大气温度具有较为明显的日变化规律，基本表现为单峰曲线。不同群落土壤温度和大气温度出现峰值的时间不同，土壤温度出现峰值的时间一般晚于大气温度。与土壤呼吸速率日变化相似，4种植被类型土壤呼吸速率的季节变化范围以及不同月份之间的土壤呼吸均存在显著差异（$P<0.05$），8月土壤呼吸速率显著高于6月、7月和9月。同期的土壤温度和大气温度与土壤呼吸的变化趋势不一致，但不

同植物群落土壤温度一般小于同期大气温度,表明该区不同植被类型土壤温度和大气温度的季节差异较大,即使是同一植物群落也是如此。可见,该区域植被对水热条件的影响较大。

不同植被相同月份土壤呼吸速率的比较研究表明,6~9月侧柏、柠条、沙棘和油松植物群落土壤呼吸的差异均达显著水平($P<0.05$)。其中,6月和8月、7月和9月不同植被类型土壤呼吸速率变化趋势一致,分别为柠条>油松>沙棘>侧柏和油松>柠条>沙棘>侧柏,表明不同植被类型在相似的气候条件下土壤呼吸的差异具有较为明显的季节变化特征。土壤呼吸和土壤温度以及大气温度的相关分析表明4个群落6~9月土壤呼吸日变化与大气温度的相关性要大于与土壤温度的相关性,但总的看来,它们与土壤温度和大气温度的变化趋势不太一致,因此可能还存在其他因子调控土壤呼吸的日变化。杨晶等[5]对农牧交错区不同植物群落土壤呼吸的日动态观测试验发现,不同植物群落土壤呼吸日动态一致性较差,规律性不明显,主要是因为土壤呼吸除受土壤温度的驱动外,同时也受到当日降水情况和云量、风速等气象因子的较大影响。土壤水分是土壤呼吸的一个重要影响因子,而且由于与土壤温度的交互作用往往使土壤呼吸与温度和水分之间的关系变得较为复杂[11, 12]。本研究中,所有植物群落土壤呼吸与不同土层土壤含水量均呈不同程度的正相关,但与气温和表层土壤温度的相关性较强,这可能是因为研究区域和植被类型不同有关。董云社等[13]研究发现不同草地类型土壤呼吸通量在植物生长季与0~10cm以及10~20cm土壤含水量均呈不同程度的正相关,而与气温和表层土壤温度的相关性较弱。黄承才等[14]对中亚热带东部3种主要木本群落土壤呼吸的研究表明,青冈常绿阔叶林、茶园的土壤呼吸速率与土壤含水率正相关,毛竹林的土壤呼吸速率与土壤含水率无明显关系。刘建军等[15]对秦岭天然油松、锐齿栎林地土壤呼吸的研究表明,土壤呼吸速率与气温和土壤温度之间具有显著的相关关系,呼吸速率与土壤温度的相关性高于和气温的相关性。本文与前人研究结果均表明同一环境因子对不同植物群落土壤呼吸的影响是不同的。

本研究中除侧柏群落土壤呼吸与20~40cm土层土壤水分的相关性达显著水平外($P<0.05$),其余群落土壤呼吸与不同土层土壤水分的季节变化的相关性并不明显,表明本研究区域土壤水分的季节变化对土壤呼吸的限制作用并不明显。本研究中土壤呼吸随不同土层深度的土壤水分的增加而增加,而Martin和Bolstad[16]对5个不同森林类型土壤呼吸的研究发现土壤呼吸随立地平均土壤水分的增加而降低。虽然研究对象都是森林群落,但是由于调控土壤呼吸不同组分(异养呼吸和根系呼吸)季节变异的因子各不相同[17],因此土壤呼吸与环境因子之间的关系发生了相应的变化,如Tang和Baldocchi[10]研究表明,土壤CO_2通量的季节变化模式是由土壤温度和土壤水分共同驱动的,土壤呼吸的日变化模式主要受树木生理的限制。对土壤呼吸的不同组分而言,异养呼吸主要由土壤温度和水分驱动,而根系呼吸则与其有关的自养呼吸的生理方面联系得更为紧密些[10],因此本研究中环境因子变化对4种植物群落自养和异养呼吸的影响作用可能不同,使得不同组分对土壤呼吸的贡献大小也不同,导致土壤呼吸的时间变异特征及其影响因素都较为复杂。综合分析可以得出4种植物群落对温度的反应要比对水分的反应敏感些,但是由于土壤呼吸对水热条件反应的复杂性,陈全胜等[8]认为研究研究土

壤呼吸对温度变化的响应有必要按照不同植被状况进行，Dave 等[4]认为土壤温度和水分模式是调控土壤呼吸的重要物理变量，为了更准确地预测生态系统的净呼吸，仍需要对它们对土壤呼吸的影响进行分类，本研究支持这些观点。

Campbell 和 Law[18]认为，由于森林土壤呼吸是全球碳循环的重要组成部分，因此理解森林土壤呼吸是如何被调控的就成为评价全球碳循环的一个重要内容，由于土壤呼吸的调控因子不仅是立地因子还有更大的空间和时间尺度上的因子，因此仅仅研究土壤温度和土壤水分是不够的。Ekblad 等[19]研究认为虽然土壤呼吸与土壤温度和土壤水分有关，但是却很少能解释土壤呼吸的季节变异。不同林分对土壤呼吸具有显著的影响，这可能与不同林分中的生物和非生物因子存在变异有关。Ma 等[12]研究发现异质森林中植被、根系、微生物群落在理解土壤呼吸变异模式方面的重要性，因此以后土壤呼吸研究应将这方面很好地结合起来。本研究中土壤呼吸与温度和水分之间的关系因研究的时间尺度和植物群落类型不同而不同，可能还存在一些其他因子（土壤凋落物等）影响土壤呼吸排放，因此需要进一步探明引起这种变化的原因以及这些因子对土壤呼吸的调控机制。另外，土壤质地、有机质含量、风速、降水、土壤 C/N 等非生物因子，生物量、叶面积指数、植被凋落物等生物因子以及人类活动等诸多环境因子均会改变土壤呼吸的排放速度[5, 20]，因此具体分析某一群落土壤呼吸发生变化时，需要考虑这些因子的单个或者综合作用。

研究土壤呼吸还有助于人们了解地下碳分配过程和生态系统的生产力[6]，因而能更好地了解不同植被类型的结构和功能，更好的预测植被对未来气候变化的响应，且土壤呼吸的时间变异模式可以通过对土壤水分、温度或者其他变量的连续观测模拟得以实现[10]，因此，为了进一步探明不同植被类型土壤呼吸的日和季节变异模式及其机制，需要对这 4 种植被类型土壤呼吸不同组分开展长期的动态研究，这对于了解其他相似植被类型的土壤呼吸可提供一定的参考依据，也可为准确的估测黄土高原地区土壤碳收支提供一定的数据积累。

参 考 文 献

[1] King J S, Hanson P J, Bernhardt E, et al. A multiyear synthesis of soil respiration responses to elevated atmospheric CO_2 from four forest FACE experiments. Global Change Biology, 2004, 10: 1027-1042.

[2] Schlesinger W H, Andrews J A. Soil respiration and the global carbon cycle. Biogeochemistry, 2000, 48: 7-20.

[3] Raich J W, Tufekcioglu A. Vegetation and soil respiration: correlations and controls. Biogeochemistry, 2000, 48: 71-90.

[4] Dave R, Kellman L, Beltrami H. Are soil respiration processes geographically invariant? Banff: Canadian Geophysical Union Meeting, 2002.

[5] 杨晶, 黄建辉, 詹学明, 等. 农牧交错区不同植物群落土壤呼吸的日动态观测与测定方法比较. 植物生态学报, 2004, 28（3）: 318-325.

[6] Davidson E A, Verchot L V, Cattânio H, et al. Effects of soil water content on soil respiration in forests and cattle pastures of eastern Amazonia. Biogeochemistry, 2000, 48（1）: 53-69.

[7] Raich J W, Schlesinger W H. The global carbon dioxide flux in soil respiration and its relationship to vegetation and climate. Tellus B：Chemical and Physical Meteorology，1992，44：81-89.

[8] 陈全胜，李凌浩，韩兴国，等. 温带草原 11 个植物群落夏秋土壤呼吸对气温变化的响应. 植物生态学报，2003，27（4）：441-447.

[9] 陈四清，崔骁勇，周广胜，等. 内蒙古锡林河流域大针茅草原土壤呼吸和凋落物分解的 CO_2 排放速率研究. 植物学报，1999，41（6）：645-650.

[10] Tang J, Baldocchi D D. Spatial-temporal variation in soil respiration in an oak-grass savanna ecosystem in California and its partitioning into autotrophic and heterotrophic components. Biogeochemistry，2005，73：183-207.

[11] 李凌浩，王其兵，白永飞，等. 锡林河流域羊草草原群落土壤呼吸及其影响因子的研究. 植物生态学报，2000，24：680-686.

[12] Ma S, Chen J, Butnor J R, et al. Biophysical controls on soil respiration in the dominant patch types of an old-growth, mixed-conifer forest. Forest Science，2005，51（3）：221-232.

[13] 董云社，齐玉春，刘纪远，等. 不同降水强度 4 种草地群落土壤呼吸通量变化特征等. 科学通报，2005，50（5）：473-480.

[14] 黄承才，葛滢，常杰，等. 中亚热带东部三种主要木本群落土壤呼吸的研究. 生态学报，1999，19（3）：324-328.

[15] 刘建军，王得祥，雷瑞德，等. 秦岭天然油松、锐齿栎林地土壤呼吸与 CO_2 释放. 林业科学，2003，39（2）：8-13.

[16] Martin J G, Bolstad P V. Annual soil respiration in broadleaf forests of northern Wisconsin: influence of moisture and site biological, chemical, and physical characteristics. Biogeochemistry，2005，73：149-182.

[17] Lee M S, Nakane K, Nakatsubo T, et al. Seasonal changes in the contribution of root respiration to total soil respiration in a cool-temperate deciduous forest. Plant and Soil，2003，255（1）：311-318.

[18] Campbell J L, Law B E. Forest soil respiration across three climatically distinct chronosequences in Oregon. Biogeochemistry，2005，73：109-125.

[19] Ekblad A, Boström B, Holm A, et al. Forest soil respiration rate and d13C is regulated by recent above ground weather conditions. Oecologia，2005，143（1）：136-142.

[20] 张东秋，石培礼，张宪洲. 土壤呼吸主要影响因素的研究进展. 地球科学进展，2005，20（7）：778-785.

不同土壤类型白榆种子贮藏物质动态变化的研究*

钱红格　陈存根　王　强　徐怀同

摘要

通过对中西部 5 个不同土壤类型，15 个地区的白榆种子贮藏物质变化的研究，旨在丰富白榆种质资源的研究，并为白榆在西部风沙区的科学种植提供理论依据。以采集的当年生成熟的白榆种子为材料，对其发芽率、发芽势和在萌发过程中的淀粉含量、可溶性糖含量和 α-淀粉酶、α+β 淀粉酶活性进行研究。不同土壤类型区白榆种子发芽率和发芽势由高到低依次为褐土>荒漠土>黄棕壤>棕钙土>黄绵土；褐土类型区种子淀粉含量最高，黄绵土类型区种子淀粉含量最低，72h 种子中可溶性糖含量的比较结果为黄绵土>棕钙土>黄棕壤>荒漠土>褐土；相同的时间，不同土壤类型区 α-淀粉酶、α+β 淀粉酶活性在大部分时间达到极显著差异水平，5 个土壤类型 α+β 淀粉酶活性的最大值比较结果为褐土>荒漠土>黄棕壤>棕钙土>黄绵土。综合各项指标得出 5 个不同土壤类型白榆种子贮藏物质变化存在显著差异，种子活力高低依次是褐土>荒漠土>黄棕壤>棕钙土>黄绵土。

关键词：土壤类型；淀粉；可溶性糖；淀粉酶活性

白榆（*Ulmus pumilal* L.）落叶乔木，树体高大，耐寒耐旱，在年降雨量不足 200mm 的地区仍能正常生长，是我国防风固沙、水土保持和盐碱地造林的重要树种[1]。种子是植物生命的延存器官，其优质特性一方面受其遗传因素的决定，另一方面也受到生态环境影响，其中气候和土壤等生态环境会使同一品种的种子品质性状的表现产生很大变异。白榆分布广泛，分布区内生态环境差异显著，在长期的系统发育、自然选择和人工选择影响下，种内群体间存在着明显的性状差异[2]。本文以我国中西部 5 个主要的土壤类型区白榆种子为研究对象，研究种子贮藏物质的动态变化，比较种子活力和保护性酶活性的高低，对从生物系统学角度了解白榆的地理变异规律具有重要的理论参考价值，同时研究结论对于白榆栽培中的种子调拨、种子处理等实践环节具有一定的指导意义。

---

* 原载于：西北农业学报，2009，18（4）：223-228.

1 材料和方法

1.1 材料

供试材料为 5 个不同土壤类型区的 15 个地区 30 年树龄白榆母树当年生成熟的白榆种子，具体分布见表 1。

表1 5个不同土壤类型区

土壤类型	编号	分布
A 褐土	A1	陕西宝鸡市扶风县
	A2	山西忻州宁武县
	A3	陕西咸阳市兴平
B 荒漠土	B1	陕西榆林市神木县
	B2	内蒙古鄂尔多斯市达拉特旗
	B3	新疆格尔木市区
C 黄绵土	C1	陕西榆林市米脂县
	C2	陕西榆林市绥德县
	C3	陕西延安市宝塔区
D 黄棕壤	D1	河南三门峡市义马市
	D2	河南偃师市内
	D3	河南郑州市巩义县
E 棕钙土	E1	新疆乌鲁木齐市内
	E2	新疆伊犁市巩留县
	E3	内蒙古准格尔旗

1.2 方法

1.2.1 材料的处理

随机选取发育良好、形态完整的 5 个不同土壤类型 15 个地区的白榆种子各 50 粒，3 个重复，用 1%次氯酸钠消毒 15min 后，均匀地排列在垫有 2 层滤纸的培养皿中，加入适量的蒸馏水，盖上培养皿盖，放置在 25℃恒温条件下培养，进行标准发芽试验，统计发芽率，并在萌发后 8h、16h、24h、32h、40h、48h、56h、64h、72h 九个不同时期取样进行各项生理生化指标的测定。

1.2.2 发芽率和发芽势的测定

发芽率的测定：参考孟祥才等[2]方法，每隔 12h 统计一次发芽个数，到 96h 时结束发芽试验，发芽率=发芽种子数/供试种子数×100%；发芽势=60h 时正常发芽种子数/

供试种子数×100%。

1.2.3 生理生化指标的测定

1）种子可溶性糖和淀粉含量的测定：采用蒽酮比色法[3]，取不同萌发时间的种子0.1g，加10ml 80%的乙醇研磨，80℃保温30min，过滤得到上清液加蒽酮试剂测定可溶性糖含量，单位用百分数表示。提取后的残渣加入2% HCl 5ml，在沸水浴中糊化30min，冷却后用蒸馏水定容至50ml，混匀后离心，测定方法同可溶性糖的测定，单位用百分数表示。

2）淀粉酶活性测定：参考谢逸萍等的3,5二硝基水杨酸还原法[4]，单位用 mg/g·5min 表示。

1.3 数据处理

所有数据均采用 Excel 软件和 SPSS 13.0 软件进行相关性分析和方差分析。

2 结果和分析

2.1 不同土壤类型白榆种子的发芽率和发芽势

由图1看出，5个不同土壤类型采集的白榆种子发芽率随着时间的变化而升高，在前12h种子开始吸胀，24h部分种子开始萌发，随后在60h时各土壤类型白榆种子发芽率迅速升高，其中，A褐土类型白榆种子发芽率最高，为71%，是其在24h时发芽率的18倍，发芽率最低的D棕钙土类型，为56%，72h种子发芽率达到一个高峰，72h后种子发芽率增加缓慢，开始逐渐趋于稳定，不同土壤类型白榆种子发芽率由高到低依次为 A>B>D>E>C，这和在60h时计算的种子发芽势趋势一致（图2）。种子的发芽势表示种子发芽时间的快慢，种子的发芽势越高，说明种子发芽越快，这样就有可能避免停留在土壤中的时间过长而造成的对种子的伤害[5]，从而保证种子较高的发芽率，5个土壤类型种子发芽势差异极显著（$P<0.01$），这从一方面反映了种子活力的高低，表明A土壤类型区种子活力最高。

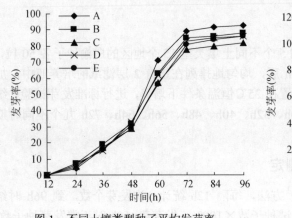

图1 不同土壤类型种子平均发芽率

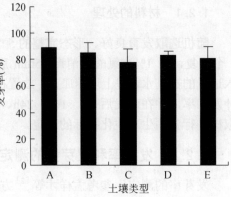

图2 不同土壤类型种子的发芽势

2.2 不同土壤类型区白榆种子萌发过程中淀粉和可溶性糖含量的变化

种子在萌发的过程中，淀粉被水解，由图3可以看出，不同土壤类型区淀粉含量随时间的增加呈下降趋势，不同土壤类型干种子淀粉含量差异极显著（$P<0.01$），淀粉含量最高的是A褐土类型区，淀粉含量在64h下降趋于平稳，到72h，下降幅度最大的是D黄棕壤类型区，比萌发前降低了8.22倍，幅度最小的是B荒漠土类型区，降低了5.03倍。72h A和B淀粉含量达到极显著差异水平（$P<0.01$），C、D和E之间无显著差异。贮藏淀粉含量的迅速下降的同时，可溶性糖含量却显著的增加，从图4可以看出，不同土壤类型区白榆种子在萌发过程中，可溶性糖含量的变化趋势是先上升后下降，其中A和B在40h时达到最大值，而其他三个类型区均在48h时才达到最大值，这说明前两个类型区的种子活力较高，萌发较快。在48h A、D和E可溶性糖含量之间达到极显著差异（$P<0.01$），这种现象反映了B和C之间差异不显著。种子萌发后可溶性糖用于胚根生长或转化合成新的物质，因此含量会逐渐下降，物体内碳水化合物的运转情况，这也是种子能够顺利萌发和生长的重要的生理基础之一[6]，72h时种子中可溶性糖含量的比较的结果是：C>E>D>B>A。

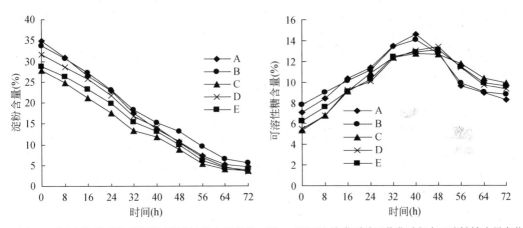

图3 不同土壤类型种子萌发过程中淀粉含量变化　　图4 不同土壤类型种子萌发过程中可溶性糖含量变化

2.3 不同土壤类型区白榆种子淀粉酶活性的比较

淀粉酶主要包括α-淀粉酶、β-淀粉酶、葡萄糖淀粉酶和R-酶，不同来源的淀粉酶，性质有所不同。植物中最重要的淀粉酶是α-淀粉酶和β-淀粉酶，淀粉酶活力等于α-淀粉酶与β-淀粉酶活性之和，活力高的种子淀粉酶的活性也高[7]。试验表明发芽时间对α-淀粉酶活力有显著影响，α-淀粉酶活力随发芽时间增加而增大，这与刘聪和安家彦[7]的研究相似。24h时A、B和D之间α-淀粉酶活力差异不显著，C和E差异极显著（$P<0.01$），32h时A和D种子α-淀粉酶活力差异不显著，B、C、E之间差异极显著（$P<0.01$），其中活性最高的是E（8.42mg/g·5min），40h时D和E种子之间α-淀粉酶活力无显著性差异（$P>0.05$），A、B和C之间差异极显著（$P<0.01$）其余各时间不同土

壤类型白榆种子α-淀粉酶活性在之间均达到极显著差异（$P<0.01$）。在72h白榆种子α-淀粉酶活性以A最高（19.34mg/g·5min），其余依次为B、D、E，最小的是C（16.72mg/g·5min）（表2）。

表2　不同土壤类型白榆种子α-淀粉酶活性的方差分析　　　（单位：mg/g·5min）

土壤类型	0h	8h	16h	24h	32h	40h	48h	56h	64h	72h
A	2.25A	6.08A	7.1D	7.66DB	8.37BB	12.13A	14.4A	17.91A	19.08A	19.34A
B	2.07B	5.04B	7.42B	7.69CB	8.27D	11.89D	14.08B	17.21B	18.41B	18.86B
C	1.01E	3.32E	6.32E	7E	7.51E	10.54E	12.6E	15.64E	16.4E	16.72E
D	1.17C	4.4C	7.43A	7.69B	8.36CB	11.38DC	13.58C	16.43C	17.65C	18.15C
E	1.09D	4.05D	7.19C	7.74A	8.42A	11.38C	13.42D	16.2D	17.09D	17.41D

注：同一列数值后的相同小写字母代表同一测定指标在0.01水平上不显著，下同。

从表3可以看出，在8h时A和B种子之间α+β淀粉酶活力无显著性差异，C、D和E种子之间差异极显著（$P<0.01$），16h时D和E种子之间α+β淀粉酶活力无显著性差异，A、B和C种子之间差异极显著（$P<0.01$），40h时A、B和E种子之间α+β淀粉酶活力无显著性差异，C和D种子之间显著极差异（$P<0.01$），其余各时间段不同土壤类型白榆种子α+β淀粉酶活力均达到极显著差异（$P<0.01$）。α+β淀粉酶活力随着发芽时间增加呈现先增大后下降的趋势，不同土壤类型种子α+β淀粉酶活力在56h均达到最高值，其中，活性最高的是A褐土（38.12mg/g·5min），最低的是黄绵土（34.56mg/g·5min）。

表3　不同土壤类型白榆种子α+β淀粉酶活性的方差分析　　　（单位：mg/g·5min）

土壤类型	0h	8h	16h	24h	32h	40h	48h	56h	64h	72h
A	10.44A	16.13A	17.57B	19.78A	22.72A	29.13A	36.96A	38.12A	37.84A	36.73A
B	10.28B	16.13BA	18.89A	19.47B	22.09C	29.01CA	36.8B	37.81B	37B	36.1B
C	7.35E	11.4E	14.79E	17.29E	21.36D	27.6E	32.92E	34.56E	32.79E	32.46E
D	8.29D	13.02D	16.27C	17.79D	21.04E	28.14D	35.57C	35.77C	34.7C	34.92C
E	8.5C	12.69C	16.24DC	18.17C	22.34B	29.07BA	35.13D	35.54D	34.01D	33.72D

2.4　不同土壤类型白榆种子生理生化指标的相关分析

由表4可以看出，5个不同土壤类型区的发芽率和α-淀粉酶、β-淀粉酶及α+β淀粉酶活力呈极显著正相关（$P<0.01$），5个土壤类型可溶性糖含量和α-淀粉酶活性呈正相关性，但不显著。A和B可溶性糖含量和β-淀粉酶活性、α+β淀粉酶活性均呈正相关，但不显著，C和E可溶性糖含量和β-淀粉酶呈极显著正相关（$P<0.01$），D可溶性糖含量和β-淀粉酶活性呈显著正相关（$P<0.05$）。C、D、E可溶性糖含量和α+β淀粉酶活性呈显著正相关（$P<0.05$），5个不同土壤类型区淀粉含量与α-淀粉酶活性、β-淀粉酶活性和α+β淀粉酶活性均呈极显著负相关（$P<0.01$）。

表4　不同土壤类型白榆种子生理生化指标的相关分析

分数	土壤类型	发芽率	可溶性糖含量	淀粉含量
α-淀粉酶活性	A	0.891**	0.043	-0.971**
	B	0.913**	0.055	-0.977**
	C	0.934**	0.627	-0.983**
	D	0.915**	0.541	-0.982**
	E	0.935**	0.561	-0.987**
β-淀粉酶活性	A	0.943**	0.375	-0.911**
	B	0.924**	0.351	-0.865**
	C	0.972**	0.824**	-0.899**
	D	0.939**	0.747*	-0.883**
	E	0.963**	0.802**	-0.873**
α+β 淀粉酶活性	A	0.938**	0.197	-0.976**
	B	0.940**	0.192	-0.960**
	C	0.979**	0.746*	-0.978**
	D	0.956**	0.663*	-0.971**
	E	0.980**	0.702*	-0.971**

*表示在 0.05 水平下线性关系显著，**表示在 0.01 水平下线性关系显著。

3　结论与讨论

种子萌发需经历从异养到自养的过程，在萌芽期间种子还不能制造足够的养分，其所需的养料和能量只能动用它储藏物质的转化和利用，这些物质在种子萌发时水解成简单的营养物质，并运转到生长部位，作为构成新组织成分和产生能量的原料，为种子萌发提供必要条件[8]。本试验中，同样的白榆品种，由于生长土壤类型的不同，其吸收的水分、营养元素必定有差异，从而造成其种子内贮藏物质含量和种子活力的差异，5个不同土壤类型白榆种子发芽率和发芽势比较的结果是：A>B>D>E>C，表明 A 褐土类型的白榆种子活力最高。

白榆种子主要贮藏物质是淀粉[9]，不同土壤类型种子淀粉含量显著不同，含量最高的是 A 褐土，最低的是 C 黄绵土，在萌发过程中淀粉经淀粉酶水解成可溶性糖，参与细胞内各种生理、生化反应，以直接或间接的方式影响种子的萌发和胚体的生长[10]，本试验中 5 个不同土壤类型可溶性糖含量变化趋势是先升后降，这和张帆等[11]对香果树种子的研究结果一致。种子在萌发前期需要积累大量的小分子糖用于萌发，当种子萌发后期，种子中可溶性糖含量反而缓慢下降，种子活力越高，可溶性糖下降得越多，这和董晓虹等[12]人的研究相似。72h 时可溶性糖含量的比较的结果是：C>E>D>B>A，这和种子的发芽率和发芽势的结果相反，表明了种子中可溶性糖作为一种能源消耗性物质在种子体内的运转和利用的快慢程度。

胡标林等[13]研究水稻作物的抗旱性时发现，淀粉酶的活性和作物的抗旱性之间有着明显的正相关，尤其是α+β淀粉酶活性。淀粉酶是一种含A^{2+}的金属酶，是水解淀粉过程中的关键酶类。淀粉酶将淀粉水解为可溶性糖，可溶性糖含量不断增加，可溶性糖在种子中的大量存在，其可溶性远远超过大分子淀粉，可增大细胞液浓度，从而提高渗透压，增加细胞吸水和保水的能力[14]。榆树种子和水稻贮藏物质相似，本试验研究了不同土壤类型的白榆种子α-淀粉酶、α+β淀粉酶活性的动态变化的过程，并进行了方差分析，结果表明：不同土壤类型白榆种子α+β淀粉酶活性是先升后降，在56h达到最大值，之后稍有下降。不同土壤类型α+β淀粉酶最大值比较结果为A>B>D>E>C，这是否能表示A褐土类型的种子抗旱性最强，C黄绵土类型种子抗旱性最弱，还要进一步做种子的抗逆性试验才能够准确地说明。各不同土壤类型α-淀粉酶活力随发芽时间增加而增大，相同的时间，α-淀粉酶和α+β淀粉酶除部分土壤类型种子无显著差异外，其余均达到极显著差异（$P<0.01$）或显著的差异（$P<0.05$），这说明同一品种的植株，由于土壤和周围环境的不同，从而产生了适应性的差异[15]。

5个不同土壤类型白榆种子的发芽率和α-淀粉酶、β-淀粉酶与α+β淀粉酶活力相互之间呈极显著正相关（$P<0.01$），说明淀粉酶活力是考察种子萌发的一个重要指标，这和陈淑芳[16]的研究结果相似。淀粉含量和α-淀粉酶、β-淀粉酶与α+β淀粉酶活力之间均呈极显著负相关（$P<0.01$），这说明α-淀粉酶、β-淀粉酶是水解淀粉的关键酶。β-淀粉酶（又名a-1,4-葡聚糖麦芽糖水解酶），广泛存在于大麦、小麦、谷物中，它是一种外切型淀粉酶，可以从淀粉或低聚糖非还原性末端顺次水解a-1,4-葡萄糖苷键而切下麦芽糖单位，水解终产物为β-型麦芽糖[17,18]。有研究表明，谷物籽粒中淀粉的降解和转运主要由β-淀粉酶参与催化，从而为萌发提供能力，但是其具体的功能还存在有不少疑问。本实验中可溶性糖含量和α-淀粉酶活力之间呈正相关，但不显著，可见白榆种子可溶性糖的积累β-淀粉酶贡献较大。A和B可溶性糖含量和β-淀粉酶活性呈正相关，但不显著，C和E可溶性糖含量和β-淀粉酶呈极显著正相关（$P<0.01$），D可溶性糖含量和β-淀粉酶活性呈显著正相关（$P<0.05$），这说明β-淀粉酶活力和可溶性糖关系较为复杂，有待进一步深入的研究。

参 考 文 献

[1] 吴光彪,孟明,张富治. 白榆种源试验研究. 山西农业大学学报（自然科学版），2001，21（3）：285-287.

[2] 孟祥才,孙晖,王喜军. 防风种子发芽特性及促进发芽的试验研究. 植物研究，2008，28（5）：627-631.

[3] 王义,赵文君,孙春玉,等. 人参体细胞胚胎发生过程中的生理变化. 东北师大学报（自然科学版），2008，40（2）：93-97.

[4] 谢逸萍,李洪民,王欣. 贮藏期甘薯块根淀粉酶活性变化趋势. 江苏农业学报，2008，24（4）：406-409.

[5] 王丽娟,李青丰. 吸湿-回干处理对披碱草属种子萌发和耐高温的影响. 种子，2008，8（27）：14-21.

[6] 陶嘉龄,郑光华. 种子活力. 北京：科学出版社，1991：35-36.

[7] 刘聪，安家彦. 小麦发芽α-淀粉酶活力及相应条件下淀粉黏度的研究. 食品研究和开发，2008，29（9）：41-44.

[8] Uniyal R C, Nautiyal A R. Seed germination and seedling extension growth in Ougeinia dal bergioides Benth under water and salinity stress. New Forests, 1998, 16: 265-272.

[9] 敦论，林新福，王铁章，等. 白榆. 北京：中国林业出版社，1984：37-38.

[10] 臧运祥，郑伟尉，孙仲序，等. 植物胚状体发生过程中主要代谢产物变化动态研究进展. 山东农业大学学报（自然科学版），2004，35（1）：131-136.

[11] 张帆，梁宏伟，熊丹，等. 香果树种子萌发中生理生化变化的研究. 种子，2007，10（26）：21-23.

[12] 董晓红，万清林，徐娜. 胡萝卜种子萌发过程中生理生化变化的研究. 生物技术，2005，15（6）：55-57.

[13] 胡标林，李名迪，万勇，等. 我国水稻抗旱性鉴定方法和指标研究进展. 江西农业学报，2005，17（2）：56-60.

[14] 徐茂，吴昊. 江苏省不同类型土壤基础供氮能力对水稻产量的影响. 南京农业大学学报，2006，29（40）：1-5.

[15] 高乐旋，陈家宽，杨继. 表型可塑性变异的生态-发育机制及其进化意义. 植物分类学报，2008，46（4）：441-451.

[16] 陈淑芳. GA诱导NaCl胁迫下黄瓜种子萌发和幼苗耐盐性效应. 西北植物学报，2008，28（7）：1429-1433.

[17] Diaz A, Sieiro C, Villa T G. Production and partial characterization of a β-amylase by Xanthophyllomyces dendrorhous. Letters in Applied Microbiology, 2003, 36: 203-207.

[18] Van Damme J M, Hu J, Barre A, et al. Purification, characterization, immunolocalization and structural-analysis of the abundant cytoplasmic β-amylase from calystegia sepium（hedge bindweed）rhizomes. European Journal of Biochemistry, 2001, 268: 6263-6273.

基于高光谱参数的枫杨叶绿素含量估算模型优化*

李文敏　魏　虹　李昌晓　陈存根

摘要　采用 ASD Fieldspec 光谱仪测定了不同土壤水分条件下枫杨叶片高光谱反射率,并给出了对应的枫杨的叶绿素含量。研究了不同土壤水分条件下枫杨幼苗叶片叶绿素含量的变化规律;综合分析了 9 个常见光谱植被指数与枫杨叶片的叶绿素含量的相关性与回归方程,利用主成分分析对光谱数据进行降维,将得到的主成分得分作为 BP 人工神经网络模型的输入变量进行了枫杨叶片叶绿素含量的估算。结果表明:不同水分处理均显著影响枫杨幼苗叶绿素含量。在所列举的 9 个常用植被指数中 VOG1 植被指数与叶绿素含量的关系最密切,相关系数 r 达到了 0.865。主成分分析-BP 神经网络模型进行叶绿素含量的估算,预测值与实测值之间的线性回归的相关系数 $r=0.934$,是一种良好的植被叶绿素含量高光谱估算模式。

关键词　高光谱;叶绿素含量;敏感光谱指数;主成分分析;BP 神经网络

1　引言

举世瞩目的三峡大坝建成以后,对原有植被和物种的生存环境产生了极大的影响。同时,季节性的裸露和淹没、流水的反复冲刷、人类的频繁活动使三峡库区消落带成为生态脆弱区,严重影响三峡库区的安全、健康和可持续发展。研究表明,筛选适生树种进行生态修复是一个行之有效的方式,其中乡土树种在维持库区的生态环境,确保库岸生态系统生态功能的正常发挥等方面具有重要意义。作为我国亚热带地区的乡土树种的枫杨(*Pterocarya stenoptera*)是胡桃科枫杨属落叶速生乔木[1],常见于河岸带和消落带[2]。有研究表明枫杨能够耐受较长时间的水淹胁迫,可以作为三峡库区库岸修复的备选物

---

* 原载于:林业科学,2014, 50 (4):55-59.

种[3]。因此,研究枫杨在不同水分下的生理特性及生长状况就具有重要的理论和实践意义。由于叶绿素含量是光合作用能力和植被发育阶段(特别是衰老阶段)的指示[4],同时也是反映植物生理功能、受害程度和其他矿质营养的重要指标[5]。为了解枫杨在不同水分下的适应性,开展其在不同水分下叶绿素含量的研究就显得十分必要了。

与传统的色素分析法需要破坏性取样,并且时间上和空间上难以满足实时、快速、无损、精确诊断的要求相比,近年来发展起来的高光谱遥感技术,可以实时、快速、无损地从叶片水平、群体水平以及生态系统等多个层面研究植物在各种环境条件下的生理生态变化。因而在植被研究方面具有巨大的应用价值和广阔的发展前景[6],并且有不少学者开展了利用高光谱遥感进行树木的叶片叶绿素含量估算的研究[7-10]。大量研究表明植被的生物物理、化学参数与光谱反射率之间的关系基本上是非线性的,而神经网络对非线性问题的拟合有着无可比拟的优势,因此已有部分研究者开始把神经网络引入到高光谱数据分析中来进行叶绿素的估算[11-13]。

目前对于枫杨的研究集中于其木材解剖学特征、水淹对枫杨生理生态的影响以及在水淹胁迫下枫杨叶片的高光谱反射率变化的研究[14-18],而对于在水淹胁迫下利用高光谱数据对枫杨幼苗叶片的叶绿素含量估算的研究至今尚未见到报道。本文应用主成分分析-BP 神经网络模型反演枫杨叶绿素含量,并对比研究不同植被指数构建的回归模型与主成分分析的 BP 神经网络模型在反演枫杨叶绿素含量变化的精度,以相关系数和模型的预测值与实测值的均方根误差(RMSE)作为模型的评价标准来寻找最优反演模型。

2 实验材料和方法

2.1 材料和地点

试验地点位于西南大学三峡库区生态环境教育部重点实验室实验基地(海拔高度 249m),2010 年 3 月播种,2010 年 6 月选取长势一致的 80 棵幼苗移栽入盆,每盆一株,进行常规田间管理,2010 年 7 月 23 日,选取 65 盆长势一致的枫杨进行水分处理。

2.2 实验设计

将试验用苗随机分为三组,即对照组 CK(不进行水淹,但保证正常供水)、T1(根部水淹-干旱)、T2(根部水淹)。CK 组保持常规管理,根部水淹处理为淹水至土壤表面 5cm。干旱处理通过 Wescor 水势仪对植株叶片进行检测(叶水势位置在-0.5MPa),并及时补充水分。实验从 2010 年 7 月 23 日开始,取样间隔分别为 0d、12d、24d、36d、48d,实验处理开始时取 5 株用于 0d 测量,其余三个处理各 20 株,每组每次取 5 株。

2.3 叶片反射光谱的测定

采用便携式光谱仪 FieldSpec 3 Hi-Res,在 350~2500nm 波长范围内进行连续测量,350~1100nm 采样间隔为 1.4nm,1000~2500nm 采样间隔为 2nm。350~1100nm 光谱分辨率为 3nm,1000~1900nm 光谱分辨率为 8.5nm,1700~2500nm 光谱分辨率为 6.5nm。

利用叶片夹持器固定叶片，使用植被探头测定叶片光谱，每一植株选取三片叶片进行测量，每次测量十次重复，平均后作为该植株叶片反射率。

2.4 叶片叶绿素含量的测定

采用浸提法[19]，用岛津 2550 分光光度计测定浸提液的浓度，再依据公式将其转为含量（mg/g）。

2.5 光谱指数

依据目前文献中比较能反映植物叶绿素含量的光谱指数，列举以下常见的光谱指数（表1）。

表 1 常用的叶绿素敏感高光谱指数

光谱指数	计算公式	参考文献
PSSRa	R_{800}/R_{680}	[20]
PSSRb	R_{800}/R_{635}	[20]
ND_{705}	$(R_{750}-R_{705})/(R_{750}+R_{705})$	[21]
GNDVI	$(R_{750}-R_{550})/(R_{750}+R_{550})$	[22]
NPCI	$(R_{680}-R_{430})/(R_{680}+R_{430})$	[23]
SRPI	R_{430}/R_{680}	[24]
GM1	R_{750}/R_{550}	[25]
VOG1	R_{740}/R_{705}	[26]
VOG2	$(R_{734}-R_{747})/(R_{715}+R_{726})$	[27]

2.6 主成分分析-BP 神经网络法

BP 神经网络是最具代表性、使用最多、应用最广、发展最迅速的误差反向传播多层前馈式网络（back-propagation network，BP 网络），80%～90% 的模型都采用这种网络或它的变化形式[28]。在应用 BP 神经网络时，若将光谱仪生成的成百上千个采样数据直接输入 BP 神经网，即使建一个只含三层结点的网，其计算量之大也是目前微机难以承受，而且这样建成的模型会对输出产生"过拟合"，其预测的适应能力反而大大下降。如果舍弃其中一些因素，势必会造成某些有用信息的丢失。而主成分分析是研究利用降维的思想，把多指标转化为少数几个综合指标的一种常用的统计分析方法。它既保留数据原有的绝大部分信息，又大大降低了数据的维数。因此，我们利用主成分分析法把所取得的高光谱数据进行降维，在保留有效的数据信息的条件下，以所得主成分的得分作为 BP 神经网络的输入数据。这样既保留原有的信息，又大大降低了 BP 神经网络的输入数据的维数。本文中以 65 株枫杨幼苗为研究对象，利用便携式光谱仪 FieldSpec 3 Hi-Res 采集得到 64 组有效高光谱数据，其中对照组 25 组，根部水淹-干旱组 20 组，根部水淹组 19 组。

3 结果与分析

3.1 土壤水分处理对枫杨幼苗叶片叶绿素含量的影响

由表2可知，随着处理时间的延长，T1、T2组均显著低于对照。T1组的总叶绿素含量先下降后上升，在24d呈最低值1.93mg/g，仅为对照的67.48%，在48d达最大值2.21mg/g；T2组的变化趋势与T1组一致，在24d达最低值2.07mg/g，仅为对照的72.38%。这可能是植物受到水淹胁迫后，各种生理过程受到干扰，造成膜系统结构破坏、有害代谢产物积累、蛋白质合成下降等后果，直接或间接地影响了叶绿素的含量。

表2 不同土壤水分处理下枫杨叶片叶绿素总浓度的动态变化

处理时间	0d (mg/g)	12d (mg/g)	24d (mg/g)	36d (mg/g)	48d (mg/g)
CK	2.19±0.03	2.74±0.10	2.86±0.15	2.85±0.05	2.86±0.04
T1	2.19±0.03	2.08±0.08	1.93±0.17	1.98±0.08	2.21±0.11
T2	2.19±0.03	2.08±0.01	2.07±0.05	2.09±0.04	2.17±0.05

注：表中数据均为平均值±标准误。

3.2 枫杨叶片叶绿素含量的高光谱估算模型

以表1所列举的常用的叶绿素敏感光谱指数为自变量，枫杨叶片叶绿素含量为因变量，进行一元线性回归，所得的结果如表3所示。

表3 枫杨的叶片叶绿素含量（y）与光谱参数（x）的回归分析（$n=64$）

光谱参数	线性回归模型		
	回归方程	相关系数 R	均方根误差 RMSE
PSSRa	$y=0.215x+0.565$	0.530	0.0773
PSSRb	$y=0.389x+0.262$	0.834	0.0522
ND_{705}	$y=8.081x-0.386$	0.841	0.0486
GNDVI	$y=8.045x-1.099$	0.761	0.0723
NPCI	$y=-5.602x+2.694$	-0.344	0.0951
SRPI	$y=3.152x-0.450$	0.383	0.0858
GM1	$y=1.189x-0.661$	0.781	0.0648
VOG1	$y=1.820x-1.252$	0.865	0.0418
VOG2	$y=-44.251x+0.784$	-0.854	0.0433

以回归模型的相关系数（r）和预测值与实测值的均方根误差（RMSE）作为模型的评价标准。从表3可知，以VOG1为自变量所得的回归模型的相关系数在这9个叶绿素敏感光谱指数中最大，为0.865，而相应的总均方根误差为0.0418，低于其余的叶绿素

敏感光谱指数总均方根误差。因此，以 VOG1 为自变量所得的回归模型来预测枫杨叶片叶绿素含量的效果为最好。

3.3 基于主成分分析-BP 神经网络的叶绿素含量估算

在本研究中，利用 matlab7.01 对原始光谱 350～2500nm 数据进行主成分分析，提取 $\lambda>1$ 的作为主成分，共 12 个，累积贡献率达到 99.70%。具体的贡献率见表 4，同时也给出相应主成分载荷矩阵（表 5），进而求出每个样本各个主成分得分（表 6）。

表 4 主成分的贡献率

主成分	贡献率（%）	累计贡献率（%）
主成分 1	70.00	70.00
主成分 2	18.95	88.95
主成分 3	6.84	95.79
⋮	⋮	⋮
主成分 12	0.05	99.70

表 5 主成分载荷矩阵

变量	$x1$	$x2$	$x3$	$x4$	…	$x2150$	$x2151$
主成分 1	0.5111	0.4749	0.5019	0.5324	…	0.6871	0.6833
主成分 2	0.5703	0.5823	0.5698	0.6124	…	0.5748	0.5762
主成分 3	−0.1317	−0.1839	−0.2096	−0.1805	…	0.2532	0.2597
⋮	⋮	⋮	⋮	⋮	…	⋮	⋮
主成分 12	0.1364	0.0407	−0.0569	0.1901	…	0.0271	0.0286

表 6 主成分得分矩阵

样本 S	$S1$	$S2$	$S3$	$S4$	…	$S63$	$S64$
因子 1	25.6422	26.9722	26.8901	28.0248	…	24.1325	25.4057
因子 2	9.5425	4.8684	5.5533	4.6950	…	4.2226	4.7712
因子 3	−2.3263	−1.2613	−1.4076	−0.9757	…	−1.8029	−1.3016
⋮	⋮	⋮	⋮	⋮	…	⋮	⋮
因子 12	0.0096	0.0016	0.0020	0.0019	…	0.0079	0.0021

本研究的 BP 神经网络是由 Matlab 的 Neural Net2 work Toolbox 提供，网络共有 3 层，依次为输入层、隐含层和输出层。将主成分得分（表 6）作为神经网络的输入层神经元，以相对应样本叶绿素含量为输出层神经元。根据 Kolmogorov 定理，本研究设定网络的中间层神经元个数为 11。隐含层的激活函数为 tansig，输出层为 purelin，网络的训练函数为 traingd，网络的学习函数 learngdm，网络的性能函数采用均方误差性能函数 rmse。

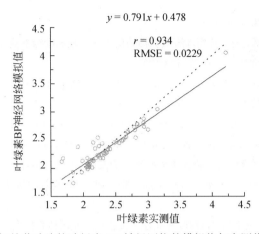

图 1 枫杨幼苗叶片的叶绿素 BP 神经网络的模拟值与实测值的效果图

注：实线表示实测值和模拟值完全一致时的直线；虚线表示实测值与利用本文的方法得出的预测值的拟合直线；圆圈表示实测值与利用本文的方法得出的预测值的拟合效果。

在所得到 64 个枫杨在不同水分条件下的叶绿素数据中随机选取 40 个数据作为 BP 神经网络的训练样本的学习目标，将训练样本的主成分的得分作为 BP 神经网络的输入数据，设置网络学习的精度为 0.001，最大迭代次数为 1000。当网络训练完毕后，将所有样本的光谱数据作为输入矢量，进行模拟所得的预测值与实测值进行拟合，其结果如图 1 所示。

由图 1 的结果可知，用主成分分析-BP 神经网络方法来拟合枫杨的叶绿素含量时，其预测值与实测值的一元线性回归模型的相关系数为 0.934，其均方根误差为 0.0229。

4 结论

本文研究了在不同水分条件下枫杨幼苗叶片的叶绿素含量的估算模型，通过对 9 个叶绿素敏感光谱指数的回归分析和基于主成分分析-BP 神经网络模拟，得出以下结论。

1）在所列举的 9 个常用叶绿素敏感光谱指数中，VOG1 指数中与枫杨幼苗的叶绿素含量的线性回归模型最好，其相关系数达到了 0.865。所以在应用光谱敏感指数对枫杨幼苗的叶绿素含量进行反演时，应优先考虑这个叶绿素敏感光谱指数。

2）将主成分分析-BP 神经网络应用到枫杨幼苗的叶绿素含量反演中，预测值与实测值之间的线性回归的相关系数 r 为 0.934，预测值与实测值均方根误差 RMSE 为 0.0229。相比较前面用叶绿素敏感光谱指数所得回归模型的结果，无论是在回归的相关系数上，还是在所得预测值与实测值的均方根误差上都是最好的。并且预测精度都大大提高了，达到了理想的效果。

参 考 文 献

[1] 彭镇华. 中国长江三峡植物大全（上卷）. 北京：科学出版社，2005：861.

[2] 徐有明，邹明宏，史玉虎，等. 枫杨的生物学特性及其资源利用的研究进展. 东北林业大学学报，

2002, 30 (3): 42-48.

[3] 贾中民, 魏虹, 田晓峰, 等. 长期水淹对枫杨幼苗光合生理和叶绿素荧光特性的影响. 西南大学学报 (自然科学版), 2009, 31 (5): 124-129.

[4] Minolta C. Chlorphyll SPAD-502 Instruction Manual. Osaka, Japan: Radiometric Instruments Operation, 1989: 17-21.

[5] Strachan I B, Pattey P, Boisvert J B. Impact of nitrogen and environmental conditions on corn as detected by hyper spectral reflectance. Remote Sensing of Environment, 2002, 80 (2): 213-224.

[6] 杨哲海, 韩建锋, 宫大鹏, 等. 高光谱遥感技术的发展与应用. 海洋测绘, 2003, 23 (6): 55-81.

[7] 刘秀英, 熊建利, 林辉. 基于高光谱特征参数的樟树叶绿素含量的估算模型研究. 广东农业科学, 2011, 5: 1-4.

[8] 张永贺, 陈文惠, 郭乔影, 等. 桉树叶片光合色素含量高光谱估算模型. 生态学报, 2013, 33 (3): 876-887.

[9] 宋晓东, 江洪, 余树全, 等. 亚热带典型常绿阔叶树种叶片叶绿素含量与其高光谱特征间的关系. 生态学报, 2008, 28 (5): 1959-1963.

[10] 伍南, 刘君昂, 闫瑞坤. 利用光谱特征参数估算病害胁迫下杉木叶绿素含量. 植物保护, 2012, 38 (4): 72-76.

[11] Combal B, Baret F, Weiss M, et al. Retrieval of canopybiophysical var iables from bi-direct ional reflectance: using prior information to solve the ill-posed inverse problem. Remote Sensing of Environment, 2003, 84 (1): 1-15.

[12] Gong P, Wang, D X, Liang S. Inverting a canopy reflectance model suing a neural network. International Journal of Remote Sensing, 1999, 20 (1): 111-122.

[13] 宋开山, 张柏, 王宗明, 等. 大豆叶绿素含量高光谱反演模型研究. 农业工程学报, 2006, 22 (8): 16-21.

[14] 汪佑宏, 曹仁忠, 徐斌, 等. 水淹程度对滩地枫杨主要力学性质的影响. 安徽农业大学学报, 2003, 31 (2): 168-172.

[15] 王振夏, 魏虹, 吕茜, 等. 枫杨幼苗对土壤水分"湿-干"交替变化的光合及叶绿素荧光响应. 生态学报, 2013, 33 (3): 888-897.

[16] 吕茜, 魏虹, 李昌晓. 土壤水分条件对枫杨幼苗光合生理的影响. 西南大学学报, 2010, 32 (3): 116-123.

[17] 衣英华, 樊大勇, 谢宗强, 等. 模拟淹水对枫杨和栓皮栎气体交换、叶绿素荧光和水势的影响. 植物生态学报, 2006, 30 (6): 960-968.

[18] 谢小红, 魏虹, 李昌晓, 等. 水淹胁迫下枫杨 (*Pterocarya stenoptera* C. DC.) 幼苗叶片高光谱特征的研究. 西南大学学报 (自然科学版), 2011, 33 (4): 93-98.

[19] 高俊凤. 植物生理学实验指导. 北京: 高等教育出版社, 2006: 74-77.

[20] Blackburn G A. Quantifying chlorophylls and carotenoids at leaf and canopy scales: an evaluation of some hyperspectral approaches. Remote Sensing of Environment, 1998, 66 (3): 273-285.

[21] Gitelson A A, Merzlyak M N. Signature analysis of leaf reflectance spectra: algorithm development for remote sensing. Journal of Plant Physiology, 1996, 148 (3-4): 494-500.

[22] Gitelson A A, Kaufman Y J, Merzlyak M N. Use of a green channel in remote sensing of global vegetation from EOS-MODIS. Remote Sensing of Environment, 1996, 58 (3): 289-298.

[23] Peñuelas J, Gamon J, Fredeen A L, et al. Reflectance indices associated with physiological changes in nitrogen and water-limited sunflower leaves. Remote Sensing of Environment, 1994, 48 (2): 135-146.

[24] Peñuelas J, Filella I, Lloret P, et al. Reflectance assessment of mite effects on apple trees. International Journal of Remote Sensing, 1995, 16 (14): 2727-2733.

[25] Gitelson A A, Merzlyak M N. Remote estimation of chlorophyll content in higher plant leaves. Internal Journal of Remote Sensing, 1997, 18 (12): 2691-2697.

[26] Vogelmann J E, Rock B N, Moss D M. Red-edge spectral measurement from SugarMaple leaves. Internal Journal of Remote Sensing, 1993, 14 (8): 1563-1575.

[27] Zarco-Tejada P J, Miller J R, Noland T L, et al. Scaling-up and model inversion methods with narrow-band optical indices for chlorophyll content estimation in closed forest canopies with hyperspectral data. IEEE Transactions on Geoscience and Remote Sensing, 2001, 39 (7): 1491-1507.

[28] 朱瑜馨, 张锦宗, 赵军. 基于人工神经网络的森林资源预测模型研究. 干旱区资源与环境, 2005, 19 (1): 101-104.

重金属铅、镉在白榆中分布规律和累积特性研究*

张永超　陈存根　钟发明　王强　钱红格

摘要

研究了重金属铅、镉在白榆中分布和积累，了解其对重金属镉、铅的耐性和抗性，为重金属严重污染地区的造林提供技术依据。通过盆栽试验，探讨了白榆在铅、镉单一及复合处理条件下，对铅、镉元素的吸收累积特性。结果表明：在铅、镉单一及复合处理下，铅、镉元素在白榆各器官中分布规律均表现为：根＞茎＞叶；在铅、镉单一处理下，转运系数均随处理浓度增加而降低，说明白榆在较高浓度的重金属危害下，通过根部对重金属的大量累积，来保证其茎和叶功能相对正常，而其地上部分、地下部分的富集系数则随着重金属处理浓度的增加均呈先增后减的趋势；在复合污染下，镉对叶、茎累积铅具有协同作用，对根累积铅无明显影响，铅对根、茎、叶累积镉具有拮抗作用。白榆对铅、镉元素不符合超富集植物条件，但对铅、镉元素表现出一种耐性，因此，白榆适用于铅、镉污染区的种植和绿化。

关键词：白榆；铅；镉；转运系数；富集系数

随着工矿业及农业的发展，受重金属污染的土壤面积迅速增加，目前我国受重金属污染的耕地面积约占总耕地面积的 1/5[1-3]，这些有毒重金属可通过食物链进入人体[4]，对人体健康造成危害，因此重金属对土壤的污染问题引起了人们的广泛关注。近年来，国内外学者对某些草本或木本植物吸收、积累、转化重金属元素进行了大量相关的研究报道[5-7]，而白榆对重金属镉、铅的吸收、积累特性缺乏报道。白榆是我国北方分布广泛的造林树种，抗逆性强，生长快。通过不同浓度的铅、镉污染及其不同浓度配比的复合污染，应用盆栽试验，探讨了白榆体内重金属铅、镉的积累、分布规律，并通过多元回归分析，进一步认识铅、镉元素在榆树体内的累积、分布及重金属复合作用的基本规律。为进一步控制白榆体内重金属含量，探讨其对白榆的毒害机理及其复合作用的特征，既可为应用白榆进行重金属污染防治提供理论根据，又可为在我国北方干旱、半干旱重金属污染地区采用白榆植树造林恢复植被提供科学指导。

* 原载于：西北林学院学报，2011，26（5）：6-11.

1 材料与方法

1.1 试验材料

盆栽实验于 2008 年 12 月至 2009 年 10 月在西北农林科技大学进行。供试树种为白榆（*Ulmus pumila*），土壤为塿土，将采回的土壤混合均匀作为供试土壤。在供试土壤中随机取多个点的土壤混合，经风干，压碎，过筛后测定其基本理化性质（表1）。

表1 供试土壤养分含量状况

pH	有机质（g/kg）	碱解氮（mg/kg）	有效磷（mg/kg）	速效钾（mg/kg）	全铅（mg/kg）	全镉（mg/kg）
7.9	17.6	58.2	157.5	144.9	30.43	0.38

1.2 试验方法

2008 年 12 月，选口径 40cm，高 60cm 的实验盆（塑料桶），每桶称取供试土壤 15kg，加入尿素（N 46%）500mg/kg 和复合肥（N 15%，P_2O_5 5%，K_2O 25%）200mg/kg 作为底肥，充分混匀。用去离子水将分析纯醋酸铅 [$Pb(CH_3COO)_2 \cdot 3H_2O$] 和硝酸镉 [$Cd(NO_3)_2 \cdot 4H_2O$] 配成母液，再次稀释配成添加液；添加铅元素浓度分别为 0mg/kg、10mg/kg、40mg/kg、100mg/kg、200mg/kg、300mg/kg 6 个处理，添加镉元素浓度分别为 0mg/kg、5mg/kg、10mg/kg、20mg/kg、100mg/kg、400mg/kg 6 个处理；铅镉复合污染选取铅 20mg/kg、100mg/kg、200mg/kg 分别与镉 10mg/kg、20mg/kg、100mg/kg 完全正交，以不加铅镉为对照（CK）处理，每个处理重复 3 次。将土壤混合均匀，钝化 2 个月。2009 年 3 月底移栽长势一致的 2 年龄白榆每盆 1 株，在自然光照下生长，定期以称重法加自来水（pH 6.8，水中未检出 Pb^{2+} 和 Cd^{2+}）浇灌，保持土壤湿度为田间持水量的 60%~70%。2009 年 10 月结束实验。

1.3 样品制备

在白榆苗移栽前选取盆内 20cm 以下土壤 50g 左右，混合均匀，将选取的土壤样品在 60℃烘干、磨细分别过筛，制备土壤样品，用以测定土壤中铅、镉本底值。2009 年 9 月取白榆的不同部位（吸收根、嫩茎、成熟叶片），用蒸馏水冲洗干净，在 80℃烘箱烘干，磨细过筛，制备成植物样品，分别测定铅、镉（全量）的含量；每盆用四分法采集土样 500g，混合均匀，将采回的土壤样品 60℃烘干，磨细过 20 目筛，制备土壤样品，用以测定土壤中铅、镉全量。

1.4 测定方法

1.4.1 土壤基本理化性质及重金属含量测定

采用电位法（水土比为 2.5∶1）测定 pH；采用半微量凯氏法测定碱解氮含量；采

用重铬酸钾容量（外加热）法测定有机质含量；采用醋酸铵（pH 7.0）交换法测定阳离子交换量（CEC）；采用 $NaHCO_3$ 浸提法测定有效磷；采用 NH_4OAc 浸提-火焰光度法测定测定速效钾[8]，土壤样品中的铅、镉含量采用王水消解法，用火焰原子吸收光谱器测定其含量[9, 10]。

1.4.2 植物样品重金属含量测定

白榆根、茎、叶样中的铅、镉元素含量采用湿法消化法（HNO_3：$HClO_4$=5：1）消解，用日立 Z-5000 型原子吸收分光光度法测定重金属含量[8]。

1.5 数据处理与分析

富集系数（BCF）和转运系数（TF）是衡量植物是否为超积累植物的重要指标，其计算式[11, 12]如下。

富集系数（BCF）=地上部器官中重金属含量/土壤中重金属含量

转运系数（TF）=地上部器官中重金属含量/根部重金属含量

数据采用 Microsoft Excel 和 SPSS 13.0 软件处理。

2 结果与讨论

2.1 不同铅处理下白榆各部位中铅的累积量

白榆的枝、叶、根中铅含量随土壤铅离子的增加而持续升高（表 2）。白榆不同部位铅的累积量依次为：根>茎>叶，反映出整个白榆体从根到叶片对铅的累积量显著降低；表明铅元素在白榆体内活性较低，迁移能力较弱，大部分运移到根部即被固定，向地上部分运输的量较低，可见白榆根部能够有效阻止土壤中的铅向白榆地上部分中转移，发挥了重要的缓冲和屏障作用。本试验研究结果与其他植物上的研究结果一致，都认为铅主要积累在植物根部，在重金属污染环境中生长的植物更大比例地积累于根部[6, 13, 14]。

表 2 不同处理下白榆各部位中铅的累积量

处理浓度（mg/kg）	白榆各部位铅累积量（mg/kg）		
	叶	枝	根
CK	3.84±0.09f	7.87±0.34f	11.42±0.14g
10	4.79±0.19e	9.67±0.16e	17.11±0.49f
40	6.74±0.22d	11.83±0.56d	38.11±0.93e
100	8.46±0.37c	14.15±0.63c	74.80±1.89d
200	9.76±0.34b	15.1±0.53b	112.79±1.96c
300	10.92±0.42a	16.43±0.59a	185.48±1.85b

注：表中数据均为平均值±标准差，小写字母分别表示用 LSD 检验中 0.05 显著水平，下同。

2.2 不同镉处理下白榆各部位镉的累积量

试验结果表明，随着土壤中镉处理浓度的提高，植物各部分镉含量均有所增加（表3）。白榆受到不同浓度镉污染后，根的累积量最大，在镉投入量≥20mg/kg 时，叶中镉的含量小于枝中镉的含量，在镉投入量＜20mg/kg 时，叶中镉的含量大于枝中镉的含量，这是因为低浓度镉处理有助于白榆的光合作用，因而叶上的镉积累相对较高，在高浓度时，重金属镉对叶片会造成伤害，白榆叶片具有较强的自我防护机制，有效阻止了重金属镉向叶片的过度转移；不同处理条件下，根部镉含量变化最大，这可能因为白榆根系分泌物和通过根系分泌物诱导的一系列变化（如 pH 的变化）有助于根对镉的吸收[15]，这与檀建新等[16]、阎雨平等[17]、石元值等[18]、陆晓怡和何池全[19]、肖昕等[20]的研究结果相一致。

表3 不同处理下白榆各部位中镉的累积量

处理浓度（mg/kg）	白榆各部位镉累积量（mg/kg）		
	叶	枝	根
CK	0.32±0.02f	0.13±0.01f	0.49±0.01f
5	2.97±0.15e	0.22±0.02f	4.08±0.13f
10	4.74±0.17d	2.42±0.16e	11.2±0.33e
20	5.98±0.25c	7.74±0.31d	26.42±0.34d
100	7.9±0.21b	12.13±0.38c	47.39±1.90c
200	9.89±0.36a	18.95±0.18b	80.34±4.86b

2.3 铅镉复合污染中重金属在白榆中的分布累积特性

铅镉交互作用对白榆不同部位累积铅镉的影响明显（表4和表5）。白榆不同部位的铅、镉累积量均为：根＞茎＞叶片。交互处理下白榆累积铅和镉的量分别以不加镉和不加铅为对照，在对照处理下，白榆中铅的含量增加缓慢，铅浓度低于 200mg/kg 时，镉的加入促进了白榆根、茎、叶中铅的吸收，但增加量较小，增加量不显著；当铅浓度不变时，随着镉浓度的增加，白榆根、茎、叶中铅的含量都逐步增加，均大于对照；由此可以说明，在试验浓度范围内，镉的加入可以促进白榆各部分对铅的吸收（表5）。

表4 不同的铅、镉浓度组合下白榆各部位中镉的累积量

处理浓度（mg/kg）	白榆各部位镉累积量（mg/kg）		
	叶	枝	根
CK	0.32±0.03h	0.13±0.01e	0.49±0.02h
$Cd_{10}Pb_{40}$	4.47±0.27e	2.58±0.14d	13.41±0.50fg
$Cd_{10}Pb_{100}$	3.93±0.16fg	2.84±0.08d	14.02±0.57f
$Cd_{10}Pb_{200}$	3.72±0.21g	2.65±0.13d	12.27±0.33g
$Cd_{20}Pb_{40}$	6.24±0.08b	7.55±0.41c	28.88±0.43b

续表

处理浓度（mg/kg）	白榆各部位镉累积量（mg/kg）		
	叶	枝	根
$Cd_{20}Pb_{100}$	6.08±0.10bc	7.61±0.22c	22.95±1.19c
$Cd_{20}Pb_{200}$	5.96±0.08c	7.54±0.38c	20.91±0.76d
$Cd_{100}Pb_{40}$	6.92±0.19a	12.04±0.32a	34.45±1.29a
$Cd_{100}Pb_{100}$	4.95±0.12d	11.72±0.12a	27.61±1.47b
$Cd_{100}Pb_{200}$	4.04±0.19f	10.03±0.50b	19.16±1.04e

在对照处理下，白榆叶片、茎和根各器官对镉的累积量随镉的添加量增加而增加（表4），在镉元素处理浓度不变时，加入铅元素后，随着加入铅元素浓度增加，白榆根、茎、叶中镉的累积量均出现下降的趋势。相对单一镉污染加入铅元素之后，根中镉元素含量增加变得缓慢；当镉元素投入量≥100mg/kg时，白榆叶、茎、根中镉含量随着铅元素的增加都显著降低，说明铅的加入可以抑制叶片、茎和根对镉的吸收，并且铅的加入量越大，对镉元素累积的影响越大。

表5 不同的铅、镉浓度组合下白榆不同部位中铅的累积量

处理浓度（mg/kg）	白榆各部位铅累积量（mg/kg）		
	叶	枝	根
CK	3.77±0.15f	8.10±0.11c	11.25±0.26h
$Cd_{10}Pb_{40}$	6.96±0.25e	13.49±0.68bc	16.57±0.84g
$Cd_{10}Pb_{100}$	8.61±0.18d	14.60±0.53ab	75.49±3.96d
$Cd_{10}Pb_{200}$	9.92±0.37bc	15.55±0.42ab	113.89±2.33b
$Cd_{20}Pb_{40}$	8.45±0.38d	14.93±0.38ab	19.77±0.58g
$Cd_{20}Pb_{100}$	9.36±0.50c	16.84±0.04ab	84.29±0.76b
$Cd_{20}Pb_{200}$	10.41±0.49ab	17.22±0.74ab	121.55±6.07a
$Cd_{100}Pb_{40}$	9.73±0.23c	15.90±0.35ab	20.20±0.59g
$Cd_{100}Pb_{100}$	9.98±0.54bc	18.05±0.56ab	36.90±1.97f
$Cd_{100}Pb_{200}$	10.63±0.34a	20.21±1.04a	44.52±2.15e

2.4 白榆对重金属铅、镉富集系数及转运系数

转运系数（TF）是指地上部元素的含量与地下部同种元素含量的比值，用来评价植物将重金属从地下向地上的运输和富集能力。转移系数越大，则重金属从根系向地上部器官转运系数越强[21]。从图1～图4可以看出在铅、镉单一和复合污染时白榆对重金属铅、镉的转运和富集系数比较低，富集系数除镉处理为10mg/kg、20mg/kg外都小于1，说明白榆对重金属镉、铅的吸收、富集能力较弱，不具备超富集植物的特征[22]。

在单一重金属镉污染下，白榆的转运系数（TF）随着处理浓度的增加明显降低，由对照中的 0.93 降到处理浓度为 200mg/kg 时的 0.38（图2），由此也可以看出，随着土壤中重金属镉的增加，根部累积的镉元素占吸收量的比重逐渐增加，从而极大防止了重金属元素对白榆地面营养器官的伤害，保证了其有较高的光和能力；在单一重金属铅污染下，白榆对铅元素的转运系数随着处理浓度的增加明显降低，由对照中的 1.03 降到处理浓度为 300mg/kg 时的 0.15（图1），说明随着土壤中重金属铅的增加，白榆通过增大根部对重金属元素的累积，降低转运系数，来保障其生理习性的正常。

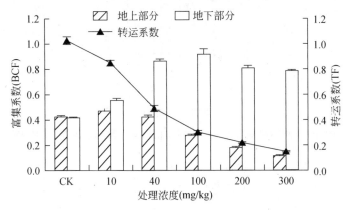

图1 不同水平铅处理对榆树富集系数和转运能力的影响

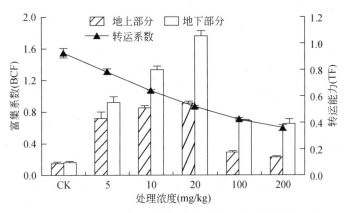

图2 不同水平镉处理对榆树富集系数和转运能力的影响

在重金属铅、镉复合污染下，铅、镉元素转移系数（TF）之间的交互作用非常明显（图3和图4）。在镉元素处理浓度为 100mg/kg 时，白榆对镉的转移系数是 0.38，而加入铅元素 40mg/kg、100mg/kg、200mg/kg 后，白榆对镉的转移系数增加到 0.55、0.60 和 0.73（图3），分别上升了 44%、58%、92%，因此铅、镉复合污染中，铅元素可促进镉元素在白榆植株内由下往上转移，当镉元素浓度相同时，植物体内镉元素转运系数随着土壤中铅浓度的增加而增加；镉对铅元素在白榆植株体内的转运系数具有促进作用，在铅投入浓度较低时（40mg/kg），镉元素的加入对铅元素的转运系数影响最大，提高两倍以上，但镉元素投入量的多少对铅的转移系数影响不明显，在铅投入浓度中等或较高时（100~200mg/kg），随着镉投入浓度的增加白榆对铅的转运系数逐步增加。

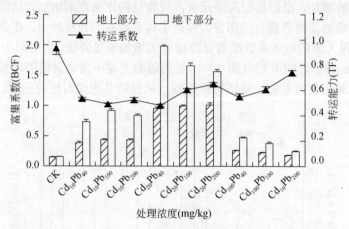

图3 不同铅镉复合处理中镉对榆树富集系数和转运能力的影响

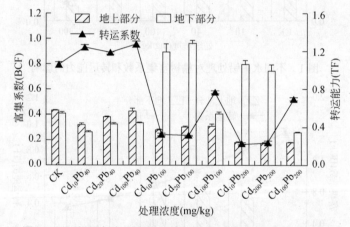

图4 不同铅镉复合处理铅对榆对富集系数和转运能力的影响

富集系数（BCF）是指植物不同器官元素含量与土壤中元素含量的比值，它表征的是元素从土壤迁移到植物各个器官的难易程度，这是反映植物将重金属吸收转移到体内能力大小的评价指标[11, 23]，富集系数越高，表明植物地上部重金属富集质量分数大。从图2可以看出，在镉元素单一污染时，地上部分与地下部分的富集系数随着投入量的增加而成先增后减的趋势，在投入量为20mg/kg时地上与地下部分的富集系数达到最大，对镉的富集能力最强；在铅元素单一污染中，地上部分的富集系数随着处理浓度的增加而减少，地下部分的富集系数随处理浓度的增加先增加到铅投入浓度为100mg/kg时的0.91最高值而后又呈降低趋势。在镉、铅复合污染中，当镉元素投入浓度相同时，白榆地上部分与地下部分对镉元素的富集能力随着铅投入量浓度增加而减小；在铅元素投入浓度较低时，镉元素的加入对地下部分铅富集能力影响不显著，对地上部分铅富集能力有一定的促进作用，在铅元素投入浓度较高时，镉元素的加入对地上部分铅富集能力具有抑制作用，对地下部分影响不显著。

3　结论

在目前重金属污染日趋严重情况下，减少和消除重金属污染的根本在于治理重金属污染源，控制工厂、矿山污染物的排放。对重金属污染区治理改造，这是当前国内外亟待解决的实际问题之一，但难度很大，其治理费用比改造污染源要大得多，目前国内外学者都在寻找超富集植物来解决重金属污染问题，这可很大程度上节省修复资金[24]，白榆作为榆林工矿区本土树种，它对重金属等污染表现为一种抗性、耐性，白榆受到不同浓度铅、镉单一污染后，各部位铅、镉的累积特性均表现为：根>茎>叶；随着土壤中重金属铅、镉的增加，转运系数逐渐降低，根部集积的铅、镉元素占吸收量的比重逐渐增加，有效保护了白榆叶，使其维持较为正常的光合生理活动；铅镉富集系数成先增后减趋势，说明在严重的铅污染处理下，白榆通过植株个体调节，可以有效防治白榆个体受铅的伤害。复合污染对白榆体内铅、镉含量有着不同的影响，在试验浓度范围内，镉的加入可以促进白榆叶和枝对铅的吸收，但对地下部分的影响则是呈先增后减的趋势，镉对叶、茎累积铅具有协同作用，对根累积铅无明显影响；在镉处理浓度一定时，铅的加入可以抑制茎、叶、根对镉的吸收，铅对根、茎、叶累积具有拮抗作用。白榆对铅、镉元素虽不是一种超富集植物，但对铅、镉元素表现出一种耐性，因此，白榆适用于铅、镉污染区的种植和绿化。

参 考 文 献

[1] 陈同斌. 重金属对土壤的污染. 金属世界，1999，(3)：10-11.

[2] 肖鹏飞，李法云，付宝莱，等. 土壤重金属污染及其植物修复研究. 辽宁大学学报（自然科学版），2004，31（3）：279-283.

[3] 韦朝阳，陈同斌. 重金属超富集植物及植物修复技术研究进展. 生态学报，2001，21（7）：1196-1203.

[4] 郭彬，李许明，陈柳燕，等. 土壤重金属污染及植物修复技术研究. 安徽农业科学，2007，35（33）：10776-10778，10781.

[5] 张连忠，路克国，杨洪强. 苹果幼树铜、镉分布特征与累积规律研究. 园艺学报，2006，33（1）：111-114.

[6] 李云，张进忠，童华荣. 茶苗对重金属 Pb、Cu、Cd 和 Cr 的吸收累积规律. 农业环境科学学报，2009，28（3）：454-459.

[7] 秦天才，吴玉树，王焕校. 镉、铅及其相互作用对小白菜生理生化特性的影响. 生态学报，1994，14（1）：46-50.

[8] 鲍士旦. 土壤农化分析. 北京：中国农业出版社，2000：22-380.

[9] 秦攀鑫，段婷婷. 测定土壤中重金属 Pb、Cd、Cu、Zn、Ni 前处理方法. 贵州师范大学学报（自然科学版），2005，23（2）：81-83.

[10] 杨艳芳，刘凤校，蔡彦明. 土壤样品的王水回流消解重金属测定方法的研究. 农业环境与发展，2005，22（4）：44-45.

[11] 栾以玲,姜志林,吴永刚. 栖霞山矿区植物对重金属元素富集能力的探讨. 南京林业大学学报(自然科学版), 2008, 32 (6): 69-72.

[12] 魏树,杨传杰,周启星. 三叶鬼针草等7种常见菊科杂草植物对重金属的超富集特征. 环境科学, 2008, 29 (10): 2912-2918.

[13] 兰海霞,夏建国. 川西蒙山茶树中铅、镉元素的吸收累积特性. 农业环境科学学报, 2008, 27 (3): 1077-1083.

[14] 张守文,呼世斌,肖璇,等. 油菜对Pb污染土壤的修复效应研究. 西北植物学报, 2009, 29 (1): 122-127.

[15] Van Assche F V, Clijster H. Effects of metal on enzyme activity in plants. Plant Cell Environment, 1990, 13: 195-206.

[16] 檀建新,尹君,王文忠,等. 镉对小麦、玉米幼苗生长和生理生化反应的影响. 河北农业大学学报, 1994, 17 (S1): 83-87.

[17] 阎雨平,蔡士悦,史艇. 广东赤红壤、红壤含镉的农作物污染效应及其临界含量研究. 环境科学研究, 1992, 5 (2): 49-53.

[18] 石元值,阮建云,马立峰,等. 茶树中镉、砷元素的吸收累积特性. 生态与农村环境学报, 2006, 22 (3): 70-75.

[19] 陆晓怡,何池全. 蓖麻对重金属Cd的耐性与吸收积累研究. 农业环境科学学报, 2005, 24 (4): 674-677.

[20] 肖昕,朱子清,王晨,等. 重金属Cd在小麦中的富集特征. 安徽农业科学, 2009, 37 (20): 9584-9585, 9605.

[21] 刘维涛,张银龙,陈喆敏,等. 矿区绿化树木对镉和锌的吸收与分布. 应用生态学报, 2008, 19 (4): 752-756.

[22] 何兰兰,角媛梅,王李鸿,等. Pb、Zn、Cu、Cd的超富集植物研究进展. 环境科学与技术, 2009, 32 (11): 120-123.

[23] 贾玉华. 三种植物对重金属Cd和Pb抗性及修复潜力的研究. 乌鲁木齐: 新疆农业大学硕士学位论文, 2008.

[24] 章家琪,吴燕玉. 镉污染与农业. 北京: 科学普及出版社, 1989: 31-42.

A coastal and an interior Douglas fir provenance exhibit different metabolic strategies to deal with drought stress[*]

Du Baoguo   Kirstin Jansen   Anita Kleiber   Monika Eiblmeier

Bernd Kammerer   Ingo Ensminger   Arthur Gessler

Heinz Rennenberg   Jürgen Kreuzwieser

Abstract

Drought is a major environmental stress affecting growth and vitality of forest ecosystems. In the present study, foliar nitrogen(N) and carbon(C) metabolism of two Douglas fir (*Pseudotsuga menziesii*) provenances with assumed different drought tolerance were investigated. We worked with one-year-old seedlings of the interior provenance Fehr Lake (FEHR) originating from a dry environment and the coastal provenance Snoqualmie (SNO) from a more humid origin. Total C and N, structural N, and the concentrations of soluble protein, total amino acids (TTAs) and individual amino acids as well as the relative abundance of polar, low-molecular-weight metabolites including antioxidants were determined in current-year needles exposed either to 42 days of drought or to 42 days drought plus 14 days of re-watering. The seedlings reacted in a provenance specific manner to drought stress. Coastal provenance SNO showed considerably increased contents of TAAs, which were caused by increased abundance of the quantitatively most important amino acids arginine, ornithine, and lysine. Additionally, the polyamine putrescine accumulated exclusively in drought stressed trees of this provenance. In contrast, the interior provenance FEHR

[*] 原载于：Tree Physiology, 2015, 36（2）：148-163.

showed the opposite response, i.e. drastically reduced concentrations of these amino acids. However, FEHR showed considerably increased contents of pyruvate derived and aromatic amino acids, but also higher drought induced levels of the antioxidants ascorbate and α-tocopherol. In response to drought, both provenances produced large amounts of carbohydrates, such as glucose and fructose, most likely as osmolytes that can readily be metabolized for protection against osmotic stress. We conclude that FEHR and SNO cope with drought stress in a provenance specific manner: The coastal provenance SNO was mainly synthesizing N based osmolytes, a reaction not observed in the interior provenance FEHR; instead, the latter increased the levels of scavengers of reactive oxygen species (ROS). Our results underline the importance of provenance-specific reactions to abiotic stress.

Keywords: Douglas fir (*Pseudotsuga menziesii*); drought; amino acid; foliar nitrogen partitioning; osmolytes; polyols; polyamines; antioxidants

1 Introduction

Global climate change will cause considerable changes in precipitation amount and patterns together with rising air temperature. In Europe, the amount of summer precipitation is projected to decrease by 9%-41% until 2070 [1]. Decreasing precipitation together with increased air temperature will have significant effects on growth, development and distribution of trees, and will alter forest ecosystem and bio-geophysical processes [2]. Climate change will therefore affect forest productivity and tree performance [3], as already observed during the extreme heat and drought period in Europe in summer 2003 [4].

Cultivation of more resistant and resilient tree species or provenances at sites characterized by increasingly occurring drought events is considered. In Central Europe, Douglas fir (*Pseudotsuga menziesii*) is discussed as an alternative to Norway spruce and Scots pine because it seems to cope better than these species with drought as indicated, for example, by the maintenance of relatively high radial growth under conditions of water deficit [5-7]. However, within its wide distribution in America's Pacific Northwest (USA, Canada) [8], Douglas fir has evolved a great variety of different provenances adapted to the specific environmental conditions in different habitats [9]. Both, genetic and physiological variations in response to drought and high temperature stresses have been documented for the two major geographic races (the coastal var. *menziesii* and the interior or Rocky Mountain var. *glauca*) and among different provenances within these races [10-15].

Soil water availability affects almost all processes underlying forest tree growth [16]. Clearly, photosynthetic CO_2 assimilation is impaired under drought, whereas stored non-structural carbohydrates have been revealed as key determinants to cope with drought stress in plants. For example, soluble sugars such as disaccharides, oligosaccharides of the raffinose family, fructans and polyols play multiple roles as substrates in energy metabolism, as cellular osmolytes, signal molecules, and scavengers of reactive oxygen species (ROS) under drought stress [17-20]. Specific responses to drought were not only observed in carbohydrates but also in nitrogen (N) containing compounds. Water availability is supposed to influence N mineralization and N uptake [21], thereby affecting tree growth and physiological processes. Foliar N metabolism may control photosynthetic responses of trees to climate change, but also may be altered by climate change [22]. Several studies showed that foliar N contents and labile N compounds of tree leaves, such as amino acids and soluble proteins, are strongly affected by drought [23-25] and population or provenance specific responses were demonstrated in some tree species [25-27]. However, compared to studies on the effects of elevated temperature and / or CO_2 contents [22,28-31] and fertilization [32-37] on N metabolism, the knowledge of drought effects on N metabolism of Douglas fir provenances with contrasting environmental origin is scarce.

Following the 2003 drought and heat wave over Europe, an increased decline and dieback of Douglas-fir has been reported. As a higher frequency of drought episodes is predicted in future, a deep understanding of the drought resistance capacity of Douglas-fir and its variation between different provenances is urgently needed [38,39]. Elucidation of C and N metabolism of Douglas fir as well as provenance specific variations under drought stress is obviously of fundamental importance to better understand physiological responses of this species to water shortage, but also to provide useful information for the selection of suitable provenances for future forest management [40].

The central objective of this study was to characterize important features of the C and N metabolism as affected by water deficit in seedlings of two contrasting provenances of Douglas fir. For this purpose, we performed an experiment under controlled conditions in which we compared the drought induced reactions of the interior provenance Fehr Lake originating from a relatively dry habitat with the coastal provenance Snoqualmie originating from a humid habitat. Seedlings of both provenances were grown under reduced soil water availability for six weeks, followed by an additional period of re-watering. In the seedlings of the two provenances, we analyzed the concentrations of major compounds of the C and N metabolism as well as the levels of the antioxidants ascorbate and glutathione in current year needles. We hypothesized that ① the six weeks of drought lead to changes in C and N partitioning and the levels of antioxidants in both provenances and that re-watering of seedlings can partially reverse or alleviate the observed drought response; ② the responses vary between the provenances, with the interior provenance Fehr Lake from the drier origin being less affected by drought than the coastal variety Snoqualmie. We further tested the hypotheses that ③ the assumed more drought tolerant interior provenance possesses a higher capacity to quench ROS and a stronger potential to maintain plant water status by accumulation of osmoprotectants compared with the coastal provenance, whereas ④the latter provenance will show indications of impaired primary metabolism due to drought.

2 Materials and methods

2.1 Douglas fir provenances and acclimation

One-year-old seedlings of two provenances of Douglas fir originating from contrasting habitats were used for this experiment, i.e., the provenance Fehr Lake (FEHR, seedlot: FDI 39481) from a relatively dry original habitat with an annual precipitation of 333 mm and Snoqualmie [SNO, seedlot: pme 07(797) 412-10] from a more humid habitat with nearly seven times higher annual precipitation (for further information see Table 1). Seedlings of FEHR and SNO were purchased from commercial nurseries, i.e., BC Timber Sales (Vernon, Canada) and Forestry Commission, Wykeham Nursery (Sawdon, Great Britain), respectively. The seedlings were planted in 3 L square-shaped pots filled with commercial substrate (Container substrate 1 medium+ GreenFibre basic, pH 5.3, Klasmann-Deilmann GmbH, Geeste, Germany) and NPK fertilizer (N170 + P200 +K230 +Mg100 +S150 mg/L) directly after arrival (FEHR: March 8, SNO: April 14, 2011) at the Leibniz Centre for Agricultural Landscape Research in Müncheberg, Germany, where the experiment took place. The seedlings were placed into two walk-in environmental chambers (VB 8018, Vötsch Industrietechnik GmbH, Balingen-Frommern, Germany). The light period lasted for 16 h from 5:00 to 21:00, and light intensity in the canopy was held at 500 μmol/(m^2/s) PPFD with metal halide lamps (Powerstar HQI-BT 400 W/D PRO Daylight, Osram GmbH, Munich, Germany). The temperature was kept at 21 ℃ during day and night at a relative humidity of 70%. Seedlings were allowed to acclimate for three months and all pots were watered daily with 100 ml mixed tap water/distilled water (1/1, v/v). On July 20, at the end of the acclimation period, half of the plants were shifted to drought conditions and the other half was used as controls. Bud swelling started on March 21 and April 16 on seedlings of FEHR and SNO, respectively. The second flush development of saplings of the two provenances was finished before the drought treatment was started.

Table 1 Description of the original sites of the two selected Douglas fir provenances

Name	Fehr Lake (FEHR)	Snoqualmie (SNO)
Province/State of origin	British Columbia (BC), Canada	Washington, USA
Region	Thompson Okanogan Valley in Southern interior BC, near Fehr Lake, interior	North Cascades west side, zone Snoqualmie, coastal

Continued

Name	Fehr Lake (FEHR)	Snoqualmie (SNO)
Geographical coordinates	N50.71,W120.86	N46.83-48.05, W121.97-120.79
Elevation (m)	800	457-610
Mean annual temperature (°C)	5.80	7.94
Sum of annual precipitation (mm)	333.0	2134.2
Mean of summer (May to Sept.) precipitation (mm)	162	364.6
AHM ①	47.3	9.04
SHM ②	105.4	52.62

Note: ① AHM, annual heat-moisture index (MAT+10)/(MAP/1000); MAT, mean annual temperature (°C); MAP, mean annual precipitation (mm); ② SHM, summer heat-moisture index (MWMT)/(MSP/1000) ; MWMT, mean warmest month temperature (°C); MSP-mean annual summer (May to Sept.) precipitation (mm) [50, 51].

2.2 Drought stress treatment and needles sampling

Control plants of each provenance were watered daily according to soil moisture measurements with tap water/distilled water (1/1) to keep a soil volumetric water content of around 37%. In contrast, the water supply of drought exposed trees was reduced stepwise and completely stopped on August 10. A time course of soil water content was measured regularly during the experiment (see Supplementary Fig. 1). After 42 days of stress exposure, current year needles of five trees of each provenance and treatment were harvested. Remaining five drought stressed trees of both provenances were re-watered fully and supplied with 200 ml water per day. After 14 days of re-watering, current year needles from recovered and control seedlings of the two provenances were harvested. During the experiment, soil moisture content was determined by EC-5 sensors (Decagon Devices, Inc., Pullman, USA). In addition, predawn twig water potential was determined with the method originally described by Scholander et al. [41]. as modified by Rennenberg et al. [42]. Needles excised from twigs were immediately shock-frozen in liquid N_2, homogenized with mortar and pestle in liquid N_2, and stored at −80℃ until analysis.

2.3 Foliar amino acids and NH_4^+ analysis

Amino compounds and ammonium were extracted and determined as described by Winter et al. [43]. Compounds were extracted from 50 mg frozen needle powder, freeze-dried and analyzed using an ultra-performance-liquid-chromatography (UPLC) system (Waters Corp., Milford, MA) at a wave length of 260 nm as described by Luo et al. [44]. Amino acid standards were analyzed under identical conditions for quantification. For the quantification of total amino acids, the same extracted solutions were analysed colorimetrically at 570 nm

(UV-DU650 Spectrophotometer, BECKMAN, USA) after derivatization with ninhydrin reagent as described by Du et al.[45]. Concentrations were calculated according to a standard curve of different glutamine concentrations.

2.4 Soluble protein analysis

Determination of total soluble protein was performed as described by Du et al.[45]. Polyvinylpolypyrrolidone (PVPP) was added at a ratio of 3∶1 to the needle powder (w/w) which was suspended in 1.5 ml 50 mM Tris–HCl extraction buffer. Soluble protein contents were calculated from the absorbance of the mixture of 5 μl extract with 200 μl Bradford reagent (Amresco Inc., Solon, Ohio, USA) measured at 595 nm on a 96-well micro plate ELISA reader (Sunrise Basic, Tecan, Austria). Bovine serum albumin (BSA; Sigma-Aldrich Chemie GmbH, Germany) was used as reference for quantification.

2.5 Total N determination and structural N calculation

Total N content in oven dried needle powder was determined with an elemental analyzer (NA 2500, CE Instruments, Milan, Italy)[45]. Structural N content in needles was calculated by subtracting the N fractions of soluble proteins, amino acids and ammonium N from total N in leaf bulk material. As chlorophyll N contents were quite low compared to other N compounds in Douglas fir seedlings[45], they were not considered in structural N calculation in this study.

3 Determination of the levels of antioxidants

3.1 Nontargeted analysis of water soluble metabolites

Polar low-molecular-weight metabolites were extracted from current year needles and derivatised according to a modified method of Kreuzwieser et al.[46]. For each sample, approximately 50 mg of homogenized frozen tissue powder were weighed into a pre-frozen 2ml Eppendorf tube and 500μl of cold 85% (v/v) methanol were added as extraction medium. 10μl 100ng/μl ribitol was added as an internal standard. Tubes were rapidly heated to 65℃ and shaken at 1400rpm for 15min. After centrifugation, 50μl supernatant were transferred to 1.5ml microfuge tubes and dried under vacuum. Dried extracts were methoximated by adding 20μl of a solution containing 20mg/ml methoxyamine hydrochloride in anhydrous pyridine (Sigma); the mixtures were incubated at 30℃ for 90min and shaken at 1400rpm. For trimethylsilylation, 70μl N-methyl-N-(trimethylsilyl) trifluoroacetamide (MSTFA; Sigma) was added to each tube and incubated at 37℃ for 30min and shaken at 1400rpm. 10μl n-alkane retention index calibration standard (n-alkane-mix, C10−C40, 50μg/ml in n-hexane; Sigma) was added to each sample. After short vortexing, reaction mixtures were centrifuged

(14 000g, 20℃) for 2 min and 80μl supernatant were transferred to amber GC-MS vials with low volume inserts and screw top seals (Agilent Technologies, Palo Alto, CA, USA) for GC-MS analysis.

The derivatised metabolite samples were analysed on an Agilent GC/MSD system consisting of an Agilent GC 7890A gas chromatograph (Agilent Technologies, Palo Alto, CA, USA) equipped with a Gerstel MultiPurpose Sampler (MPS2-XL, Gerstel, Mülheim, Germany) and connected to a 5975C Inert XL EI/CI MSD quadrupole MS detector (Agilent Technologies, Palo Alto, CA, USA). Separation occurred on a capillary column (HP-5MS 5% Phenyl Methyl Silox, length: 30m, diameter: 0.25mm, film thickness: 0.25μm; Agilent Technologies, Palo Alto, CA, USA) at a helium flow of 1ml/min. GC-MS run conditions were set up according to Erxleben et al.[47] with some changes. The initial oven temperature was held at 80℃ for 2min, and then increased to 325℃ at 5℃/min before being held at 325℃ for 10min. Total run time was 61min. Transfer line and MS source temperatures were adjusted to 280℃ and 230℃, respectively. Electron impact ionisation energy was 70 eV and the MS detector was operated in full scan mode in the range of 83 to 500m/z at 3 scans per second.

The raw data files were processed with freely available AMDIS (automated mass spectral deconvolution and identification system, version 2.69) software supplied by NIST (National Institute of Standards and Technology, Gaithersburg, MD, USA). Mass spectra were searched against the Golm Metabolome Database[48] and compounds were identified based on spectrum similarity match and retention index. For relative quantification the peak areas for each metabolite were first standardized for sample weight, then normalized using the internal ribitol standard, and finally corrected for blank values. Metabolites detected in less than 3 of all 5 replicates of each provenance were discarded. Treatment specific differences of the corrected and normalized peak areas are given as ratios of treatments versus controls.

3.2　Needle dry mass determination

Frozen needle powder was dried in 1.5ml reaction tubes at 60℃ until constant weight was achieved. All biological parameters were based on dry mass unless indicated otherwise.

3.3　Statistical analysis

The software package SigmaPlot 11.0 (Systat Software GmbH, Erkrath, Germany) was applied for statistical analysis and figure generating. Student's t-test was performed when only two groups were compared, i.e., the differences between two provenances within treatment and the differences between treatments within provenance. To ensure that data of metabolite concentrations matched normal distribution in order to perform t-tests, data were transferred by denary logarithm. For data still failing normality test after normalization, the t-test on Ranks was employed to determine the significance of differences. Partial least square discriminant analysis (PLS-DA) was conducted using the Multibase Excel Add-in for PCA (http：//www.numericaldynamics.com/)[49]. Data shown in figures and tables represent means ± standard error (*n*=3-5).

4 Results

4.1 Plant water status

After 42 days of reduced water supply, soil volumetric water content was below 2.3% in representative pots of both provenances (see Supplementary Fig. 1), which was well below the permanent wilting point; re-watering quickly recovered soil water content and 14 days later, control levels (around 38%) were reached. Pre-dawn water potential of control plants amounted to −0.98MPa and did not change during the course of the experiment. Reduced water supply caused significantly altered twig water potentials at the end of the drought period of −2.7±0.12MPa and −2.4±0.32 in SNO and FEHR, respectively, indicating strong drought stress in both provenances. Re-watering for 14 days completely recovered plant water potential reaching values of −0.7±0.08MPa and −0.6±0.04MPa in SNO and FEHR, respectively. Provenance specific differences in plant water potential during the course of the drought period were not observed (data not shown).

4.2 Drought effects on total C and N contents

There was a trend to higher total C contents under drought stress, which was significant in FEHR (Fig. 1(a)), see Supplementary Table 1 for statistical results); this effect was lost after re-watering the trees. No significant effects of drought were, however, observed on total N contents (Fig. 1(b)) and C/N ratios (Fig. 1(c)). Drought stress caused higher total amino acid (TAA) contents in SNO, as values increased from 16.8 μmol/g DW to 120.6μmol/g DW (Fig. 1(d)). No consistent trends of drought and re-watering effects were observed for the contents of soluble protein and structural N (Fig. 1(e), (f)).

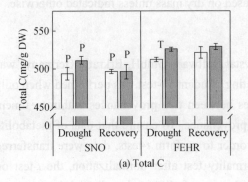
(a) Total C

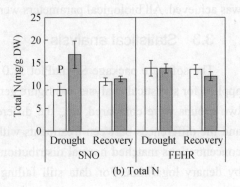
(b) Total N

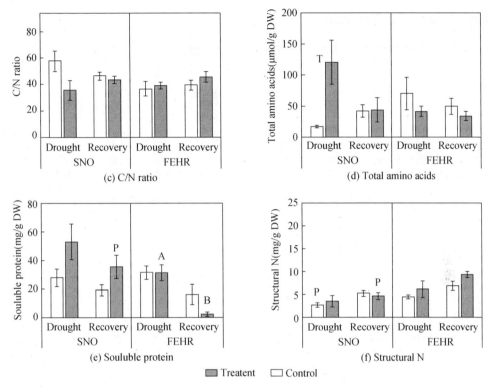

Fig. 1 Effect of drought and re-watering on total carbon (C), total nitrogen (N), C/N ratio, total amino acids, soluble protein and structural N contents in needles of the Douglas fir provenances SNO (right panels) and FEHR (left panels)

Note: Means ± standard error of three to five replicates are shown. P, significant provenance specific differences at same treatments; T, treatment specific differences in same provenances. Different upper case letters indicate significant differences ($P<0.05$) between drought stressed and re-watered trees of one provenance.

4.3 Provenance specific effects of drought on the concentrations of individual amino acids

Irrespective of provenance, arginine, ornithine, lysine, glutamate, proline and aspartate were quantitatively the most important amino acids (Fig. 2). Together these amino acids contributed ca. 85% to TAA-N of the trees. Among these amino acids, arginine dominated in all treatments of both provenances, representing up to 58% of the TAA-N pool (Fig. 2(a)). Drought stress significantly affected the concentrations of arginine, ornithine and lysine in the trees' needles, but in a provenance dependent way (Fig.2(a)-(c), Fig.3). Whereas significantly increased arginine, ornithine and lysine contents were found in drought stressed trees of SNO, the concentrations of these amino acids decreased in FEHR. Re-watering completely removed the differences between previously stressed trees and controls (Fig. 2(a)-(c), Fig. 3). In SNO, re-watering even caused lower concentrations of arginine, ornithine, lysine and aspartate than in control plants (Fig. 2(a)-(c), (f)).

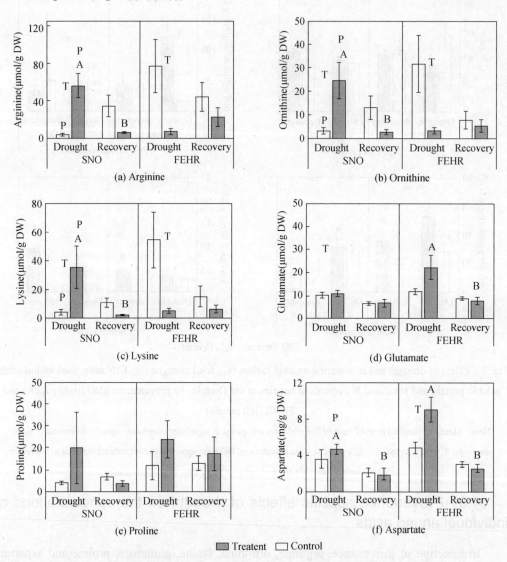

Fig. 2 Effect of drought and re-watering on amino acid concentrations in needles of the Douglas fir provenances SNO (right panels) and FEHR (left panels)

Note: Means ± standard error of three to five replicates are shown. P, significant provenance specific differences at same treatments; T, treatment specific differences in same provenances. Different upper case letters indicate significant differences ($P<0.05$) between drought stressed and re-watered trees of one provenance.

Moreover, only in FEHR, drought caused considerably increased levels of the aromatic amino acids tryptophan, tyrosine and phenylalanine as well as the pyruvate derived amino acids valine, leucine and isoleucine (Fig.3, Fig.4(a)-(f)). Ammonium concentrations also significantly increased from 8.7 to 29.5μmol/g DW in FEHR due to drought (Table 2). After re-watering the drought induced differences of these amino acid and ammonium concentrations to control levels disappeared (Fig. 3).

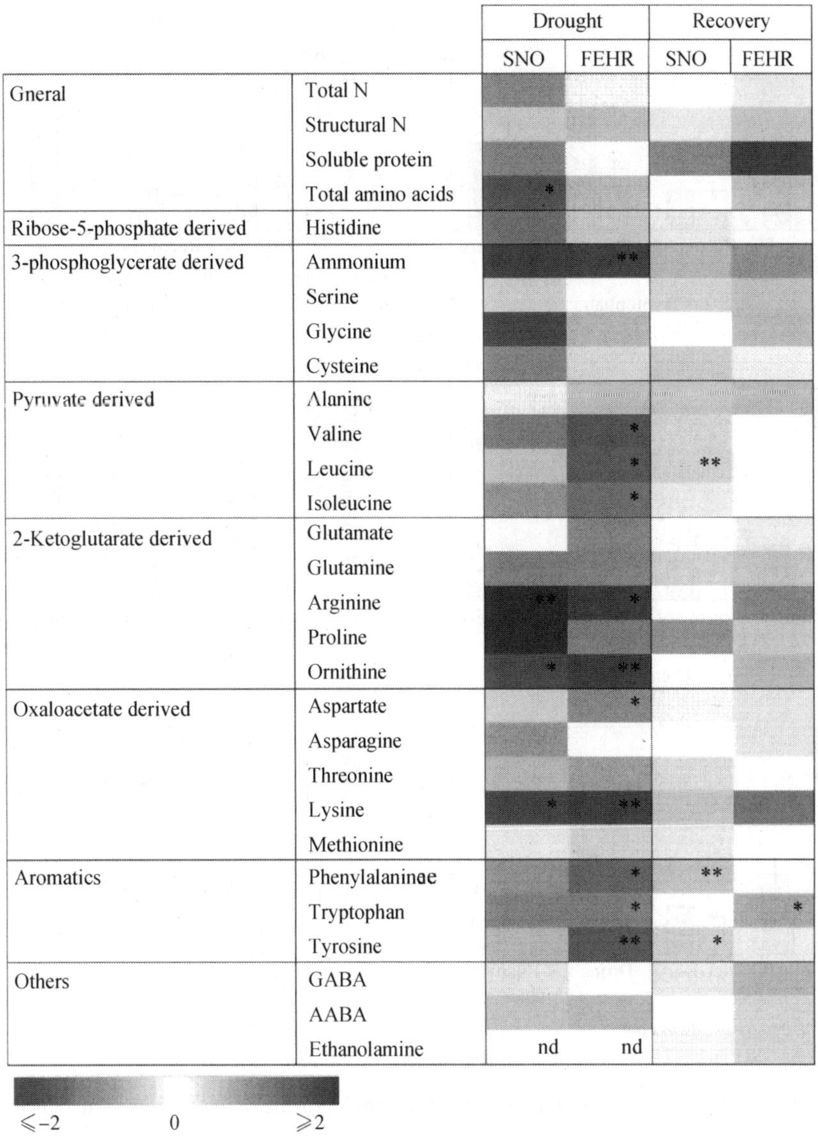

		Drought		Recovery	
		SNO	FEHR	SNO	FEHR
Gneral	Total N				
	Structural N				
	Soluble protein				
	Total amino acids	*			
Ribose-5-phosphate derived	Histidine				
3-phosphoglycerate derived	Ammonium		**		
	Serine				
	Glycine				
	Cysteine				
Pyruvate derived	Alanine				
	Valine		*		
	Leucine		*	**	
	Isoleucine		*		
2-Ketoglutarate derived	Glutamate				
	Glutamine				
	Arginine	**	*		
	Proline				
	Ornithine	*	**		
Oxaloacetate derived	Aspartate		*		
	Asparagine				
	Threonine				
	Lysine	*	**		
	Methionine				
Aromatics	Phenylalaninae		*	**	
	Tryptophan		*		*
	Tyrosine		**	*	
Others	GABA				
	AABA				
	Ethanolamine	nd	nd		

⩽-2　　0　　⩾2

Fig. 3　Effect of drought and re-watering on fold-changes of foliar N metabolites of the two Douglas fir provenances SNO and FEHR

Note: The colour code indicates the $\log_2$ of fold-changes of the metabolite concentrations of treated trees vs. control trees. Blue and red colours indicate increased and decreased concentrations compared to controls, respectively. Significant differences between treatments and controls at $P<0.05$ and 0.01 are indicated by * and ** respectively; nd, value below detection limit.

(见彩图 A)

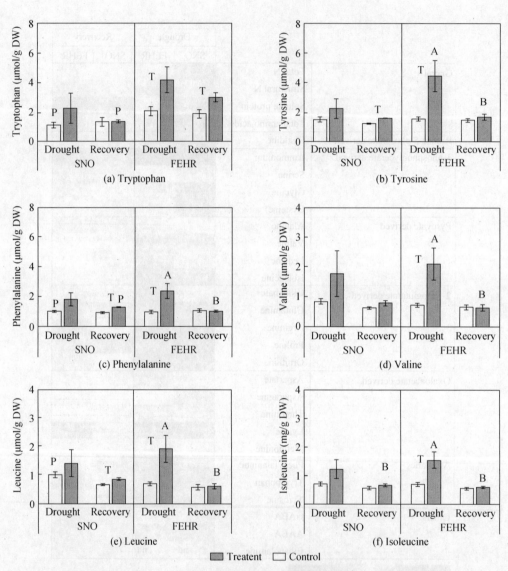

Fig. 4 Effect of drought and re-watering on aromatic and pyruvate derived amino acid concentrations in needles of the Douglas fir provenances SNO (right panels) and FEHR (left panels)

Noet: Means ± standard error of three to five replicates are shown. P, significant provenance specific differences at same treatments; T, treatment specific differences in same provenances. Different upper case letters indicate significant differences ($P<0.05$) between drought stressed and re-watered trees of one provenance.

Table 2 Effect of drought and re-watering on the content of N compounds in needles of the Douglas fir provenances SNO and FEHR

Compounds		SNO (μmol/g · DW)		FEHR (μmol/g · DW)	
		Control	Treatment	Control	Treatment
Ammonium	D	16.63±4.71 a	**58.17±27.48**	8.66±0.55 b	**29.50±7.08**
	R	9.20±1.09	**6.25±2.43**	10.22±1.53	**5.55±1.40**
Serine	D	2.43±0.54	1.97±0.27	2.38±0.26	2.40±0.28
	R	1.75±0.12	2.19±0.21	2.67±0.33	1.94±0.24
Glycine	D	1.27±0.16	4.16±2.48	2.70±0.82	1.78±0.25
	R	1.23±0.32 a	1.20±0.39	1.61±0.92 b	1.10±0.23
Histidine	D	2.16±0.11	4.83±1.72	5.40±1.56	**4.21±0.72**
	R	1.91±0.32	2.18±0.65	2.09±0.59	**2.26±0.41**
Alanine	D	2.48±0.39	2.25±0.49	2.24±0.13	**3.16±0.37**
	R	2.03±0.30	1.54±0.03	2.36±0.47	**1.65±0.32**
Glutamine	D	1.50±0.29	3.03±0.61	1.86±0.40	3.54±1.10
	R	1.36±0.09	1.09±0.14	1.90±0.25	1.43±0.36
Asparagine	D	0.77±0.06	1.35±0.36	0.84±0.19	0.83±0.06
	R	1.17±0.28	1.17±0.21	1.03±0.42	0.89±0.21
Threonine	D	**1.44±0.14**	2.10±0.71	1.28±0.11	**2.10±0.35**
	R	**1.05±0.07**	1.29±0.14	1.20±0.12	**1.13±0.13**
Methionine	D	0.98±0.10	1.12±0.20	0.93±0.07	1.16±0.17
	R	0.91±0.07	1.08±0.07	0.86±0.09	0.81±0.07
GABA	D	**1.10±0.11**	0.94±0.25	**1.18±0.22**	1.21±0.12
	R	**0.31±0.07 a**	0.27±0.05 A	**0.60±0.06 b**	**0.92±0.27 B**
AABA	D	1.62±0.44	2.29±0.71	0.95±0.13	**1.45±0.56**
	R	0.77±0.16 a	0.83±0.27 A	0.56±0.14 b	**0.45±0.06 B**

Note: Shown are Means ± standard error of three to five replicates. Different lower case or upper case letters indicate significant provenance specific differences ($P < 0.05$). Asterisks indicate significant differences ($P < 0.05$) between control and treatments of each provenance. Bold indicates significant differences ($P < 0.05$) between drought stressed (D) and re-watered (R) trees. GABA, γ-amino butyric acid; AABA, α-aminobutyric acid.

4.4 Provenance specific effects of drought on the concentrations of antioxidants

There was a distinct trend for higher concentrations of glutathione and its precursors cysteine and γ-glutamyl-cysteine in drought stressed SNO compared to control trees (Fig. 5(a)-(c), (e)). This effect was not observed in FEHR. In this provenance, however, total ascorbate and reduced ascorbate concentrations increased in response to drought (Fig. 5(f), (g)). This pattern was not seen in the coastal provenance SNO. Interestingly, FEHR contained significantly higher amounts of both total and reduced ascorbate than SNO. Neither in SNO nor FEHR, the ratios of oxidized to reduced glutathione and dehydroascorbate were affected by drought (detailed statistical analysis results see Supplementary Table 2).

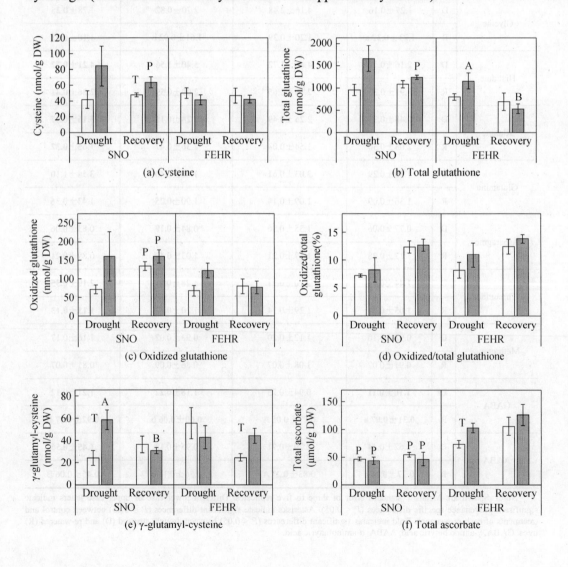

(a) Cysteine
(b) Total glutathione
(c) Oxidized glutathione
(d) Oxidized/total glutathione
(e) γ-glutamyl-cysteine
(f) Total ascorbate

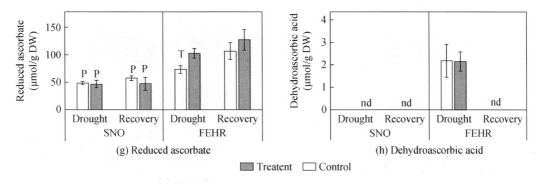

Fig. 5 Effect of drought and re-watering on thiols (cysteine, γ-glutamyl-cysteine, glutathione) and ascorbate concentrations in needles of the Douglas fir provenances SNO (right panels) and FEHR (left panels)

Note: Means ± standard error of three to five replicates are shown. P, significant provenance specific differences at same treatments; T, treatment specific differences in same provenances. Different upper case letters indicate significant differences ($P<0.05$) between drought stressed and re-watered trees of one provenance. nd, value below detection limit.

4.5 Effects of drought stress on other metabolites

Besides the amino acids mentioned above, we studied the relative contents of organic acids, carbohydrates, polyols, phosphates, polyamines and some other metabolites in the foliage of the two Douglas fir provenances (Fig. 6). Among these compounds, sucrose, quinic acid, shikimic acid and D-pinitol were most abundant. However, the most pronounced drought effects were on the carbohydrates lyxose, fructose, galactose, glucose and raffinose, which showed considerably increased concentrations compared to the controls. Similar to the observation for amino acids, the provenance FEHR reacted much stronger to water deficit than SNO (Fig. 6). The concentrations of the organic acids shikimic acid, quinic acid and gallic acid slightly increased in the needles of drought stressed plants of both provenances. Citrate, malate, benzoate and threonic acid contents were increased in needles of SNO, but decreased in FEHR. Increased concentrations of polyols (D-pinitol, threitol and sequoyitol in FEHR) and glycerol-3-phosphate, putrescine, 2-hydroxy-pyridine, β-sitosterol and three unidentified peaks (NP, not identified peak; NP4, NP10 and NP12) were also found in plants subjected to drought stress (Fig. 6). Moreover, other compounds (i.e., threonic acid, kestose, threitol, glycerol-3-phosphate, mannose-6-phosphate and the unknown compound NP11) were detected only in needles of stressed plants while gulonic acid was exclusively found in controls (Fig. 6). PLS-DA analysis based on all metabolite data plus total C and N contents showed clear clustering in a treatment (Supplementary Fig. 2(a), (e)) and a provenance (Supplementary Fig. 2 (c), (e)) specific manner. As suggested by the respective loading plots, fructose, glucose, unknown compound NP4 and isoleucine were mainly responsible for the treatment specific separation (Supplementary Fig. 2(b), (f)), whereas total and reduced ascorbate, phosphoric acid, β-sitosterol, myo-inositol, structural N and some unknown compounds were mainly responsible for the provenance specific pattern (Supplementary Fig. 2 (d)).

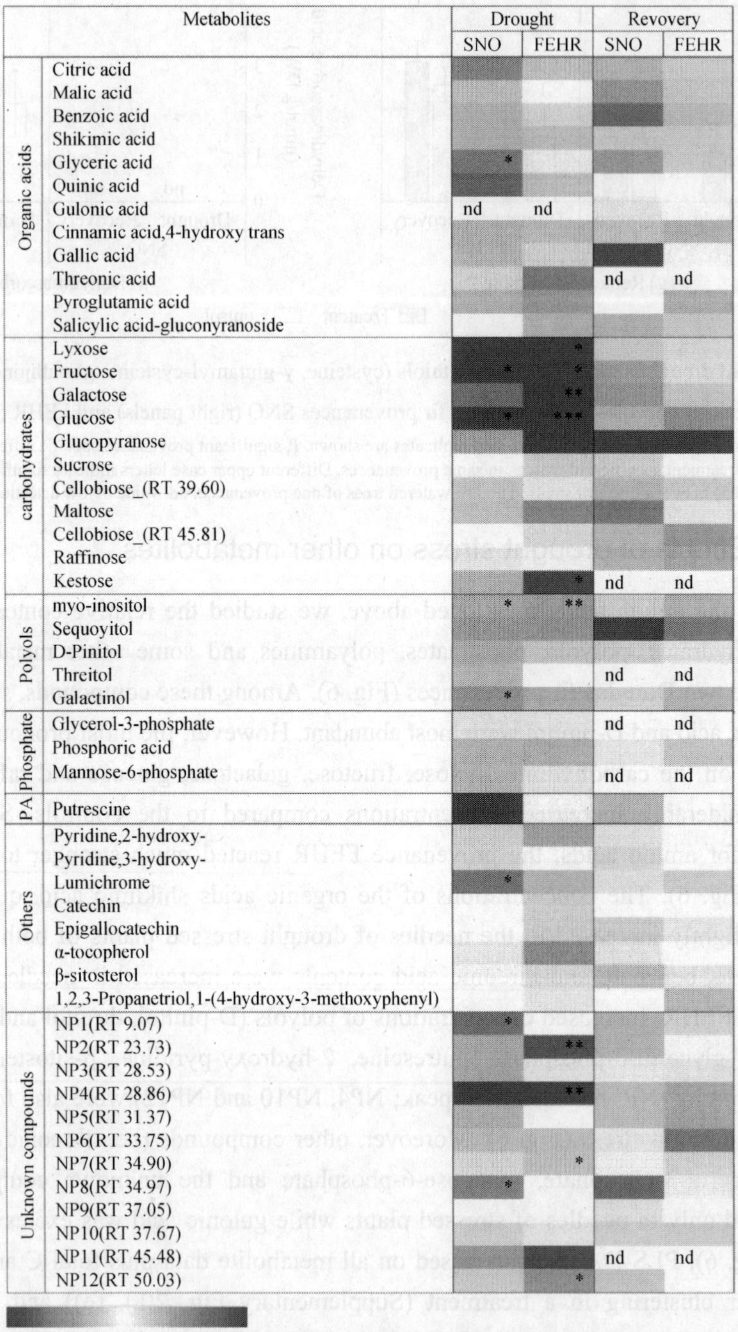

	Metabolites	Drought		Revovery	
		SNO	FEHR	SNO	FEHR
Organic acids	Citric acid				
	Malic acid				
	Benzoic acid				
	Shikimic acid				
	Glyceric acid	*			
	Quinic acid				
	Gulonic acid	nd	nd		
	Cinnamic acid,4-hydroxy trans				
	Gallic acid				
	Threonic acid			nd	nd
	Pyroglutamic acid				
	Salicylic acid-gluconyranoside				
carbohydrates	Lyxose	*	*		
	Fructose		*		
	Galactose		**		
	Glucose	*	***		
	Glucopyranose				
	Sucrose				
	Cellobiose_(RT 39.60)				
	Maltose				
	Cellobiose_(RT 45.81)				
	Raffinose				
	Kestose		*	nd	nd
Polyols	myo-inositol	*	**		
	Sequoyitol				
	D-Pinitol				
	Threitol			nd	nd
	Galactinol	*			
PA Phosphate	Glycerol-3-phosphate			nd	nd
	Phosphoric acid				
	Mannose-6-phosphate			nd	nd
PA	Putrescine				
Others	Pyridine,2-hydroxy-				
	Pyridine,3-hydroxy-				
	Lumichrome	*			
	Catechin				
	Epigallocatechin				
	α-tocopherol				
	β-sitosterol				
	1,2,3-Propanetriol,1-(4-hydroxy-3-methoxyphenyl)				
Unknown compounds	NP1(RT 9.07)	*			
	NP2(RT 23.73)		**		
	NP3(RT 28.53)				
	NP4(RT 28.86)		**		
	NP5(RT 31.37)				
	NP6(RT 33.75)				
	NP7(RT 34.90)			*	
	NP8(RT 34.97)	*			
	NP9(RT 37.05)				
	NP10(RT 37.67)				
	NP11(RT 45.48)			nd	nd
	NP12(RT 50.03)		*		

≤-2   0   ≥2

Fig. 6　Effect of drought and re-watering on fold-changes of foliar low-molecular-weight metabolite in needles of the two Douglas fir provenances SNO and FEHR

Note: The colour code indicates the log$_2$ of fold-changes of the metabolite concentrations of treated trees vs. control trees. Blue and red colours indicate increased and decreased concentrations compared to controls, respectively. Significant differences between treatments and controls at $P<0.05$, 0.01 and 0.001 are indicated by *, **, and *** respectively. PA, polyamines; NP, not identified peak; nd, value below detection limit; RT in parentheses, retention time in minute.

(见彩图 B)

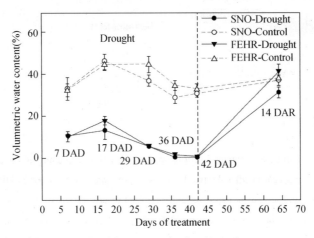

Supplementary Fig. 1 The time course of soil volumetric water contents during the experiment for controls and stressed individuals of the two Douglas fir provenances SNO and FEHR at different days after drought (DAD) and after re-watering (DAR).

Note: Means ± standard error of three replicates are shown.

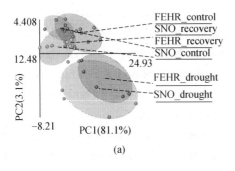

(a)

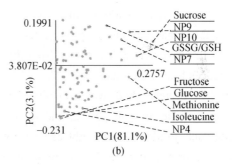

(b)

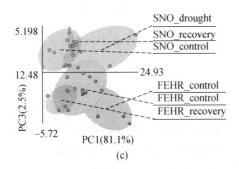

(c)

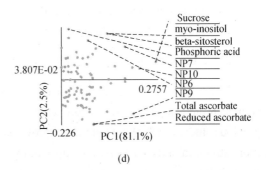

(d)

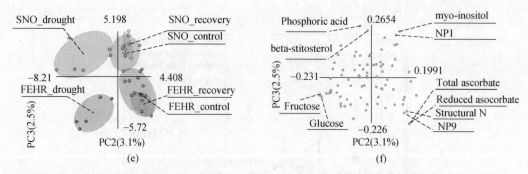

Supplementary Fig. 2 Partial least square discriminant analysis (PLS-DA) using all metabolites determined in current year needles the two Douglas fir provenances SNO and FEHR.

Note: Spots indicate individual plants of the respective treatment (control, drought and recovery). Clustering of samples from each provenance within treatment categories is color-coded. (a), (c), (e): Bi-plots of principal components 1 and 2 representing 84.2 %, principal components 1 and 3 representing 83.6 % and principal components 2 and 3 representing 5.6 % of the total variance observed in the metabolite profiles of Douglas fir needles, respectively.(b), (d), (f): Loadings of individual identified components showing the highest variability under the studied treatments.

5 Discussion

In this study, we worked with provenance SNO originating from a coastal and therefore wetter habitat than the interior provenance FEHR (Table 1). It is often proposed that interior Douglas fir provenances originating from dry habitats are more drought tolerant than coastal provenances from wetter habitats[10,14,15,39]. However, this assumption was not fully supported by our data, because shoot water potential was affected in the same way in both provenances at the end of the drought period, and recovery of the water status also occurred similarly. Despite such similarities, drought caused distinct changes in the C and N partitioning and concentrations in many metabolites including antioxidants, which was widely reversible after re-watering, thus supporting our hypothesis 1. Many of such effects were provenance specific as stated in our hypothesis 2, however, in contrast to our assumption, we did not observe more dramatic changes in the coastal than in the interior provenance.

5.1 Provenance specific characteristics

Generally, needles of the interior provenance FEHR contained more structural N but less soluble protein than the coastal provenance SNO (Fig. 1). This finding is consistent with a previous study, in which a Douglas fir provenance from a drier habitat contained considerably more structural N and less soluble protein in previous year needles than a provenance from a mesic habitat[45]. Similar results were also obtained for other species such as oak (*Quercus robur*, *Q. petraea*), where provenances from drier origins showed lower soluble protein concentrations but higher structural N compared to provenances from wetter areas [25]. We assume that such N allocation pattern might be related to leaf structural characteristics, as plants from drier habitats often possess thicker, more robust leaves with higher leaf mass per area and consequently with a higher portion of structural compounds than plants from more humid habitats [52].

5.2 Drought does not affect total N and protein abundance

Our study indicated that foliar total N and soluble protein contents in both provenances were not affected by drought (Fig. 1). This finding is consistent with work on *Pinus taeda* [53], *Picea abies* [54], different *Fagus sylvatica* ecotypes [55] and several oak species (i.e. *Quercus robur, Q. petraea, Q. pubescens*) [25] suggesting little impact of drought on foliar N content. Some other studies, however, demonstrated that drought stress decreases foliar N in *F. sylvatica* [23,56] and *Picea asperata* [57] most probably as a consequence of reduced N uptake under water shortage [58-60]. In contrast, it was also observed that total N content in current year needles of *Pinus halepensis* [61] and *Pinus sylvestris* [54] significantly increased due to growth decline under drought.

5.3 The abundance of individual amino acids is strongly affected by water deficit in a provenance specific manner

The coastal provenance SNO exhibited an increased TAA content due to drought, whereas in the interior provenance FEHR, TAA concentrations rather decreased. Such patterns were obviously driven by the contents of the three most abundant amino acids, i.e., arginine, ornithine and lysine (Fig. 2). Given the assumed different drought tolerances of the two provenances such difference in amino acid concentrations might give a hint to metabolic acclimation to drought stress. Amino acids are likely to accumulate in drought-stressed leaves due to the proteolytic breakdown of leaf proteins or reduction of protein synthesis [62]. For example, drought significantly reduced the soluble protein contents in current-year needles of 2-year-old *Pinus halepensis* seedlings [61]. However, as we did not observe decreased soluble protein concentrations in SNO, the elevated levels of arginine, ornithine and lysine are more likely the result of *de novo* biosynthesis or degradation of structural N. Together with these main amino compounds, drought stressed trees of the provenance SNO showed also considerably increased levels of proline and the polyamine putrescine. Similar patterns have been described for other conifers [63]. Apparently, there is a tight metabolic link between these N compounds in plants. Putrescine can be derived via decarboxylation of ornithine; the latter amino acid is the product of the activity of arginase, an enzyme converting arginine into ornithine [64]. Notably, ornithine can also be used for the biosynthesis of proline that also showed increased levels in drought stressed trees of the provenance SNO. Most of these N compounds, i.e. arginine, proline and putrescine, are well known osmolytes being produced either during periods of reduced water availability or increased salt content [65,66]. Similar to our work with Douglas fir seedlings, drought induced proline accumulation also occurred in an eco-type specific manner in beech [67]. Glycine was another amino acid accumulating in drought stressed SNO but not in FEHR. This amino acid is an intermediate in the formation of the osmolyte glycine-betaine [68]. However, the occurrence of this compound has not yet been shown in Douglas fir and further studies are required to reveal its role in osmoprotection of conifers. It seems that the coastal provenance SNO reacted to drought with an intense biosynthesis of such N containing osmolytes, a response which was in contrast to our hypothesis 3, where we expected such changes in the interior provenance. However, FEHR did not reveal increased production of N based osmolytes when exposed to drought. In contrast to SNO, drought stressed interior FEHR produced significantly higher amounts of many aromatic and pyruvate derived amino acids as well as aspartate and ammonium, and there was a general trend of elevated abundances of numerous other amino acids. This general increase seemed to occur at the expense of the main amino acids arginine, ornithine and lysine. The significance of such rearrangement of N compounds remains unclear, but might be linked to the biosynthesis of stress-induced proteins.

5.4 Drought causes biosynthesis of readily metabolizable osmolytes in both provenances

Both provenances accumulated considerable amounts of carbohydrates and to smaller degree sugar alcohols (Fig. 6). The production of such compounds is a well-understood reaction of plants to osmotic stress like drought [18-20,68-70]. From a study with eucalypts, it seems that species from mesic sites rather produce readily usable compounds such as monosaccharides as osmolytes, whereas species from arid sites produce more stable compounds, most likely because they "expect" long-term drought events [71]. Similarly, *Quercus robur* produced readily usable carbohydrates as osmolytes if drought stressed on a short term, but formed stable osmolytes (mainly quercitol and mannitol) under long-term drought periods [69]. It remains unclear if Douglas fir would accumulate more stable osmolytes if drought stressed for longer time, or if strong production of such osmoprotectants does not occur in this species. Noteworthy, both provenances accumulated at least small amounts of the stable D-pinitol, a polyol formed under abiotic stress [72]. In consistence with our work, drought stress led to increased D-pinitol levels also in other conifers such as black spruce [73] and maritime pine [17]. Interestingly, myo-inositol contents dropped during drought in Douglas fir (Fig. 6). Because myo-inositol is a precursor of D-pinitol, this decrease might be related to an enhanced consumption for biosynthesis of this osmoprotectant [68,74]. Similarly, the galactinol levels decreased during drought stress which could also be an effect of enhanced consumption for the production of osmolytes of the raffinose-family which are also known as antioxidants and putative signaling compounds [75]. In general, the production of osmolytes in

Douglas fir seemed to occur at the expense of other carbohydrates because the contents of several sugars such as sucrose and maltose decreased during the drought period. In consistence, Aranjuelo et al[19] found increased D-pinitol levels during drought but decreased amounts of fructose and glucose in leaves of alfalfa. Nevertheless, it cannot be excluded that such drop in carbohydrate levels is due to consumption in energy metabolism as suggested from studies with drought stressed conifers[76-78].

5.5 Provenance specific effects on antioxidant levels

Besides the biosynthesis of osmolytes as osmoprotectants, the ascorbate-glutathione cycle plays a crucial role in detoxification of reactive oxygen species (ROS) formed in drought exposed plants[79]. Trees of the coastal provenance SNO accumulated glutathione and its precursors cysteine, γ-glutamyl-cysteine and glycine under drought stress, but did not change the ascorbate levels. The opposite effect was observed in the interior provenance FEHR. Notably, controls of this provenance contained significantly more ascorbate than controls of SNO. Similar results were observed in European beech (*Fagus sylvatica*) seedlings from different provenances[80]. Apparently, the redox status of both provenances does not seem to be impaired by drought stress, as suggested from the unchanged ratios of oxidized to reduced glutathione and ascorbate. Needles of drought stressed FEHR trees contained higher concentrations of the ROS scavenger α-tocopherol than control trees. This anti-oxidant is assumed to play a crucial role as singlet oxygen scavenger in chloroplasts of plants exposed to oxidative stress[81]. By removing oxygen radicals, α-tocopherol contributes to maintain the functioning of the photosynthetic apparatus under stress conditions[81]. Because regeneration of α-tocopherol after reaction with the oxygen radical needs the abundance of ascorbate, there is a tight metabolic link between the two antioxidants, which might explain their common pattern in provenance FEHR in the present study. However, protection against ROS also involves other antioxidant defences including carotenoids, isoprenoids, flavonoids and enzymatic regeneration of antioxidants that were not determined in the present study focused on plant N metabolism. Therefore, differences in antioxidative defences of drought treated Douglas fir provenances require attention in further studies.

5.6 The ability of recovery of the two provenances

The ability to recover from stress is a crucial feature of plant stress tolerance, because this resilience capacity determines post-stress performance and competitiveness in its environment. Compared to the great number of reports on plant responses to drought stress, little is known about the capacity to recover[82]. In the present study, the concentrations of most compounds, which were affected by drought, recovered completely to the levels of well-watered control trees within the two weeks lasting recovery period. This pattern indicates the high drought tolerance and resilience of both Douglas fir provenances tested.

6 Conclusion

Reduced water supply caused clear effects on the trees' water potential. Although the interior provenance FEHR was expected to be more drought tolerant than the coastal provenance SNO, differences in the water potential of both provenances were not observed indicating similarly impaired water relations. We also assumed stronger impairment of the coastal provenance's primary metabolism (hypothesis 4), which was not the case. Nevertheless, both provenances responded to drought in different ways. Needles of SNO accumulated high amounts of N containing osmolytes and increased its levels of glutathione and related compounds, a finding not seen in FEHR. The latter provenance, in contrast, showed a general but moderate increase in many amino acids mainly at the expense of arginine. This observation suggests that FEHR has a high potential to synthesize protective proteins. FEHR also accumulated α-tocopherol and ascorbate, two antioxidants with known benefit for the photosynthetic apparatus, which at least partially supported our hypothesis 3. The production of readily utilizable carbohydrates as osmolytes was seen in both provenances and might be crucial for survival of Douglas fir under drought conditions. The present study underlines that provenances of a given species can respond in very different manner to environmental factors, highlighting the need to carefully test reactions of provenances planned to be used in forestry.

Conflict of interest

None declared.

Funding

This work was financially supported by the German Science Foundation (DFG, grants KR 2010/4-1, EN829/5-1 and GE1090/7-1), the Forest Research Institute of the German State Baden-Württemberg (FVA) and the Natural Sciences and Engineering Research Council of Canada (NSERC). B.D. obtained funding from the China Scholarship Council, the Ministerium für Wissenschaft, Forschung und Kunst Baden Württemberg of Germany and the University of Freiburg.

References

[1] Frei C. Eidgenössische Technische Hochschule(Zürich) Institut für Atmosphäre und Klima, Meteo-Schweiz. Klimazukunft der Schweiz - Eine probabilistische Projektion. Zürich: MeteoSchweiz, 2004.

[2] European Environment Agency. Climate Change, Impacts and Vulnerability in Europe 2012. (An

indicator-based report). Copenhagen: European Environment Agency, 2012.
[3] IPCC. Climate Change 2007-The Physical Science Basis: Working Group I Contribution to the Fourth Assessment Report of the Intergovernmental Panel on Climate Change. Cambridge: Cambridge University Press, 2007.
[4] Ciais P, Reichstein M, Viovy N, et al. Europe-wide reduction in primary productivity caused by the heat and drought in 2003. Nature, 2005, 437: 529-533.
[5] Eilmann B, Rigling A. Tree-growth analyses to estimate tree species' drought tolerance. Tree physiology, 2012, 32: 178-187.
[6] Lévesque M, Saurer M, Siegwolf R, et al. Drought response of five conifer species under contrasting water availability suggests high vulnerability of Norway spruce and European larch. Global Change Biology, 2013, 19: 3184-3199.
[7] Nadezdhina N, Urban J, Čermák J, et al. Comparative study of long-term water uptake of Norway spruce and Douglas-fir in Moravian upland. Journal of Hydrology and Hydromechanics, 2014, 62: 1-6.
[8] Smith W B, Miles P D, Perry C H, et al. Forest Resources of the United States, 2007: A Technical Document Supporting the Forest Service 2010 RPA Assessment. Washington D.C.: USDA Forest Service, 2009.
[9] Perić S, Jazbec A, Tijardović M, et al. Provenance studies of Douglas fir in the locality of Kontija (Istria). Periodicum Biologorum, 2009, 111: 487-493.
[10] Ferrell W K, Woodard E S. Effects of seed origin on drought resistance of Douglas-fir (*Pseudotsuga menziesii*) (Mirb.) Franco. Ecology, 1966, 47: 499-503.
[11] Pharis R P, Ferrell W K. Differences in drought resistance between coastal and inland sources of Douglas fir. Canadian Journal of Forest Research, 1966, 44: 1651-1659.
[12] Franklin J F, Waring R H. Distinctive features of the northwestern coniferous forest: development, structure, and function. Oregon: The 40th Biology Colloquium Corvallis, 1980.
[13] White T L.Drought tolerance of southwestern Oregon Douglas-fir. Forest Science, 1987, 33: 283-293.
[14] Joly R J, Adams W, Stafford S G. Phenological and morphological responses of mesic and dry site sources of coastal Douglas-fir to water deficit. Forest Science, 1989, 35: 987-1005.
[15] Eilmann B, de Vries SM, den Ouden J, et al. Origin matters! Difference in drought tolerance and productivity of coastal Douglas-fir (*Pseudotsuga menziesii* (Mirb.)) provenances. Forest Ecology & Management, 2013, 302: 133-143.
[16] Lucier A, Ayres M, Karnosky D, et al. Forest responses and vulnerabilities to recent climate change. IUFRO World Series, 2009, 22: 29-52.
[17] Nguyen A, Lamant A. Pinitol and myo-inositol accumulation in water-stressed seedlings of maritime pine. Phytochemistry, 1988, 27: 3423-3427.
[18] Sudachkova N E, Milyutina I L, Semenova G P. Influence of water deficit on contents of carbohydrates and nitrogenous compounds in *Pinus sylvestris* L. and *Larix sibirica* Ledeb. tissues. Eurasian Journal of Forest Research, 2002, 4: 1-11.
[19] Aranjuelo I, Molero G, Erice G, et al. Plant physiology and proteomics reveals the leaf response to

drought in alfalfa (*Medicago sativa* L.). Journal of Experimental Botany, 2011, 62: 111-123.

[20] Keunen E, Peshev D, Vangronsveld J, et al. Plant sugars are crucial players in the oxidative challenge during abiotic stress: extending the traditional concept. Plant Cell and Environment, 2013, 36: 1242-1255.

[21] Rennenberg H, Loreto F, Polle A, et al. Physiological responses of forest trees to heat and drought. Plant Biology, 2006, 8: 556-571.

[22] Lewis J, Lucash M, Olszyk D, et al. Relationships between needle nitrogen concentration and photosynthetic responses of Douglas-fir seedlings to elevated CO_2 and temperature. New Phytologist, 2004, 162: 355-364.

[23] Fotelli M, Rennenberg H, Geβler A. Effects of drought on the competitive interference of an early successional species (*Rubus fruticosus*) on *Fagus sylvatica* L. seedlings: ^{15}N uptake and partitioning, responses of amino acids and other N compounds. Plant Biology, 2002, 4: 311-320.

[24] Guo J, Yang Y, Wang G, et al. Ecophysiological responses of *Abies fabri* seedlings to drought stress and nitrogen supply. Physiologia Plantarum, 2010, 139: 335-347.

[25] Hu B, Simon J, Kuster T, et al. Nitrogen partitioning in oak leaves depends on species, provenance, climate conditions and soil type. Plant Biology, 2013,15: 198-209.

[26] Raitio H, Sarjala T. Effect of provenance on free amino acid and chemical composition of Scots pine needles. Plant and Soil, 2000, 221: 231-238.

[27] Duan B, Lu Y, Yin C, et al. Physiological responses to drought and shade in two contrasting *Picea asperata* populations. Physiologia Plantarum, 2005, 124: 476-484.

[28] Ormrod D P, Lesser V M, Olszyk D M, et al. Elevated temperature and carbon dioxide affect chlorophylls and carotenoids in Douglas-fir seedlings. International Journal of Plant Sciences, 1999, 160: 529-534.

[29] Hobbie E, Olszyk D, Rygiewicz P, et al. Foliar nitrogen concentrations and natural abundance of ^{15}N suggest nitrogen allocation patterns of Douglas-fir and mycorrhizal fungi during development in elevated carbon dioxide concentration and temperature. Tree Physiology, 2001, 21: 1113-1122.

[30] Tingey D T, Mckane R B, Olszyk D M, et al. Elevated CO_2 and temperature alter nitrogen allocation in Douglas-fir. Global Change Biology, 2003, 9: 1038-1050.

[31] Boczulak S, Hawkins B, Roy R. Temperature effects on nitrogen form uptake by seedling roots of three contrasting conifers. Tree Physiology, 2014, 34: 513-523.

[32] Van den Driessche R, Webber J. Total and soluble nitrogen in Douglas fir in relation to plant nitrogen status. Canadian Journal of Forest Research, 1975, 5: 580-585.

[33] Van den Driessche R, Webber J. Variation in total and soluble nitrogen concentrations in response to fertilization of Douglas-fir. Forest Science, 1977, 23: 134-142.

[34] Margolis H, Waring R. Carbon and nitrogen allocation patterns of Douglas-fir seedlings fertilized with nitrogen in autumn. I. Overwinter metabolism. Canadian Journal of Forest Research, 1986, 16: 897-902.

[35] Margolis H, Waring R. Carbon and nitrogen allocation patterns of Douglas-fir seedlings fertilized with

nitrogen in autumn. II. Field performance. Canadian Journal of Forest Research, 1986, 16: 903-909.

[36] Gijsman A J. Soil water content as a key factor determining the source of nitrogen (NH_4^+ or NO_3^-) absorbed by Douglas-fir (*Pseudotsuga menziesii*) and the pattern of rhizosphere pH along its roots. Canadian Journal of Forest Research,1991, 21: 616-625.

[37] Pérez-Soba M, de Visser P H. Nitrogen metabolism of Douglas fir and Scots pine as affected by optimal nutrition and water supply under conditions of relatively high atmospheric nitrogen deposition. Trees, 1994, 9: 19-25.

[38] Sergent A S, Rozenberg P, Bréda N. Douglas-fir is vulnerable to exceptional and recurrent drought episodes and recovers less well on less fertile sites. Annals of Forest Science, 2014, 71: 697-708.

[39] Sergent A S, Bréda N, Sanchez L, et al. Coastal and interior Douglas-fir provenances differ in growth performance and response to drought episodes at adult age. Annals of Forest Science, 2014, 71: 709-720.

[40] Knoke T, Hahn A. Global change and the role of forests in future land use systems. Developments in Environmental Science, 2013, 13: 569-588.

[41] Scholander P F, Bradstreet E D, Hemmingsen E, et al. Sap pressure in vascular plants: negative hydrostatic pressure can be measured in plants. Science, 1965, 148: 339-346.

[42] Rennenberg H, Schneider S, Weber P. Analysis of uptake and allocation of nitrogen and sulphur compounds by trees in the field. Journal of Experimental Botany, 1996, 47: 1491-1498.

[43] Winter H, Lohaus G, Heldt H W. Phloem transport of amino acids in relation to their cytosolic levels in barley leaves. Plant Physiology, 1992, 99: 996-1004.

[44] Luo Z B, Janz D, Jiang X, et al. Upgrading root physiology for stress tolerance by ectomycorrhizas: insights from metabolite and transcriptional profiling into reprogramming for stress anticipation. Plant Physiology, 2009, 151: 1902-1917.

[45] Du B, Jansen K, Junker L V, et al. Elevated temperature differently affects foliar nitrogen partitioning in seedlings of diverse Douglas fir provenances. Tree Physiology, 2014, 34: 1090-1101.

[46] Kreuzwieser J, Hauberg J, Howell KA, et al. Differential response of gray poplar leaves and roots underpins stress adaptation during hypoxia. Plant Physiology, 2009,149: 461-473.

[47] Erxleben A, Gessler A, Vervliet-Scheebaum M, et al. Metabolite profiling of the moss *Physcomitrella patens* reveals evolutionary conservation of osmoprotective substances. Plant Cell Reports, 2012, 31(2): 427-436.

[48] Kopka J, Schauer N, Krueger S, et al. GMD@ CSB.DB: the Golm Metabolome Database. Bioinformatics, 2005, 21: 1635-1638.

[49] Pavli O I, Vlachos C E, Kalloniati C,et al. Metabolite profiling reveals the effect of drought on sorghum (*Sorghum bicolor* L. Moench) metabolism. Plant Omics Journal, 2013, 6: 371-376.

[50] Tuhkanen S. Climatic parameters and indices in plant geography. Acta Phytogeographica Suecica, 1980, 67: 1-105.

[51] Wang T, Hamann A, Spittlehouse D, et al. Development of scale-free climate data for Western Canada for use in resource management. International Journal of Climatology, 2006, 26: 383-397.

[52] Abrams M D, Kubiske M E, Mostoller S A. Relating wet and dry year ecophysiology to leaf structure in contrasting temperate tree species. Ecology, 1994, 75: 123-133.

[53] Green T H, Mitchell R J, Gjerstad D H. Effects of nitrogen on the response of loblolly pine to drought. New Phytologist, 1994,128: 145-152.

[54] Turtola S, Manninen A M, Rikala R, et al. Drought stress alters the concentration of wood terpenoids in Scots pine and Norway spruce seedlings. Journal of Chemical Ecology, 2003, 29: 1981-1995.

[55] Peuke A, Rennenberg H. Carbon, nitrogen, phosphorus, and sulphur concentration and partitioning in beech ecotypes (*Fagus sylvatica* L.): phosphorus most affected by drought. Trees, 2004, 18: 639-648.

[56] Nahm M, Matzarakis A, Rennenberg H, et al. Seasonal courses of key parameters of nitrogen, carbon and water balance in European beech (*Fagus sylvatica* L.) grown on four different study sites along a European North–South climate gradient during the 2003 drought. Trees, 2007, 21: 79-92.

[57] Yang Y, Han C, Liu Q, et al. Effect of drought and low light on growth and enzymatic antioxidant system of *Picea asperata* seedlings. Acta Physiologiae Plantarum, 2008, 30: 433-440.

[58] Geβler A, Keitel C, Nahm M, et al. Water shortage affects the water and nitrogen balance in central European beech forests. Plant Biology, 2004, 6: 289-298.

[59] Geβler A, Jung K, Gasche R, et al. Climate and forest management influence nitrogen balance of European beech forests: microbial N transformations and inorganic N net uptake capacity of mycorrhizal roots. European Journal of Forest Research, 2005, 124: 95-111.

[60] Rennenberg H, Dannenmann M, Gessler A, et al. Nitrogen balance in forest soils: nutritional limitation of plants under climate change stresses. Plant Biology, 2009, 11: 4-23.

[61] Wellburn F A, Lau K-K, Milling P M, et al.Drought and air pollution affect nitrogen cycling and free radical scavenging in *Pinus halepensis* (Mill.). Journal of Experimental Botany, 1996, 47: 1361-1367.

[62] Good A G, Zaplachinski S T. The effects of drought stress on free amino acid accumulation and protein synthesis in *Brassica napus*. Physiologia Plantarum, 1994, 90: 9-14.

[63] Cyr D, Buxton G, Webb D, et al. Accumulation of free amino acids in the shoots and roots of three northern conifers during drought. Tree Physiology, 1990, 6: 293-303.

[64] Bitrián M, Zarza X, Altabella T, et al. Polyamines under abiotic stress: metabolic Crossroads and hormonal crosstalks in plants. Metabolites, 2012, 2: 516-528.

[65] Zhu J K. Plant salt tolerance.Trends in Plant Science, 2001, 6: 66-71.

[66] Shao H B, Chu L Y, Shao M A, et al. Higher plant antioxidants and redox signaling under environmental stresses. Comptes Rendus Biologies, 2008, 331: 433-441.

[67] Schraml C, Rennenberg H. Sensitivity of different ecotypes of beech (*Fagus sylvatica*) to drought stress. Forstwissenschaftliches Centralblatt vereinigt mit Tharandter forstliches Jahrbuch, 2000, 119: 51-61.

[68] Krasensky J, Jonak C. Drought, salt, and temperature stress-induced metabolic rearrangements and regulatory networks. Journal of Experimental Botany, 2012, 63: 1593-1608.

[69] Spieß N, Oufir M, Matušíková I, et al. Ecophysiological and transcriptomic responses of oak (*Quercus robur*) to long-term drought exposure and rewatering. Environmental and Experimental Botany, 2012, 77: 117-126.

[70] Harfouche A, Meilan R, Altman A. Molecular and physiological responses to abiotic stress in forest trees and their relevance to tree improvement. Tree physiology, 2014, 34(11): 1181-1198.

[71] Merchant A, Tausz M, Arndt S K, et al.Cyclitols and carbohydrates in leaves and roots of 13 Eucalyptus species suggest contrasting physiological responses to water deficit. Plant Cell and Environment, 2006, 29: 2017-2029.

[72] Orthen B, Popp M, Smirnoff N. Hydroxyl radical scavenging properties of cyclitols. Proceedings of the Royal Society of Edinburgh Section B Biological Sciences, 1994, 102: 269-272.

[73] Deslauriers A, Beaulieu M, Balducci L, et al. Impact of warming and drought on carbon balance related to wood formation in black spruce. Annals of Botany, 2014, 114: 335-345.

[74] Jansen K, Du B, Kayler Z, et al. Douglas-fir seedlings exhibit metabolic responses to increased temperature and atmospheric drought. PloS One, 2014, 9: e114165.

[75] Valluru R, Van den Ende W. *Myo*-inositol and beyond–Emerging networks under stress. Plant Science, 2011, 181: 387-400.

[76] Gruber A, Pirkebner D, Florian C, et al. No evidence for depletion of carbohydrate pools in Scots pine (*Pinus sylvestris* L.) under drought stress. Plant Biology, 2012, 14: 142-148.

[77] Adams H D, Germino M J, Breshears D D, et al. Nonstructural leaf carbohydrate dynamics of *Pinus edulis* during drought-induced tree mortality reveal role for carbon metabolism in mortality mechanism. New Phytologist, 2013, 197: 1142-1151.

[78] Sevanto S, McDowell N G, Dickman L T, et al. How do trees die? A test of the hydraulic failure and carbon starvation hypotheses. Plant Cell Environment, 2014, 37: 153-161.

[79] Foyer C H, Noctor G. Ascorbate and glutathione: the heart of the redox hub. Plant Physiology, 2011, 155: 2-18.

[80] García-Plazaola J I, Becerril J M. Effects of drought on photoprotective mechanisms in European beech (*Fagus sylvatica* L.) seedlings from different provenances. Trees, 2000, 8: 485-490.

[81] Munné-Bosch S. The role of α-tocopherol in plant stress tolerance.Journal of Plant Physiology, 2005, 162: 743-748.

[82] Gallé A, Haldimann P, Feller U. Photosynthetic performance and water relations in young pubescent oak (*Quercus pubescens*) trees during drought stress and recovery. New Phytologist, 2007, 174: 799-810.

Elevated temperature differently affects foliar nitrogen partitioning in seedlings of diverse Douglas fir provenances*

Du Baoguo　　Kirstin Jansen　　Laura Verena Junker

Monika Eiblmeier　　Jürgen Kreuzwieser　　Arthur Gessler

Ingo Ensminger　　Heinz Rennenberg

Abstract

Global climate change causes an increase in ambient air temperature, a major environmental factor influencing plant physiology and growth that already has been perceived at the regional scale and is supposed to become even more severe in the future. In the present study, we investigated the effect of elevated ambient air temperature on nitrogen metabolism of two interior provenances of Douglas fir (*Pseudotsuga menziesii* var. *glauca*) originating from contrasting habitats, namely the provenances Monte Creek (MC) from a drier environment and Pend Oreille (PO) from a more humid environment. Three to four years old seedlings of the two provenances were grown for three months in controlled environments under either control temperature (CT, day 20℃, night 15℃) or under high temperature conditions (HT, 30℃/25℃). Total nitrogen (N), soluble protein, chlorophyll and total amino acid(TAA) contents as well as individual amino acid concentrations were determined in both current year and previous-year needles. Our results show that the foliar total N contents of the two provenances were unaffected by HT. Arginine, lysine, proline, glutamate, and glutamine were the most abundant amino acids, which together contributed 88% to the TAA pool of current- and previous-year needles. HT decreased the contents of most amino acids of the glutamate family (i.e., arginine, proline, ornithine and

---

* 原载于：Tree Physiology，2014，34（10）：1090-1101.

glutamine) in current-year needles. However, HT did not affect the concentrations of metabolites related to the photorespiratory pathway, such as NH_4^+, glycine, and serine. In general, current year needles were considerably more sensitive to HT than previous year needles. Moreover, provenance PO originating from a mesic environment showed stronger responses to HT than provenance MC. Our results indicate provenance-specific plasticity in the response of Douglas fir to growth temperature. Provenance-specific effects of elevated temperature on nitrogen use efficiency suggest that origin might determine the sensitivity and growth potential of Douglas fir trees in a future warmer climate.

Keywords: Douglas fir (*Pseudotsuga menziesii*); amino acid; arginine

1 Introduction

Douglas fir (*Pseudotsuga menziesii*) is native to America's Pacific Northwest (USA, Canada) where it covers more than 20 million hectares of natural forests [1]. Its longitudinal distribution ranges from northern British Columbia, CA, to north of New Mexico, USA. From West to east it spans over 1000km from the Pacific coast to the eastern borders of the Rocky Mountains. The altitudinal distribution of Douglas fir ranges from sea level up to 2700m above sea level in New Mexico [2]. Consequently, Douglas fir has evolved a great variety of different provenances adapted to the contrasting environmental conditions in different habitats [3].

Due to its outstanding growth and great economic potential, Douglas fir was introduced to Central Europe at the end of the 19th century [4,5]. In southwestern Germany, Douglas fir currently contributes 3% of the current stocking volume of forests. Forest management practice aiming to address increasing air temperature and changes in precipitation pattern are expected to further increase the share of Douglas fir to managed forests in southwestern Germany [6,7]. This is owing to the fact, that besides its high growth potential and a more effective use of soil nutrients [8], many Douglas fir provenances exhibit better tolerance to drought [9,10] and increased adaptive potential to warming, than other economically important conifer forest trees in Central Europe, such as Norway spruce or Scots pine [6].

Anthropogenic climate change already caused a 1.4℃ increase of the average land surface air temperatures in Central Europe since 1850 [11]. Climate models estimate a further temperature rise of 1.1–6.4℃ by 2100 [12]. Therefore, it is obvious that temperature will become one of the most important factors influencing plant performance, growth, and physiology [13]. The responses of plants to high temperature stress depend on the plasticity of physiological characteristics of each species [14]. In addition, intraspecific variability further contributes to the specific responses and plasticity within provenances[15,16] and between individuals within a provenance or population[17,18].

Numerous studies have reported that warming enhances the growth of trees (reviewed by Way and Oren[19]). In natural stands of coastal Douglas fir, for example, growth rates correlated positively with higher temperatures during autumn, winter and spring; however, they negatively correlated with high summer temperatures[20]. Net rates of photosynthesis in Douglas fir seedlings are generally increased by elevated temperatures but the temperature optimum is assumed to be not higher than 20℃[21]. Besides temperature, photosynthetic performance is strongly linked to foliar nitrogen (N) content[22], since N is an important constituent of compounds of the photosynthetic apparatus such as chlorophylls, Rubisco and other Calvin cycle enzymes[23]; Rubisco alone can bind 10%–30% of total leaf N[24]. Therefore, needle N contents is an important parameter determining photosynthetic responses of Douglas fir seedlings to climate change[25]. Moreover, N is often the primary factor limiting plant growth and development in natural forest ecosystems with low anthropogenic N input[26].

Many studies investigated how climate change related parameters[20,22,25,27-31] and N availability[23,32-34] affect the N metabolism of Douglas fir. There are, however, very few investigations on the effect of elevated ambient air temperature on Douglas fir provenances, and, to our knowledge, temperature effects on N partitioning in the needles of this species have not been reported.

In the present study, we examined the effects of elevated temperature on leaf N partitioning of two interior Douglas fir provenances, Pend Oreille (Po), originating from a humid environment in North Western Washington, USA, and Monte Creek, originating from a drier habitat in the Rocky Mountains near Kamloops, British Columbia, CA. In light of future warmer climate, we addressed the questions ① if growth under elevated air temperature affects the quantity and quality of foliar N compounds in Douglas fir trees, and ② if provenance specific reactions and differences in N partitioning between the two interior provenances can be established in the response to elevated air temperatures. For this purpose, in an experiment under controlled conditions we exposed seedlings of the two Douglas fir provenances for three months either to air temperatures, which were close the optimum for photosynthesis, or to higher air temperature clearly exceeding this optimum level[21]. We characterized foliar N partitioning by the analysis of total N, soluble protein, chlorophyll N, amino acid concentrations as well as structural N in previous and current year needles.

2 Materials and methods

2.1 Plant cultivation and experimental design

The present study was performed with three to four years old seedlings of two interior

Douglas fir (*Pseudotsuga menziesii* var. *glauca*) provenances, i.e., the provenance PO from a relatively humid original environment with the mean annual precipitation of 736mm and Monte Creek (MC) from a drier habitat, where the mean annual precipitation was 362mm (Table 1). These provenances are presently not commercially used in Germany and are not available there; the trees were therefore purchased from nurseries in Canada (MC, Nursery Services Interior BC Timber Sales, Vernon, BC) and the USA (PO, Webster Forest Nursery, Washington State Department of Natural Resources, Olympia, WA, USA). The trees were grown in 4-l pots filled with a substrate consisting of commercial potting soil (Anzucht-und Pikiererde, Ökohum, Herbertingen, Germany), perlite and sand (v/v/v, 1/1/1). This substrate was supplemented with long-term fertilizer (3g/L substrate) (Osmocote Exact high-end 5-6, 15N+9P+12K+2Mg, The Scotts Company, LLC, Marysville, OH, USA). All pots were placed into two walk-in environmental chambers (KTLK 20000-IV, Nema Industrietechnik GmbH, Netzschkau, Germany). The plants were watered with tap water every second day and, when exposed to high temperature, watered daily in order to avoid any water stress for the trees. The chambers were illuminated by a mixture of sodium-vapor lamps (NC 1000-00, -01 and -62, Narva, Plauen, Germany, with 6%, 8% and 10%-11% red light, respectively).

Table 1 Description of the habitats of origin of the two selected Douglas fir provenances

Name		MC	PO
Province/State of origin		British Columbia, Canada	Washington State, USA
Region		Southern Interior Seed planting zone "Thompson Okanogan arid"	Okanogan Highlands near PO, Coey, Deer Park, Fruitland, Leadpoint, Metaline Falls
Geographic co-ordinates		W119, N50	W117, N48
Elevation (m)		800-900	850-1000
Mean temperature (°C)	annual	5.2	6.5
	vegetation period	13.6	14.8
Mean precipitation (mm)	annual	362	736
	vegetation period	171	220

Over a period of 78 days from February to May 2010, all plants were stepwise acclimatized to experimental conditions (Table 2). During the first ten days, the air temperature in both chambers was raised from 10℃/10℃ to 20℃/15℃ (light/dark period), the photoperiod was extended from 12 h to 16 h and the light intensity was raised from 250 μmol/(m^2 · s) to 600 μmol/(m^2 · s) photosynthetic photon flux density (PPFD). During the last four days of acclimatization, the temperature in the HT chamber was raised to 30℃/25℃ (light/dark period) in two steps, while the CT chamber remained at 20℃/15℃. During the treatment phase, the vapor pressure deficit (VPD) during the light period was held at 0.3 kPa in the CT chamber and at 1.9 kPa in the HT chamber, which we presumed to represent humid and arid conditions, respectively. The plants had developed needles of three age-classes at the beginning of the experiment. New needles appeared in the course of the acclimatization and were fully expanded before the HT treatment started. After 76 days of treatment (end of July 2010), five to six trees of each treatment and provenance were harvested from which current and previous year needles were excised from the twigs, immediately shock-frozen in liquid nitrogen and stored at −80 ℃ until analyses.

Table 2 Environmental conditions in the climate chambers during the acclimatization and treatment phase

Date	Days before/after treatment	PPFD [μmol/(m² · s)]	Photoperiod (h/day)	Temperature (℃) (light/dark period)	
				CT	HT
17 Feb.	−78	250	12	10/10	10/10
23 Feb.	−72	250	14	15/13	15/13
27 Feb.	−68	250	16	20/15	20/15
30 Apr	−6	250	16	20/15	20/15
3 May	−3	400	16	20/15	25/20
6 May–6 Aug.	0–92	600	16	20/15	30/25

Note: Trees were placed into the chamber and acclimatized to the conditions until 5 May. The temperature treatment, CT and HT started from 6 May.

2.2 Dry mass determination

Needles were homogenized to a fine powder and aliquots of 100 mg were dried at 60°C to weight constancy for dry mass determination. Needle water content was calculated as the difference between fresh and dry weight.

2.3 Total N determination

Total N concentration in oven-dried (24 h, 105 ℃) needle powder was determined after Kjeldahl digestion [35]. For this purpose, 0.2g needle powder was solubilized in 2.4ml H_2O_2 and 2.4ml selenium sulfuric acid at 380 ℃ for 90min, cooled down and filled up to 100ml with H_2O. The solution was analyzed for total N using the photometer AT200 (Beckman Coulter, Brea, California, USA / Olympus, Tokyo, Japan).

2.4 Determination of amino acids and ammonium

Amino compounds and ammonium were extracted as described by Winter et al.[36]. For this purpose, 50mg frozen needle powder were added to 0.2ml extraction buffer (20mmol HEPES, 5mmol EGTA, 10mmol NaF, pH 7.0) and 1ml chloroform/methanol (1.5/3.5, v/v). The homogenate was shaken for 30min on ice. After centrifugation, water soluble amino compounds were extracted twice with 600μl H_2O_{demin}. The aqueous phases were combined, centrifuged for 10min and freeze-dried (Alpha 2-4, Christ, Osterode, Germany) for 3 days. Amino acids and ammonium were then analyzed using an ultra-performance-liquid-chromatography (UPLC) system (Waters Corp., Milford, MA, USA) as described by Luo et al.[37]. Briefly, freeze-dried extracts were re-suspended in 1ml 0.02mol HCl. Subsequently, 5μl of the re-suspended extract was mixed with 35μl AccQ-Tag Ultra Borate buffer and 10μl AccQ-Tag Reagent (Waters) and incubated for 1min at room temperature followed by 10min derivatization at 55℃. Norvaline (10μmol in total) was added to the AccQ-Tag Ultra Borate buffer to account for the efficiency of derivatization. Sample aliquots of 1μl were injected into the UPLC system; separation was achieved on an AccQ-Tag Ultra column (2.1×100mm; Waters) using a gradient of AccQ-Tag Ultra eluents A and B (Waters) at a flow rate of 0.7ml/min (for details see Luo et al.[37]). The absorbance was measured at 260nm. For quantification amino acid standards were analyzed under identical conditions. Total amino acid content was calculated as the sum of amino acids determined by UPLC.

2.5 Soluble protein analysis

Determination of total soluble protein was performed as previously described by Dannenmann et al.[38] with several adaptations. For extraction, 35mg frozen needle powder was added to 1.5ml buffer (50mmol Tris–HCl, 1mmol EDTA, 15% glycerol (v/v), 1mmol PMSF, 5mmol DTT, 0.1% Triton X-100, 150mg PVPP; pH 8.0). After shaking for 30min at 4℃ and centrifugation, the pellet was extracted again with 0.5ml buffer; supernatants were combined. The same volume of 10% (w/v) trichloroacetic acid was added to the combined supernatants and the mixture was incubated for 10 min at 4℃. The upper layer was discarded after centrifugation, and the pellets were dissolved in 0.5ml 1mol KOH. Five microliter aliquots of the protein extract were mixed with 200μl Bradford reagent (Amresco Inc., Solon, Ohio, USA) and the absorbance was measured at 595nm. Quantification was achieved using bovine serum albumin (Sigma-Aldrich, Taufkirchen, Germany) as a standard.

2.6 Chlorophyll determination

Chlorophyll a and b were extracted from 50 mg of frozen needle powder, using 700μl 98% methanol buffered with 0.5mol ammonium acetate (pH 7.1) following the method of Mantoura et al.[39]. After 2 h at 4℃ in darkness, the supernatant containing the pigments was separated from the pellet. The pellet was re-suspended in methanol twice. Pigments were analyzed from the combined extracts using a Genesys 10S spectrophotometer (Fisher Scientific, Waltham, MA, USA) at 470.0nm, 652.4nm and 665.2nm according to Lichtenthaler et al. [40].

2.7 Structural N calculation

Structural N content in needles was calculated by subtracting the N fractions in amino acids, soluble proteins, chlorophyll a and b, and ammonium N from total N.

2.8 Statistical analysis

Data were statistically analyzed by one-way analysis of variance (ANOVA) and Student's t-test (when only two groups were considered) using the software package SigmaPlot 11.0 (Systat Software GmbH, Erkrath, Germany); raw data were transferred by denary logarithm if required. Least significant difference (LSD) was employed for detecting the differences between treatments followed by one-way ANOVA. For data failing the normality test, the Kruskal-Wallis one-way analysis of variance on ranks and the Mann-Whitney rank sum test were used to determine the significance of differences. Concentrations of all parameters were calculated based on needle dry weight.

3 Results

Higher air temperature did not affect total N content in needles of both, MC and PO irrespective of needle age (Fig. 1(a)). Generally, previous year needles contained slightly more total N than current year needles; this difference was significant for MC under HT (17.9mg/g DW in previous year needles compared to 12.8mg/g DW in current year needles). Similarly, HT had no effect on structural N content in both types of needles (Fig. 1(b)). However, we observed a provenance specific effect in the previous year needles with higher structural N contents in provenance MC than in PO originating from a more humid environment. Like total N, structural N contents in previous year needles were generally slightly higher than in current year needles.

HT did not affect TAA contents of previous-year needles of both provenances, but caused significantly lower TAA contents in current year needles of provenance PO (34μmol/g DW at HT compared to 175μmol/g DW at CT) (Fig. 1(c)). Also soluble protein concentrations showed clear provenance specific effects in previous year needles with considerably higher values in PO than in MC, which originates from a drier environment (Fig. 1(d)). The current year needles contained generally less protein; this effect was highly significant for PO (25mg/g DW and 9.5mg/g DW in previous and current year needles, respectively). A very similar trend with higher concentrations in the older needles became obvious for chlorophyll a and chlorophyll b concentrations (Fig. 1(e) and (f)). For provenance PO, both pigments showed higher concentrations if the trees were grown at 30℃ than under CT conditions.

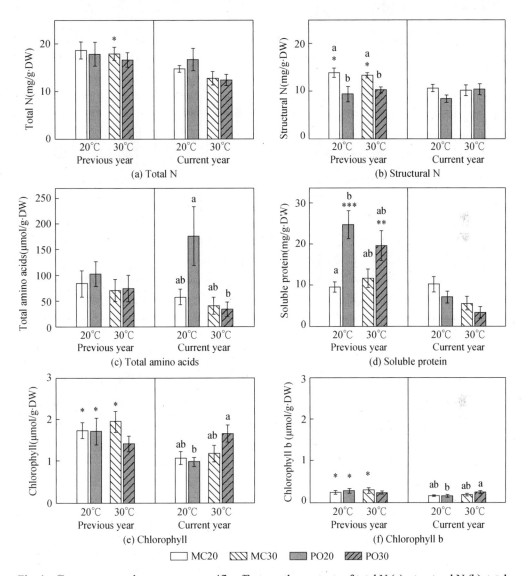

Fig. 1 Temperature and provenance specific effects on the contents of total N (a), structural N (b), total amino acid (c), soluble protein (d) and chlorophyll a (e) and b (f) of current and previous year needles of two Douglas fir provenances

Note: Provenances Monte Creek (MC, white bars) and Pend Oreille (PO, grey bars) were grown for three months at 20℃ (blank bars) and 30℃ (hatched bars). Means ± standard error of 5 to 6 independent replicates are shown. Different letters indicate statistically significant differences at $P<0.05$ within current year or previous year needles as calculated by ANOVA. Asterisks indicate significant difference between current and previous needles at the same temperature as calculated by Student's t-test (*$P<0.05$, ** $P<0.01$, *** $P<0.001$).

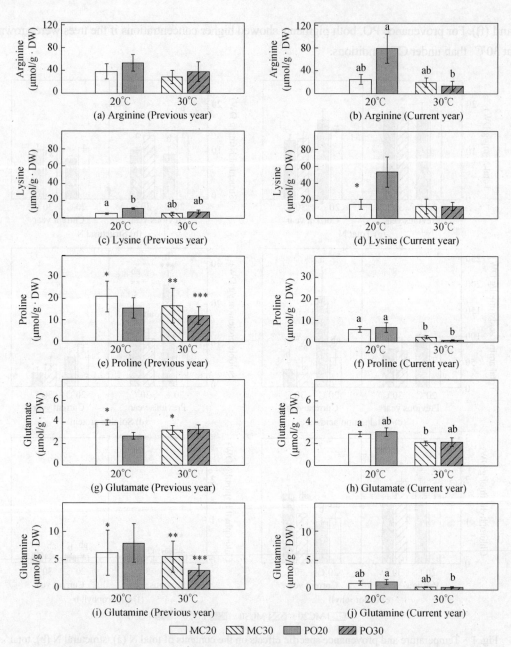

Fig. 2 Temperature and provenance specific effects on the contents of arginine (a, b), Lysine (c, d), proline (e, f), glutamate (g, h) and glutamine (i, j) of current and previous year needles of two Douglas fir provenances.

Note: Provenances Monte Creek (MC, white bars) and Pend Oreille (PO, grey bars) were grown for three months at 20℃ (blank bars) and 30℃ (hatched bars). Means ± standard error of 5 to 6 independent replicates are shown. Different letters indicate statistically significant differences at $P<0.05$ within current year or previous year needles as calculated by ANOVA. Asterisks indicate significant difference between current and previous needles at the same temperature as calculated by Student's t-test (*$P<0.05$, ** $P<0.01$, *** $P<0.001$).

Irrespective of provenance, arginine, lysine, proline, glutamate and glutamine were quantitatively the most important amino acids in all needle types (Fig. 2). Together they contributed around 90% to the amino acid N of both, current and previous year needles. Among the amino acids, arginine showed highest abundances in all treatments of both provenances, representing approximately 70% of the TAA-N pool of current year and previous year needles. There was a clear tendency for higher amino acid concentrations in previous year than in current year needles, which was significant in MC for proline, glutamate and glutamine. Temperature caused significant effects mainly in current year needles and only for provenance PO. Concentrations of arginine, proline, and glutamine significantly decreased at HT (Fig. 2(b), (f) and (j)). The only significant effect of HT in provenance MC was a lower proline concentration (Fig. 2(f)). The concentrations of N compounds associated with photorespiration did not show consistent patterns (Fig. 3). The only significant temperature effect were lower glycine concentrations in current year needles of provenance PO in response to HT. Other significant provenance and temperature specific effects on photorespiratory metabolite abundances were not observed; older needles tended to contain more serine and glycine than the current year needles.

Besides the quantitatively most important amino acids (Fig. 2) and N compounds related to photorespiration (Fig. 3), 16 other amino compounds were abundant in Douglas fir needles (Table 3). In general, there were no clear effects of HT, needle age, or provenance on the abundance of these amino compounds. Nevertheless, major trends became visible when amino compounds were grouped considering the corresponding amino acid families (Fig. 4). Amino acids derived from oxaloacetate ("aspartate family": aspartate, methionine, threonine) showed a clear tendency to lower amino acid concentrations in younger than older needles (Table 3). In addition, current-year needles of PO showed slightly lower concentrations of amino acids derived from oxaloacetate (Table 3) and 2-ketoglutarate (Fig. 4) when grown at HT. The aromatic amino acids tryptophan, phenylalanine and tyrosine showed lower concentrations in current-year needles compared with previous-year needles. The amino acids derived from pyruvate (alanine, valine, leucine, isoleucine) also showed lower abundances in younger than older needles.

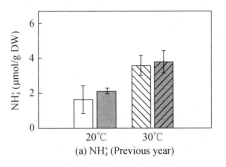

(a) NH_4^+ (Previous year)

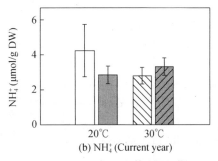

(b) NH_4^+ (Current year)

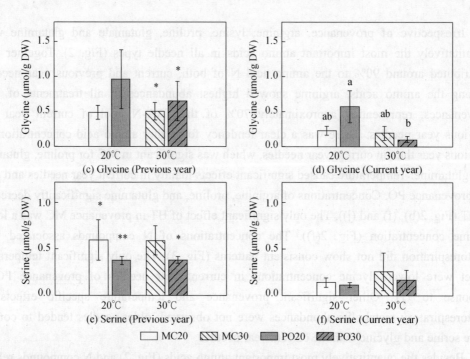

Fig. 3 Temperature and provenance specific effects on the contents of ammonium (a and b), glycine (c and d) and serine (e and f) of current (right panel) and previous (left panel) year needles of two Douglas fir provenances

Note: Provenances Monte Creek (MC, white bars) and Pend Oreille (PO, grey bars) were grown for three months at 20℃ (blank bars) and 30℃ (hatched bars). Means ± standard error of 5 to 6 independent replicates are shown. Different letters indicate statistically significant differences at $P<0.05$ within current year or previous year needles as calculated by ANOVA. Asterisks indicate significant difference between current and previous needles at the same temperature as calculated by Student's t-test (*$P<0.05$, ** $P<0.01$, *** $P<0.001$).

Table 3 Amino acid contents of MC and PO at 20℃ and 30℃

Groups	Amino compounds	Needle age	20℃		30℃	
			MC(μmol/g DW)	PO(μmol/g DW)	MC(μmol/g DW)	PO(μmol/g DW)
Ribose 5-phosphate derived	Histidine	1	0.83±0.27	1.20±0.37	1.21±0.29	0.87±0.16
		0	0.42±0.17	1.30±0.41	0.57±0.21	0.31±0.10
3-Phospho-glycerate derived	Cysteine	1	0.15±0.03	0.11±0.03	0.37±0.14	0.21±0.06
		0	NA	NA	NA	NA
Oxaloacetate-derived	Aspartate	1	**1.53±0.17**	1.18±0.11	**1.31±0.22**	1.24±0.17
		0	**0.95±0.14**	1.23±0.19	**0.68±0.13**	0.90±0.23
	Asparagine	1	0.71±0.31	1.48±0.56	1.22±0.39	0.96±0.46
		0	0.24±0.04	1.03±0.50	1.40±0.39	0.38±0.12
Oxaloacetate-derived	Methionine	1	0.10±0.01 b	0.12±0.01ab	**0.19±0.05 a**	NA
		0	0.03±0.01	0.05±0.01	**0.03±0.01**	0.04±0.01

Continued

Groups	Amino compounds	Needle age	20℃		30℃	
			MC(μmol/g DW)	PO(μmol/g DW)	MC(μmol/g DW)	PO(μmol/g DW)
Oxaloacetate-derived	Threonine	1	**0.44±0.22**	**0.40±0.06**	0.36±0.09	0.39±0.04
		0	**0.06±0.03ab**	**0.13±0.03a**	0.13±0.06ab	0.04±0.03b
Pyruvate derived	Alanine	1	0.81±0.12	0.67±0.06	**1.15±0.16**	1.03±0.14
		0	0.51±0.13	0.52±0.06	**0.55±0.08**	0.84±0.19
	Valine	1	0.20±0.10	0.16±0.02	0.38±0.08	0.20±0.04
		0	NA	0.16±0.02	NA	NA
	Leucine	1	0.17±0.09	0.20±0.04	0.15±0.04	0.14±0.02
		0	NA	NA	NA	NA
	Isoleucine	1	0.18±0.04	0.16±0.06	0.28±0.06	0.23±0.06
		0	NA	0.05±0.01	0.11±0.06	NA
Aromatics	Tryptophan	1	0.91±0.28	1.32±0.31	1.13±0.22	1.23±0.17
		0	NA	NA	NA	NA
	Phenylalanine	1	**0.33±0.07**	0.25±0.04	**0.36±0.11**	0.24±0.07
		0	**0.11±0.02**	0.12±0.03	**0.13±0.06**	0.21±0.08
	Tyrosine	1	0.30±0.03a	0.38±0.06ab	**0.52±0.10b**	0.34±0.05ab
		0	NA	NA	0.15±0.03	0.49±0.27
Others	Ornithine	1	1.13±0.32	**2.74±0.35**	1.37±0.36	1.50±0.56
		0	4.29±1.44b	**31.4±10.5a**	NA	2.63±0.74b
	GABA	1	0.37±0.11	0.22±0.08	0.42±0.02	0.29±0.05
		0	0.10±0.06	NA	NA	NA
	AABA	1	0.18±0.02	0.16±0.02	0.22±0.03	0.17±0.01
		0	0.31±0.08	0.28±0.12	0.19±0.07	0.24±0.05

Note: 0, 1: current and previous year needles, respectively. Different letters indicate significant differences between values in one row as calculated by ANOVA. Bold indicates significant difference between current and previous year needles as calculated by t-test. NA, value below detection limit; GABA, γ-aminobutyric acid; AABA, α-aminobutyric acid.

4 Discussion

4.1 Provenance specific effects

In the present study, we investigated two Douglas fir provenances originating from contrasting environment, with provenance MC originated from a drier environment than provenance PO (Table 1). There were some clear provenance-specific differences in N partitioning in the response to higher temperature. Provenance PO contained considerably more soluble protein in previous year needles and less structural N compared to provenance MC. All other components studied, except lysine in previous year needles, showed very similar levels in the two provenances under control conditions. In response to higher growth temperature, however, only provenance MC originating from a drier environment showed high-temperature related changes in previous year needles, i.e., higher tyrosine, methionine and NH_4^+ contents. In PO, elevated temperature caused higher chlorophyll a and b contents, and lower TAA, arginine, lysine, glutamine, glycine, ornithine, threonine, proline and glutamate contents (Fig. 4). Only the latter two compounds were also affected in the provenance MC. These findings support the view that the provenance originating from the humid environment (PO) reacted more sensitively to high growth temperature. Other studies on provenance specific features of Douglas fir N partitioning are rare and data on the effects of elevated temperature have not been reported.

4.2 Effects of high temperature

Temperature is an important environmental determinant of plant metabolism, because most physiological processes are strongly temperature dependent[13,41]. In the present study, the young Douglas fir trees were exposed to elevated air temperature which certainly also led to higher temperature in the soil. The observed physiological effects therefore have to be considered a consequence of the combination of higher shoot and root temperatures. Our study showed that seedlings growth at elevated air temperature did not affect foliar N contents of both interior Douglas fir provenances (Fig. 1(a)). This finding is consistent with results of van den Driessche and Webber[27] with Douglas fir seedlings. However, a range of studies report an opposite effect, i.e., higher foliar N content, due to growth at higher ambient air temperatures[22,25]. For example, 2 years old Douglas fir seedlings grown at air temperatures 3.5℃ above ambient contained 26% more needle N than trees grown under ambient conditions[25]. It is likely that such higher needle N contents are a result of improved soil N availability due to increased N mineralization rates as shown in several studies[29,42-44]. In contrast to the studies mentioned above, which were conducted under moderately elevated temperatures (3.5 ℃ above ambient), seedlings in the present study were grown at considerably higher temperatures (daytime temperature of 30℃, which was 10℃ above the ambient control temperature). Such HT might reflect a stress for the trees, which may be responsible for the contrasting effects observed.

Besides structural N, largely reflecting lignin and non-soluble protein[22], the main N containing compounds in Douglas fir needles were soluble proteins (Fig. 1). Elevated temperature did not significantly affect the needles' protein content. Only in current year needles, a trend to decreased protein concentrations became visible in both provenances. A similar negative correlation of soluble protein contents with growth temperature was demonstrated in herbaceous plants and conifers[45-47]. This tendency is in good agreement with modelling results of Dewar et al.[48] predicting lower protein levels if plants are grown at elevated temperature. This effect might result from a more pronounced stimulation of respiration compared to photosynthesis, leading to an increased ratio of CO_2 production by respiration to CO_2 consumption by photosynthesis as well as a reduced abundance of Calvin cycle proteins.

The observed temperature effects on protein concentrations were mirrored by TAA contents. Whereas elevated temperature did not affect protein and TAA contents in previous year needles, it considerably lowered the TAA content in current year needles of provenance PO (Fig. 1(c)). This decline was caused by lower abundance of most amino acids of the glutamate family (arginine, proline, glutamine) as well as lysine and ornithine in current year needles (Fig. 2 and Fig. 4, Table 3). In current and previous year needles of both provenances, arginine, proline, lysine, glutamate and glutamine were the most abundant amino acids, accounting for more than 90% of the total amino acid N. This results is in good agreement with the amino acid content of, needles of 38 years old field grown Douglas fir trees[28]. Fertilization led to increased levels of these compounds relative to other amino acids. Among these main amino acids, notably arginine dominated the TAA pool under all conditions. This finding is consistent with previous studies demonstrating that arginine is the most abundant foliar free amino acid not only in Douglas fir [28,32,49,50], but also in other conifers [51-58]. Arginine is considered a marker of the tree N status [5-9]. Consequently, field grown trees from sites with high N availability contained high arginine concentrations in the needles whereas this amino compound was less abundant in trees from N limited field sites [17,60-63]. It is a matter of debate if arginine can be considered an N storage. Because arginine cannot be reused or withdrawn from conifer needles[64], its accumulation represents sequestration of N rather than N storage [65]. Because elevated temperature decreased the content of some of the main amino acids specifically in provenance PO, it appears that the N status of this provenance was negatively affected by elevated temperatures. This will be of particular significance if Douglas fir provenances with similar properties grow under N limited conditions.

4.3 Previous year needles contain more N than current year needles

In evergreen conifers, a great portion (32%–40%) of the N needed for new needle development is derived from remobilization of stored N [60-70]. The most important storage pool of this remobilized N are the older needles of the trees [70]; a major portion of this N in Douglas fir trees is stored in RuBisCo [71]. In Douglas fir seedlings, N containing compounds, particularly RuBisCo, from 1-year-old needles are largely responsible for N supply of developing needles as well as the maintenance of bud and shoot extension[65,71]. In the present study, previous year needles of both provenances contained slightly more N than current year needles, which is in good agreement with the functional role of older needle generations for N supply of developing needles. This effect was most pronounced in soluble protein concentrations of provenance PO (Fig. 1(d)) and in the content of several amino acids such as proline, glutamate or glutamine which may have derived from degradation of RuBisCo as supported from findings of Camm[71]. The latter amino acids are typical transport compounds in trees[72-75] and might be translocated from previous to current year needles.

Previous and current year needles also differed in chlorophyll a and b content, with higher pigment contents observed in the older needle generation (Fig. 1(e) and (f)). Higher chlorophyll a and b contents in older needle generations of Douglas fir were also reported by Thomson et al.[76]. This finding together with the lower protein concentrations in current year needles compared to older needles may be interpreted as a developmental effect. We assume that the photosynthetic apparatus of current year needles was not completely build-up during the time of harvest (end of July), although the needles were already fully expanded. This view is partially supported by the observation that the development of current year needles of field-grown Douglas fir trees was completed just around end of July / beginning of August[71]; the development of our greenhouse grown plants most likely occurred somewhat slower.

In general, current year needles reacted more sensitive to elevated temperature than previous year needles. This is concluded from the significant temperature effects on the contents of TAA, chlorophylls and many individual amino acids such as arginine, proline, glutamate, glutamine, glycine, threonine, ornithine and tyrosine (Fig. 4). We assume that these temperature effects were stronger in the younger needles because they—even though fully expanded at the beginning of the treatment—still might have been subjected to some further physiological development directly under the different temperature regimes whereas the older needles were already fully developed at beginning of the experiment.

4.4 Metabolites related to photosynthesis and photorespiration

The present results clearly indicate higher chlorophyll a and b contents in current year needles of provenance PO if grown at elevated temperature (Fig. 1(e) and (f)). Work with Douglas fir seedlings of coastal origin, grown at 3.5 ℃ above ambient air temperature, exhibited similar changes in chlorophyll contents together with increased CO_2 assimilation rates[20]. However, in the present study we observed these changes only in provenance PO, originating from a more humid environment. We hypothesize that different developmental velocities under elevated temperature caused provenance specific differences in chlorophyll contents; needles of PO may have developed the photosynthetic apparatus faster at high temperature than provenance MC originating from a drier environment. On the other hand, we cannot exclude that the photosynthetic apparatus of provenance PO was better adapted to elevated temperatures than provenance MC. Growth at 30 ℃ daytime temperature may have caused stress for MC only, with the consequence of a slower development of the photosynthetic apparatus. In good agreement, combined carbon and oxygen stable isotope signatures of several plant tissues indicated a decrease in assimilation and stomatal conductance in response to HT and vapor pressure deficit, as well as lowered biomass accumulation (Jansen and Gessler, unpublished data).

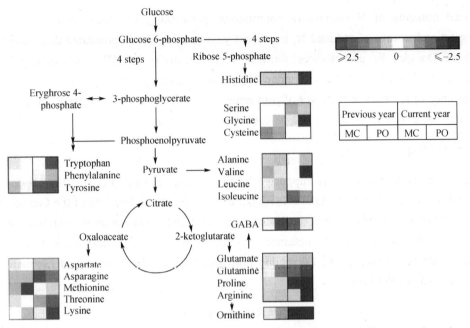

Fig. 4 Temperature and provenance specific effects on the amino acid contents of current and previous year needles of two Douglas fir provenances

Note: Provenances Monte Creek (MC) and Pend Oreille (PO) were grown for three months at 20 ℃ and 30 ℃. Color code indicating $\log_2$ of fold-changes of amino acid concentrations of trees grown at 30 ℃ and 20 ℃; blue and red colors indicate increased and decreased concentrations at 30 ℃, respectively. order of the four squares besides each amino acid from left to right: fold-changes of MC and PO of previous year and current year needles, respectively. Values of concentrations below detection limit were replaced by the smallest positive value in matrix for fold-change calculation.

(见彩图 C)

The expected higher temperatures in a future climate are most likely to increase the loss of carbon by enhanced photorespiration [77-80] and leaf dark respiration [81]. We therefore focused on N compounds related to photorespiration. However, in needles of both Douglas fir provenances, NH_4^+, glycine and serine concentrations were not changed significantly by elevated temperature. However, metabolite contents alone may not be sufficient to detect changes in photorespiration rates. Metabolite flux analyses or gas exchange measurements separating photosynthesis, dark respiration and photorespiration would be required for a better understanding of the contribution of each of these processes to net CO_2 exchange between Douglas fir needles and the atmosphere.

5 Conclusion

The present study showed that HT affects Douglas fir N partitioning in a provenance specific manner. Whereas provenance MC from a drier origin did not react strongly to HT treatment, provenance PO originating from a more humid environment showed clearly reduced contents of N containing compounds, particularly of amino acids involved in transport and storage of reduced N, in current year needles. We hypothesize that particularly trees of Douglas fir provenances showing high sensitivity to HT may experience even stronger N deficiency if grown under N limiting conditions. Our results underline the importance of research at the level of provenances in order to improve present knowledge on climate change related effects on the performance of forest trees.

Funding

This work was financially supported by the German Science Foundation (DFG, grants KR 2010/4-1, EN829/5-1 and GE1090/7-1), the Forest Research Institute of the German State Baden-Württemberg (FVA), and the Natural Sciences and Engineering Research Council of Canada (NSERC). Baoguo Du obtained funding from the China Scholarship Council, the Ministerium für Wissenschaft, Forschung und Kunst Baden-Württemberg of Germany, and the University of Freiburg.

References

[1] Smith W B, Miles P D, Perry C H, et al. Forest Resources of the United States, 2007: A Technical Document Supporting the Forest Service 2010 RPA Assessment. Washington D.C.: USDA Forest Service, 2009.

[2] Hermann R K. The genus *Pseudotsuga*: Ancestral History and Past Distribution. Oregon: Forest Research Laboratory, College of Forestry, Oregon State University, 1985.

[3] Perić S, Jazbec A, Tijardović M, et al. Provenance studies of Douglas fir in the locality of "Kontija" (Istria). Periodicum Biologorum, 2009, 111: 487-493.

[4] Brandl H. Ergänzende Untersuchungen zur Ertragslage der Baumarten Fichte, Kiefer, Buche und Eiche in Baden-Württemberg. Allg Forst Jagdztg, 1989, 160: 91-98.

[5] Hermann R K, Lavender D P. Douglas-fir planted forests. New Forests, 1999, 17: 53-70.

[6] Hein S, Weiskittel A R, Kohnle U. Effect of wide spacing on tree growth, branch and sapwood properties of young Douglas-fir [*Pseudotsuga menziesii* (Mirb.) Franco] in south-western Germany. European Journal of Pharmacology, 2008, 127: 481-493.

[7] Spielmann M, Bücking W, Quadt V, et al. Integration of nature protection in forest policy in Baden-Württemberg (Germany). Freiburg: INTEGRATE Country Report EFICENT-OEF, 2013.

[8] Ranger J, Marques R, Colin-Belgrand M, et al. The dynamics of biomass and nutrient accumulation in a Douglas-fir (*Pseudotsuga menziesii* Franco) stand studied using a chronosequence approach. Forest Ecology and Management, 1995, 2: 167-183.

[9] Keyes M R, Grier C C. Above-and below-ground net production in 40-year-old Douglas-fir stands on low and high productivity sites. Canadian Journal of Forest Research, 1981, 11: 599-605.

[10] Kantor P. Production potential of Douglas fir at mesotrophic sites of Křtiny Training Forest Enterprise. Journal of Forest Science, 2008, 54: 321-332.

[11] Hobbie E, Olszyk D, Rygiewicz P, et al. Foliar nitrogen concentrations and natural abundance of ^{15}N suggest nitrogen allocation patterns of Douglas-fir and mycorrhizal fungi during development in elevated carbon dioxide concentration and temperature. Tree Physiology, 2001, 21: 1113-1122.

[12] Christensen J H, Hewitson B, Busuioc A, et al.Regional climate projections//Solomon S, Qin D, Manning M, et al. Climate Change 2007: The Physical Science Basis Contribution of Working group I to the Fourth Assessment Report of the Intergovernmental Panel on Climate Change. Cambridge: Cambridge University Press, 2007: 847–943.

[13] Wahid A, Gelani S, Ashraf M, et al. Heat tolerance in plants: an overview. Environmental and Experimental Botany, 2007, 61: 199-223.

[14] Weston D J, Bauerle W L. Inhibition and acclimation of C_3 photosynthesis to moderate heat: a perspective from thermally contrasting genotypes of *Acer rubrum* (red maple). Tree Physiology, 2007, 27: 1083-1092.

[15] Raitio H, Sarjala T. Effect of provenance on free amino acid and chemical composition of Scots pine needles. Plant and Soil, 2000, 221: 231-238.

[16] Jansen K, Sohrt J, Kohnle U, et al. Tree ring isotopic composition, radial increment and height growth reveal provenance-specific reactions of Douglas-fir towards environmental parameters. Trees, 2012, 27: 37-52.

[17] Edfast A B, Näsholm T, Ericsson A. Free amino acid concentrations in needles of Norway spruce and Scots pine trees on different sites in areas with two levels of nitrogen deposition. Canadian Journal of Forest Research, 1990, 20: 1132-1136.

[18] Schmeink B, Wild A. Studies on the content of free amino acids in needles of undamaged and damaged spruce trees at a natural habitat. Journal of Plant Physiology, 1990, 136: 66-71.

[19] Way D A, Oren R. Differential responses to changes in growth temperature between trees from different functional groups and biomes: a review and synthesis of data. Tree Physiology, 2010, 30: 669-688.

[20] Ormrod D P, Lesser V M, Olszyk D M, et al. Elevated temperature and carbon dioxide affect chlorophylls and carotenoids in Douglas-fir seedlings. International Journal of Plant Sciences, 1999, 160: 529-534.

[21] Lewis J, Lucash M, Olszyk D, et al. Seasonal patterns of photosynthesis in Douglas fir seedlings during the third and fourth year of exposure to elevated CO_2 and temperature. Plant Cell and Environment, 2001, 24: 539-548.

[22] Tingey D T, Mckane R B, Olszyk D M, et al. Elevated CO_2 and temperature alter nitrogen allocation in Douglas-fir. Global Change Biology, 2003, 9: 1038-1050.

[23] Margolis H, Waring R. Carbon and nitrogen allocation patterns of Douglas-fir seedlings fertilized with nitrogen in autumn. I. Overwinter metabolism. Canadian Journal of Forest Research, 1986, 16: 897-902.

[24] Evans J R. Photosynthesis and nitrogen relationships in leaves of C_3 plants. Oecologia,1989, 78: 9-19.

[25] Lewis J, Lucash M, Olszyk D, et al. Relationships between needle nitrogen concentration and photosynthetic responses of Douglas‐fir seedlings to elevated CO_2 and temperature. New Phytologist ,

2004, 162: 355-364.

[26] Rennenberg H, Kreutzer K, Papen H, et al. Consequences of high loads of nitrogen for spruce (*Picea abies*) and beech (*Fagus sylvatica*) forests. New Phytologist, 1998, 139: 71-86.

[27] Van den Driessche R, Webber J. Total and soluble nitrogen in Douglas fir in relation to plant nitrogen status. Canadian Journal of Forest Research, 1975, 5: 580-585.

[28] Pérez-Soba M, de Visser P H. Nitrogen metabolism of Douglas fir and Scots pine as affected by optimal nutrition and water supply under conditions of relatively high atmospheric nitrogen deposition. Trees, 1994, 9: 19-25.

[29] IPCC. The Fourth Assessment Report of the Intergovernmental Panel on Climate Change. Geneva, Switzerland: IPCC, 2007.

[30] Bassman J H, Edwards G E, Robberecht R. Long-term exposure to enhanced UV-B radiation is not detrimental to growth and photosynthesis in Douglas-fir. New Phytologist, 2002, 154: 107-120.

[31] Olszyk D M, Johnson M G, Tingey D T, et al. Whole-seedling biomass allocation, leaf area, and tissue chemistry for Douglas-fir exposed to elevated CO_2 and temperature for 4 years. Canadian Journal of Forest Research, 2003, 33: 269-278.

[32] Van den Driessche R, Webber J. Variation in total and soluble nitrogen concentrations in response to fertilization of Douglas-fir. Forest Science, 1977, 23: 134-142.

[33] Margolis H, Waring R. Carbon and nitrogen allocation patterns of Douglas-fir seedlings fertilized with nitrogen in autumn. II. Field performance. Canadian Journal of Forest Research, 1986, 16: 903-909.

[34] Bedell J, Chalot M, Garnier A, et al. Effects of nitrogen source on growth and activity of nitrogen-assimilating enzymes in Douglas-fir seedlings. Tree Physiology, 1999, 9: 205-210.

[35] Kjeldahl J. Neue methode zur bestimmung des stickstoffs in organischen körpern. Fresenius Journal of Analytical Chemistry, 1883, 22: 366-382.

[36] Winter H, Lohaus G, Heldt H W. Phloem transport of amino acids in relation to their cytosolic levels in barley leaves. Plant Physiology, 1992, 99: 996-1004.

[37] Luo Z B, Janz D, Jiang X, et al. Upgrading root physiology for stress tolerance by ectomycorrhizas: insights from metabolite and transcriptional profiling into reprogramming for stress anticipation. Plant Physiology, 2009, 151: 1902-1917.

[38] Dannenmann M, Simon J, Gasche R, et al. Tree girdling provides insight on the role of labile carbon in nitrogen partitioning between soil microorganisms and adult European beech. Soil Biology and Biochemistry, 2009, 41: 1622-1631.

[39] Mantoura R F C, Jeffrey S W, Llewellyn C A, et al. Comparison between spectrophotometric, fluorometric and HPLC methods for chlorophyll analysis//Jeffrey S W, Mantoura R F C, Wright S W. Phytoplankton Pigments in Oceanography: Guidelines to Modern Methods. Paris: UNESCO, 1997: 361-380.

[40] Lichtenthaler H K. Chlorophylls and carotenoids: pigments of photosynthetic biomembranes. Methods in Enzymology, 1987, 148: 350-382.

[41] Kramer P, Kozlowski T. Physiology of Woody Plants. New York: Academic Press Inc., 1979.

[42] Cleve K V, Oechel W C, Hom J L. Response of black spruce (*Picea mariana*) ecosystems to soil temperature modification in interior Alaska. Canadian Journal of Forest Research, 1990, 20: 1530-1535.

[43] van Breemen N, Jenkins A, Wright R F, et al. Impacts of elevated carbon dioxide and temperature on a boreal forest ecosystem (CLIMEX project). Ecosystems, 1998, 1: 345-351.

[44] Rygiewicz P T, Martin K J, Tuininga A R. Morphotype community structure of ectomycorrhizas on Douglas fir (*Pseudotsuga menziesii* Mirb. Franco) seedlings grown under elevated atmospheric CO_2 and temperature. Oecologia, 2000, 124: 299-308.

[45] Faw W F, Shih S C, Jung G A. Extractant influence on the relationship between extractable proteins and cold tolerance of alfalfa. Plant Physiology, 1976, 57: 720-723.

[46] Kandler O, Dover C, Ziegler P. Frost hardiness in spruce. 1. Control of frost hardiness, carbohydrate-metabolism and protein-metabolism by photoperiod and temperature. Berichte Der Deutschen Botanischen Gesellschaft, 1979, 92: 225-241.

[47] Cloutier Y. Changes in the electrophoretic patterns of the soluble proteins of winter wheat and rye following cold acclimation and desiccation stress. Plant Physiology, 1983, 71: 400-403.

[48] Dewar R C, Medlyn B E, Mcmurtrie R. Acclimation of the respiration/photosynthesis ratio to temperature: insights from a model. Global Change Biology, 1999, 5: 615-622.

[49] Ebell L, McMullan E E. Nitrogenous substances associated with differential cone production responses of Douglas fir to ammonium and nitrate fertilization. Canadian Journal of Forest Research, 1970, 48: 2169-2177.

[50] Chalot M, Botton B, Banvoy J. Seasonal fluctuations of growth and nitrogen metabolism in the ectomycorrhizal association Douglas fir-Laccaria laccata. Agriculture Ecosystems and Environment, 1990, 28: 59-64.

[51] Durzan D. Nitrogen metabolism of *Picea glauca*. I. Seasonal changes of free amino acids in buds, shoot apices, and leaves, and the metabolism of uniformly labelled ^{14}C-L-arginine by buds during the onset of dormancy. Canadian Journal of Forest Research, 1968, 46: 909-919.

[52] Carrow J R. Free amino acids in grand fir needles and the effects of different forms of foliar-applied nitrogen. Canadian Journal of Forest Research, 1973, 3: 466-471.

[53] Lazarus W. Purification of plant extracts for ion-exchange chromatography of free amino acids. Journal of Chromatography A, 1973, 87: 169-178.

[54] Krupa S, Branstrom G. Studies on the nitrogen metabolism in ectomycorrhizae. II. Free and bound amino acids in the mycorrhizal fungus *Boletus variegatus*, in the root systems of *Pinus sylvestris* and during their association. Physiologia Plantarum, 1974, 31: 279-283.

[55] Durzan D J, Steward F C. Nitrogen metabolism // Steward F C. Plant Physiology, An Advanced Treatise. New York: Academic Press Inc., 1983: 55-265.

[56] Gezelius K. Free amino acids and total nitrogen during shoot development in Scots pine seedlings. Physiologia Plantarum, 1986, 67: 435-441.

[57] Näsholm T, Ericsson A. Seasonal changes in amino acids, protein and total nitrogen in needles of

fertilized Scots pine trees. Tree Physiology, 1990, 6: 267-281.

[58] Gezelius K, Näsholm T. Free amino acids and protein in Scots pine seedlings cultivated at different nutrient availabilities. Tree Physiology, 1993, 13: 71-86.

[59] Funck D, Stadelhofer B, Koch W. Ornithine-δ-aminotransferase is essential for arginine catabolism but not for proline biosynthesis. BMC Plant Biology, 2008, 8: 40.

[60] Huhn G, Schulz H. Contents of free amino acids in Scots pine needles from field sites with different levels of nitrogen deposition. New Phytologist, 1996, 134: 95-101.

[61] Stoermer H, Seith B, Hanemann U, et al. Nitrogen distribution in young Norway spruce (*Picea abies*) trees as affected by pedospheric nitrogen supply. Physiologia Plantarum, 1997, 101: 764-769.

[62] Weber P, Stoermer H, GEßLER A, et al. Metabolic responses of Norway spruce (*Picea abies*) trees to long-term forest management practices and acute $(NH_4)_2SO_4$ fertilization: transport of soluble non-protein nitrogen compounds in xylem and phloem. New Phytologist, 1998, 40: 461-475.

[63] Nordin A, Uggla C, Näsholm T. Nitrogen forms in bark, wood and foliage of nitrogen-fertilized *Pinus sylvestris*. Tree Physiology, 2001, 21: 59-64.

[64] Näsholm T. Removal of nitrogen during needle senescence in Scots pine (*Pinus sylvestris* L.). Oecologia, 1994, 99: 290-296.

[65] Millard P, Grelet G A. Nitrogen storage and remobilization by trees: ecophysiological relevance in a changing world. Tree Physiology, 2010, 30: 1083-1095.

[66] Krueger K W. Nitrogen, phosphorus, and carbohydrate in expanding and year-old Douglas-fir shoots. Forest Science, 1967, 13: 352-356.

[67] Splittstoesser W E, Meyer M M. Evergreen foliage contributions to the spring growth of *Taxus*. Physiologia Plantarum, 1971, 24: 528-533.

[68] van den Driessche R. Nutrient storage, retranslocation and relationship of stress to nutrition// Bowen G D, Nambiar E K S. Nutrition of Plantation Forests Nambiar. Michigan: Academic Press, 1984: 181-209.

[69] Manderscheid R, Jäger H J. Seasonal changes in nitrogen metabolism of Spruce needles (*Picea abies* (L.) Karst.) as affected by water stress and ambient air pollutants. Journal of Plant Physiology, 1993, 141(4): 494-501.

[70] Millard P. Ecophysiology of the internal cycling of nitrogen for tree growth. Journal of Plant Nutrition and Soil Science, 1996, 159: 1-10.

[71] Camm E. Photosynthetic responses in developing and year-old Douglas-fir needles during new shoot development. Trees, 1993, 8: 61-66.

[72] Palfi G, Koves E, Bito M, et al. The role of amino acids during water-stress in species accumulating proline. Phyton Argentina, 1974, 32: 121-127.

[73] Schneider S, Gessler A, Weber P, et al. Soluble N compounds in trees exposed to high loads of N: a comparison of spruce (*Picea abies*) and beech (*Fagus sylvatica*) grown under field conditions. New Phytologist, 1996, 134: 103-114.

[74] Gessler A, Schneider S, Weber P, et al. Soluble N compounds in trees exposed to high loads of N: a comparison between the roots of Norway spruce (*Picea abies*) and beech (*Fagus sylvatica*) trees grown

under field conditions. New Phytologist, 1998, 138: 385-399.

[75] Fotelli M N, Nahm M, Heidenfelder A, et al. Soluble nonprotein nitrogen compounds indicate changes in the nitrogen status of beech seedlings due to climate and thinning. New Phytologist, 2002, 154: 85-97.

[76] Thomson A, Goodenough D, Barclay H, et al. Effects of laminated root rot (*Phellinus weirii*) on Douglas-fir foliar chemistry. Canadian Journal of Forest Research, 1996, 26: 1440-1445.

[77] Clark R, Menary R. Environmental effects on peppermint (*Mentha piperita* L.). II. Effects of temperature on photosynthesis, photorespiration and dark respiration in peppermint with reference to oil composition. Functional Plant Biology, 1980, 7: 693-697.

[78] Long S. Modification of the response of photosynthetic productivity to rising temperature by atmospheric CO_2 concentrations: has its importance been underestimated? Plant Cell and Environment, 1991, 14: 729-739.

[79] Kozaki A, Takeba G. Photorespiration protects C_3 plants from photooxidation. Nature, 1996, 384: 557-560.

[80] Peñuelas J, Llusià J. Linking photorespiration, monoterpenes and thermotolerance in *Quercus*. New Phytologist, 2002, 155: 227-237.

[81] Lewis J, Olszyk D, Tingey D. Seasonal patterns of photosynthetic light response in Douglas-fir seedlings subjected to elevated atmospheric CO_2 and temperature. Tree Physiology, 1999, 19: 243-252.

Effects of copper-carbon core-shell nanoparticles on the physiology of bald cypress (*Taxodium distichum*) seedlings

Li Yongsheng Chen Cungen Qi Yadong

Abstract

Nanotechnology and the development of new nanomaterials opened up potential and novel applications in agriculture and biotechnology. The potential risk and safety concern caused by the releasing nanoparticles into environmental compartments and ecosystems has to be taken into consideration. As a new generation of renewable composite nanomaterial, the Copper-Carbon Core-Shell Nanoparticles (CCCSNs) possess many unique properties. The interaction between CCCSNs and living organisms and the impacts of CCCSNs on living organisms needed to be understand before this product can be commercialized. The effects of four levels of CCCSNs (with equivalent pure Cu concentrations of 0ppm, 750ppm, 1500ppm, 3000ppm) soil treatments on plant gas exchange rates, growth, and development of one-year-old bald cypress seedlings was quantified. Results demonstrated that there were no negative effects of a series of high level of CCCSNs on the physiology of cypress seedlings during a 3-month greenhouse experiment. The CCCSNs through soil treatments can be reported to be low toxic to the growth and development of one-year-old bald cypress seedlings.

Keywords: Nanoparticle; Copper; Bald cypress; Physiology

The role of nanomaterials in modern technologies is becoming increasingly important because of the feasibility and ease of adding new functions to the existing products. Many nanoparticles have been used for commercial purposes such as fillers, semiconductors, cosmetics, microelectronics, and drug carriers [1]. With the production, use, and disposal, nanomaterials inevitably are released into air, water, and soil. The potential risk and safety concern caused by the releasing nanoparticles into environmental compartments and ecosystems has to be taken into consideration.

Several studies have been conducted on the interactions between nanomaterials and plants. Study reported that iron-carbon nanoparticles were able to penetrate living plant tissues and migrate to different regions of the plant with short distances [2]. It showed that the alfalfa had the ability to actively uptake gold from a gold rich agar system [3]. It was possible to use alfalfa plants to fabricate nanoparticles in an $AuCl_4$ rich environment. However, another study by [4] indicated that the no translocation of cerium dioxide was detected in maize plants, which indicated the resistance of biological barriers of plants to translocation of the nanoparticles. Recently, some research has been done on toxicity of nanoparticles to plants, and most work has been focused on the seed germination and the root elongation tests [5-7]. The plant species used in these studies were annual agricultural and garden crops, such as spinach, radish, rape, ryegrass, lettuce, and corn. Only one research was found to study the toxicity of nanoparticles on woody plant by [8], which indicated that willow trees were not sensitive to titanium dioxide nanoparticles during a short-term exposure, due to the instability of the nanoparticles. Studies on mechanism of toxicity of nanoparticles have been carried out. It was reported that silver nanoparticles had effects on reducing ability of photosynthesis in algae through both the nanoparticles and silver ions released from the nanoparticles [9]. conducted a plant agar test, which showed that Cu nanoparticles were able to penetrate through the cell membrane and might agglomerate in the cells of the test species, causing possible toxicity[10]. However, the mechanisms of nanoparticles toxicity in plants are not fully understood.

As a new generation of renewable composite nanomaterials, Copper-Carbon Core-Shell Nanoparticles (CCCSNs) were produced using natural fibers through a technology patent [11]. CCCSNs are generally spherical in shape, and each particle contains a copper core that is encased in a carbon shell. The growth of the carbon shell structure follows a heteroepitaxial pattern (Fig. 1). Because of this core-shell structure, the CCCSNs possess many unique properties. Since the carbon shell of CCCSNs protects copper core from being rapidly oxidized when exposed to air, these particles show extreme chemical and physical stabilities in aqueous environments within a wide range of pH solution with pH values from 1 to 11 [11]. Preliminary tests by Lian and Wu [11] have been conducted on the interactions between CCCSNs and plants. One test showed that the Cu uptake was enhanced in freshly cut loblolly pine branches immersed in the CCCSNs slurry. Another test showed that the Cu uptake was increased in the stem of rose bush, which was grown in a container containing nursery soil applied with CCCSNs slurry. However, no prior research has been conducted on the interaction between CCCSNs and live trees of forest species.

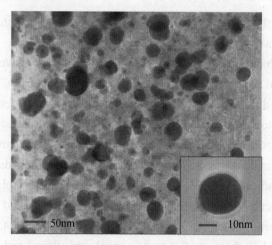

Fig. 1 Transmission electron micrographs of CCCSNs

Note: The average diameter of CCCSNs is about 30-50nm. The inset shows an individual CCCSN in a format of copper core covered by a porous carbon shell [12].

As a gymnosperm species, bald cypress (*Taxodium distichum*) is a large, deciduous conifer tree typically growing in swamps, where it forms extensive pure stands. Its location is primarily in the southeastern United States [12, 13]. The natural distribution of bald cypress is from Delaware Bay south, extends along the Atlantic Coastal Plain to Florida and westward along the Gulf of Mexico to Texas and southeastern Oklahoma, and also inland up to the Mississippi and Ohio River as far north as southern Illinois and southwestern Indiana [14]. Bald cypress is known as the dominant tree in the Gulf of Mexico coastal wetlands of Louisiana, USA. It is of great ecological importance and apparent vulnerability to current environmental changes, including contaminants such as heavy metals, salt water intrusion, and agricultural chemicals such as nitrates, phosphates and pesticides [15]. Many studies have been focusing on bald cypress' tolerance to environmental factors. It has been described that bald cypress is very tolerant to flooding [16]. A few studies have demonstrated variations to salt tolerance in morphological and physiological characteristics in bald cypress. The responses of bald cypress to flooding and salinity [17-19], as well as to both the stresses combined [20] were also reported.

In the present study, we evaluated the effects of four levels of CCCSNs on the physiology and growth of one-year-old bald cypress seedlings. The primary objectives were to quantify the effects of CCCSNs applications on ① uptake of Cu, ② gas exchange rates, and ③ above- and below-ground growth and biomass in one-year-old bald cypress seedlings.

1 Materials and Methods

1.1 Plant materials

One-year-old bare root bald cypress seedlings were purchased in January, 2011 from Louisiana Department of Agriculture and Forestry and planted thereafter in containers and grown in a greenhouse at Southern University Horticulture Farm. The seedlings were potted individually with 125 grams of oven dried soil (at 60 ℃ for 72 hours), in a 6.8-cm-diameter by 25.5-cm-deep plastic Cone-tainers (Ray Leach Cone-tainer, Hummert International, Earth City, MO) with slotted bottom for drainage. Soil used was Miracle-Gro Garden Soil for Trees & Shrubs with NPK analysis of 0.15-0.05-0.10 (Table 1).

Table 1 Nutrients contents of soil from The Scotts Miracle-Gro Company Verti-Gro Guaranteed Nutrient Analysis

Nutrient	Contents(%)
Total nitrogen (N)+	0.15
ammoniacal nitrogen	0.05
nitrate nitrogen	0.05
water insoluble nitrogen	0.05
Available phosphate (P_2O_5)+	0.05
Soluble potash (K_2O)+	0.10
iron (Fe)	0.10
water soluble iron (Fe)	0.10

Each potted seedling was watered as needed to maintain moist soil conditions and fertilized with a 5N-10P-25K hydrosol hydroponic fertilizer (applied at 3.0 g/L) (Verti-gro, Summerfield, FL), was applied once every month. This concentration was used throughout the experiment and water was applied by hand. The analysis of the hydroponic fertilizer is shown in Table 2.

Table 2 Nutrients contents of hydroponic fertilizer from Verti-Gro

Nutrient	Contents(%)
Total Nitrogen (N)	5
Nitrate Nitrogen (min)	15
Ammoniacal Nitrogen (max)	0.00
Urea Nitrogen (N)	0.00
Avaliable Phosphoric Acid (P_2O_5)	10

Nutrient	Continued Contents(%)
Soluble Pstash (K$_2$O)	25
Total Magnesium (Mg)	3
Water Soluble Magnesium (Mg)	3
Boron (B)	0.04
Copper (Cu)	0.02
Iron (Chelated) (Fe)	0.02
Total Manganese (M)	0.02
Soluble Manganese (Mn)	0.04
Molybdenum (Mo)	0.01
Zinc (Zn)	0.02

1.2 Preparation of CCCSNs

Copper-Carbon Core-Shell Nanoparticles (CCCSNs) were developed through a technology patent[11]. Before used in the application to the soil, this batch of synthesized CCCSNs went through in-house quality control which includes washing, grinding, and chemical analysis.

In the course of washing, 100 grams of the synthesized product was soaked in 700ml deionized water in a sealable wide-mouth Mason jar. Water was deionized and purified by a Direct-Q water purification system from Millipore SA (Molsheim, France). After 24hours of sedimentation, 500ml of top clear solution was taken out for checking pH level, and then 500ml of deionized water was added in the jar. After the jar tightly sealed, the mixture was vigorously shaken for a few minutes. The CCCSNs were washed and checked daily until the pH of top clear solution achieved a constant level. After washing, the CCCSNs residuals were baked in a Precision oven (Precision Scientific Co., Chicago, Illinois) at 120℃ for 48 hours until moisture was removed. Then, the CCCSNs were ready for grinding.

The aim of grinding the CCCSNs relates to a process for breaking down the soil-block-like agglomerates of CCCSNs to produce superfine particles. The equipment used in grinding was a single speed Waring commercial blender model 51BL32 (Waring Commercial, Torrington, CT). In each course of grinding, 5 grams agglomerates of CCCSNs was transferred into the 40-oz. glass container with stainless steel blade. The CCCSNs were homogenized for 5 minutes at room temperature and repeated a second time; after the materials settled, the blended product was ready for chemical analysis.

This method was simple to operate, fast, energy efficient, and requires no major equipment. The blending method was essentially free from typical contaminants, such as chemicals or residual solvents; since the container was sealed, this method also did not release any airborne dust during blending.

The CCCSNs was analyzed chemically after grinding. Total Al, B, Ca, C, Cu, Fe, Mg, Mn, N, P, K, Na, S, and Zn contents in the sample were determined by Inductively Coupled Plasma Atomic Emission Spectrometry (ICP-AES, Arcos Spectro Ltd., Germany). Chemical analysis result of CCCSNs used in one-year-old bald cypress seedlings is in Table 3.

Table 3 Chemical analysis results of CCCSNs by ICP-AES

Element	Content
Carbon(%)	16.2
Copper(%)	56.86
Sulfur(%)	0.39
Nitrogen(%)	0.06
Phosphorus(%)	0.891
Potassium(%)	0.001
Calcium(%)	0.02
Magnesium(%)	0.006
Sodium(ppm)	16.24
Zinc(ppm)	204.93
Aluminum(ppm)	268.41
Iron(ppm)	616.05
Boron(ppm)	6.89
Manganese(ppm)	4.37

The seedlings were grown for about 8 months in greenhouse. On August 22–23, 2011, 40 seedlings per species were randomly selected for treatment. The CCCSNs were suspended in deionized water. Magnetic stirring bars (L=1/2″, OD=1/8″, Fisher Scientific) and a Thermix stirring hot-plate (Model 610T, Fisher Scientific) were utilized to stir the suspension continuously at room temperature (18℃) for 20 minutes. Then 25 ml suspension was applied to the soil surface of each pot, by using the same method in the Chapter II. The plants were divided at random into four CCCSNs concentrations in order to create a factorial set-up of treatments based on pure copper content in CCCSNs (0 ppm, 750 ppm, 1500 ppm, 3000 ppm in soil).

1.3 Gas exchange measurements

A PAR of 1500μmol/(m^2·s) was used as an appropriate light intensity for measuring the maximum light-saturated photosynthetic rate (A_{max}) on a leaf-area basis. All measurements were taken with a portable Licor-6400 photosynthesis system (LI-6400, Li-Cor Inc., Lincoln, NE, USA)) equipped with a red/blue light source (6400–02B LED) on 3days, 13days, 23days, and 33 days after application of CCCSNs. Individually, unshaded, fully expanded bald cypress leaves from the most top five branches were selected for data collection. The reference CO_2 concentration in the leaf cuvette was maintained at 370 μmol/mol, and all measurements were taken between 09:00 am and 11:30 am on Central Standard Time. During the measurement, leaf cuvette temperature was maintained at 22–28℃, depending on the external temperature, and relative humidity inside the leaf cuvette was kept above 50%. Four seedlings in each treatment were selected as replicates. Actual leaf area of bald cypress seedling was re-calculated by using the same leaf area process in the Chapter II.

1.4 Chemical analysis and biomass comparison

In late November 2011, four seedlings from each treatment were collected. Each seedling was separated into above ground part (shoot) and below ground part (root) at the root collar (Fig. 2). The shoot was equally cut to two sections based on the total height of the shoots. For each shoot section, the main stem, lateral branches and leaves were separated. The root was rinsed with deionized water, dried with paper towels, and equally cut into two sections based on the length of the main root. For each root section, the lateral roots were separated from the main root. Totally, samples from eight sections (top leaves (LV-TOP), top lateral branches (LB-TOP), top main stem (MS-TOP), bottom main stem (MS-BOT), overall above ground, top fibrous roots (FR-TOP), top main roots (MR-TOP), bottom fibrous roots (FR-BOT), bottom main roots (MR-BOT)) of each seedling were collected.

The biomass dry weights of roots, main stems, lateral branches and leaves were obtained by weighing the plant material after drying at 60℃ for 48 hours. The dry samples of the plants were separately ground using a Wiley Mini-Mill (Thomas Scientific) with a 40 mesh screen for elemental analysis. Three seedlings from each treatment were used as replicates. Chemical contents in the plants were determined by an inductively coupled plasma atomic emission spectrometry (ICP-AES, Arcos Spectro Ltd., Germany).

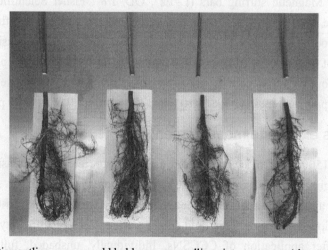

Fig. 2　Dismantling one-year-old bald cypress seedlings into two parts (shoots and roots)

1.5 Data analysis

The experiments were conducted in a total randomized design, to compare the effects of four concentrations of CCCSNs applied in soil on one-year-old bald cypress seedlings. Statistical tests were run with SAS statistical software 9.2 (SAS Institute). Variables between treatments including A_{max}, g_s, T_r, biomass and chemical contents of each section, as well as total shoot and total root, were analyzed with one-way ANOVA followed by Duncan's multiple range test at a significance level of $P=0.05$.

2 Results and Analysis

2.1 Leaf gas exchange

The presence of CCCSNs in soil did not cause differences in photosynthesis, stomatal conductance, or transpiration rates of the bald cypress seedlings. Mean value of A_{max} ranged from approximately 7.58μmol/(m² · s) to 13.33 μmol/(m² · s); g_s ranged from 0.090mol/(m² · s) to 0.207mol/(m² · s); and T_r ranged from 1.39mmol/(m² · s) to 3.72mmol/(m² · s)(Table 4).

Table 4 Maximum photosynthesis (A_{max}), stomatal conductance (g_s) and transpiration (T_r) of one-year-old bald cypress seedlings with four CCCSNs treatments (0 ppm, 750 ppm, 1500 ppm and 3000 ppm) at four intervals (day 3, 13, 23, and 33)

Day	Item	CCCSNs concentrations(ppm)				P-value
		0	750	1500	3000	
3	A_{max} [μmol/(m² · s)]	10.44	11.23	9.35	9.58	0.7667
	g_s [mol/(m² · s)]	0.12	0.12	0.1	0.1	0.8492
	T_r [mmol/(m² · s)]	3.11	3.03	2.75	2.53	0.7809
13	A_{max} [μmol/(m² · s)]	7.58	8.96	8.67	7.94	0.9015
	g_s [mol/(m² · s)]	0.11	0.13	0.12	0.09	0.8289
	T_r [mmol/(m² · s)]	1.59	1.91	1.8	1.39	0.8154
23	A_{max} [μmol/(m² · s)]	11.9	11.69	13.33	12.69	0.9332
	g_s [mol/(m² · s)]	0.17	0.17	0.21	0.17	0.4412
	T_r [mmol/(m² · s)]	2.79	2.89	3.43	2.87	0.4153
33	A_{max} [μmol/(m² · s)]	11.07	10.08	10.47	10.84	0.6939
	g_s [mol/(m² · s)]	0.17	0.17	0.17	0.17	0.9929
	T_r [mmol/(m² · s)]	3.7	3.69	3.72	3.71	0.9995

Note: The mean is an average of four samples ($n=4$).

2.2 Biomass

The presence of CCCSNs (0ppm, 750ppm, 1500ppm, and 3000ppm) did not affect the dry weights of the shoots and the roots of the plants. The values of dry root to shoot ratio were maintained constant independent of increasing concentrations of CCCSNs in the soil (Table 5).

Table 5 Dry shoot weight, dry root weight, and dry root to shoot ratio of one-year-old bald cypress seedlings

Item	CCCSNs concentrations (ppm)				P-value
	0	750	1500	3000	
dry shoot weight (gram)	5.18	5.69	4.28	4.95	0.0506
dry root weight (gram)	5.55	6.35	5.61	5.47	0.568
dry root to shoot ratio	1.08	1.12	1.32	1.12	0.5345

Note: The mean is an average of three samples (n=3).

2.3 Chemical composition

Although there were no significant differences in the dry biomass weights or physiological behavior, it was observed that, after the seedlings were exposed to increasing CCCSNs concentrations in soil, copper (Cu) contents in the total dry shoot and root tissues also increased (Table 6). The Cu content in shoots was 6.82ppm in seedlings treated with 1500ppm CCCSNs; and the Cu content was only 4.47ppm in the shoots of the control plants. This indicated a 153% rise in Cu content. The Cu content of the roots of the seedlings treated with 3000ppm CCCSNs was 424.45ppm, and the Cu content of the roots of the control plants was 32.59ppm. This indicated a 1302% rise in Cu level. This result showed that, the Cu level increased in roots was much greater than in shoots of bald cypress seedlings treated with CCCSNs (Fig. 3.(a)).

Table 6 Cu contents of one-year-old bald cypress seedlings treated with 0ppm, 750ppm, 1500 ppm and 3000ppm of CCCSNs

Section	CCCSNs concentrations (ppm)								P-value
	0		750		1500		3000		
LV-TOP	2.56	b	2.31	c	2.95	a	3.06	a	0.0002
LB-TOP	5.44	c	6.17	bc	6.60	b	7.64	a	0.0057
MS-TOP	5.17	d	6.49	c	5.58	b	6.89	a	<0.0001
MS-BOT	4.69	c	7.27	b	6.65	b	9.68	a	0.0002
FR-TOP	82.05	d	780.04	b	599.93	c	1450.80	a	<0.0001
MR-TOP	6.06	b	143.68	a	69.23	ab	102.40	a	0.0436

Continued

Section	CCCSNs concentrations (ppm)								P-value
	0		750		1500		3000		
FR-BOT	34.67	b	140.68	a	84.97	ab	136.77	a	0.0256
MR-BOT	5.38		12.53		9.46		7.85		0.0903

Note: Within each row, means with the same letter are not statistically significant according to Duncan's multiple range test at $P=0.05$. The mean is an average of three samples ($n=3$).

In shoots, Cu contents of top leaves, top lateral branches, top main stem, and bottom main stem of seedlings treated with 3000ppm CCCSNs increased 19%, 41%, 33%, and 106%, respectively, compared with the control. Cu contents of top fibrous root, top main roots, bottom fibrous roots, and bottom main roots increased 1668%, 1589%, 294%, and 46%, respectively, compared to the control (Fig. 3(b)). The results indicated Cu contents of the top sections of the roots (top fibrous root, and top main roots) increased most.

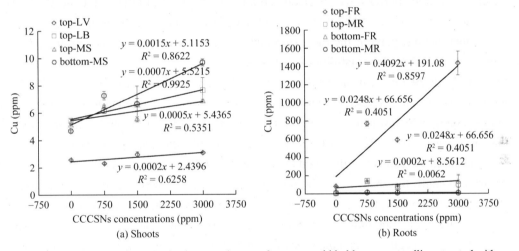

Fig. 3 Variation in Cu contents in shoots and roots of one-year-old bald cypress seedlings treated with CCCSNs (0 ppm, 750 ppm, 1500 ppm, and 3000 ppm)

Note: The mean is an average of three samples ($n=3$).

3 Discussion

The elevated concentrations of CCCSNs in the soil did not affect the efficiency of photosynthesis, or stomatal conductance of the one-year-old bald cypress seedlings. The CCCSNs treatments did not affect the transpiration rate of the plants as well. Roots serve to regulate the flow of water and nutrients into the plant, and to regulate gas exchange in leaf, thus, roots and shoots behavior plays an important role in exhibiting physiological status of a plant. There were no significant differences in roots and shoots biomass between the seedlings treated with CCCSNs and the control. It indicated that the bald cypress seedlings were undamaged from CCCSNs treatments. This was also in conjunction with non-adverse impacts of CCCSNs on the leaf gas exchange.

It has been suggested that 12mg/kg is a global value for Cu concentration in plant biomass in unpolluted conditions [21] and the values higher than 20ppm lead to toxic responses in plants [22]. Cu contents in the fibrous roots of the seedlings treated with 3000ppm CCCSNs was quite high. It was 1450 ppm in the top fibrous roots, which was 72.5-fold of the toxic threshold of Cu. At such high Cu content in plant tissue, Cu severely reduce plant growth, decrease enzyme activity of metabolism in plants, and even causes plant death [23]. However, in our research, one-year-old bald cypress seedlings treated with CCCSNs showed stable photosynthetic capacity and no change in plant biomass. This may prove that the high Cu contents in roots were caused by the attaching CCCSNs on the surface of the root system. It was reported that Cu contents were higher in roots than in shoots, due to the bounding of Cu of on cell walls in roots, which accounted for more than 60% of the total content [24]. Due to the application of CCCSNs in the top soil, the Cu content of top fibrous roots was much higher than the rest parts of the roots (top main root, bottom fibrous roots, and bottom main root) further confirmed the stability of CCCSNs.

Although the one-year-old bald cypress seedlings treated with CCCSNs had significant increase Cu content in shoots, especially in the bottom main stem, the highest Cu content was only 9.68ppm, which was lower than the toxicity threshold level (20ppm). This may explain that there was no toxic effect of CCCSNs on the shoots of bald cypress seedlings.

Increased Cu content in the plants treated with CCCSNs may affect the N, P, K content in plants mainly by changing their net uptake rate, which was suggested by Lidon and Henriques [25]. In our research, N, P, and K contents showed fluctuation patterns in different parts of the plant, but did not reveal any correlation with increasing concentrations of CCCSNs. It was suggested that the variations of N, P, and K, triggered by excess Cu, might be explained based on the interaction among Cu-uptake mechanism(s), plasma membrane-H^+ ATPase and root membrane permeability [25]. Increased N and K contents in the treated plants could also be related to the application of CCCSNs, since this batch of CCCSNs contained 0.06% nitrogen, and 0.891% phosphorus.

2000mg/L has been used as the highest concentration to investigate phytotoxicity of the selected nanoparticles in some researches. In our research, the concentrations of CCCSNs were much higher than 2000mg/L. The CCCSNs can be reported with minimal toxicity on test plants because they had no negative effects on the growth and development of one-year-old seedlings at such a high concentration. It is also suggested that the high concentration of nanoparticles in soil may enhance agglomeration, which in turn lowers the concentration of "free" nanoparticle [26]. This may also exist in our research.

4 Conclusion

Our results demonstrated that there were no negative effects of a series of high level of CCCSNs (with equivalent pure Cu concentration of 0ppm, 750ppm, 1500ppm, and 3000ppm) soil treatments on plant gas exchange, or biomass of one-year-old bald cypress seedlings during a 3-month greenhouse experiment. CCCSNs applied in the soil led to obvious Cu uptake in shoot and root tissues, but caused variations in nutrient uptake of nitrogen, phosphorus, and potassium, which were not correlated with increasing CCCSNs treatments. The CCCSNs through soil treatments can be reported to be low toxic to the growth and development of one-year-old bald cypress seedlings.

Reference

[1] Biswas, P, Wu C Y. anoparticles and the Environment. Journal of the Air and Waste Management Association, 2005, 55(6): 708-746.

[2] Corredor E, Testillano P S, Coronado M J, et al. Nanoparticle penetration and transport in living pumpkin plants: in situ subcellular identification. BMC Plant Biology, 2009, 9(1): 45.

[3] Gardea-Torresdey J, Parsons J G, Gomez E, et al. Formation and growth of Au nanoparticles inside live alfalfa plants. Nano Letters, 2002, 2(4): 397-401.

[4] Birbaum K, Brogioli R, Schellenberg M, et al. No evidence for cerium dioxide nanoparticle translocation in maize plants. Environmental Science and Technology, 2010, 44(22): 8718-8723.

[5] Zheng L, Hong F, Lu S, et al. Effect of nano-TiO_2, on strength of naturally aged seeds and growth of spinach. Biological Trace Element Research, 2005, 104(1): 83-91.

[6] Yang F, Hong F, You W, et al. Influence of nano-anatase TiO_2, on the nitrogen metabolism of growing spinach. Biological Trace Element Research, 2006, 110(2): 179-190.

[7] Lin D, Xing B. Phytotoxicity of nanoparticles: inhibition of seed germination and root growth. Environmental Pollution, 2007, 150(2): 243-250.

[8] Seeger E M, Baun A, Kästneret M, et al. Insignificant acute toxicity of TiO_2 nanoparticles to willow trees. Journal of Soils and Sediments, 2009, 9(1): 46-53.

[9] Lubick N. Nanosilver toxicity: ions, nanoparticles or both? Environmental Science and Technology, 2008, 42(23): 8617-8617.

[10] Lee W M, An Y J, Yoon H, et al. Toxicity and bioavailability of copper nanoparticles to the terrestrial plants mung bean (*Phaseolus radiatus*) and wheat (*Triticum aestivum*): plant agar test for water insoluble nanoparticles. Environmental Toxicology and Chemistry, 2008, 27(9): 1915-1921.

[11] Lian K, Wu Q. Carbon-Encased Metal Nano-particles and Sponges, Methods of Synthesis, and Methods of Use. US Patent Application13/249,558. (WIPO Patent Application WO/2007/095454)

[12] Hartzell H. The yew tree: a thousand whispers: biography of a species. Taxon, 1991, 41(2): 396.

[13] McNulty S G. Hurricane impacts on US forest carbon sequestration. Environmental Pollution, 2002,

116: S17-S24.

[14] Fowells H A.The Silvics of Forest Trees of the United States. Washington D.C.: U.S. Department of Agriculture, Forest Service, 1965.

[15] Dunbabin J S, Bowmer K H. Potential use of constructed wetlands for treatment of industrial wastewaters containing metals. Science of the Total Environment, 1992, 111(2-3): 151-168.

[16] Teskey R O, Hinckley T M. Impact of Water Level Changes on Woody Riparian and Wetland Communities. Volume II: The Southern Forest Region. Washington D.C.: Fish and Wildlife Service, Office of Biological Services, 1977.

[17] Pezeshki S R, Delaune R D, Patrick W H. Effect of Salinity on Leaf Ionic Content and Photosynthesis of *Taxodium distichum* L.. American Midland Naturalist, 1988, 119(1): 185-192.

[18] Conner W H, Askew G R. Response of baldcypress and loblolly pine seedlings to short-term saltwater flooding. Wetlands, 1992, 12(3): 230-233.

[19] Pezeshki S R. Response of *Pinus taeda* L. to soil flooding and salinity. Annales des Sciences Forestières, 1992, 49(2): 149-159.

[20] Allen J A, Pezeshki S R, Chambers J L. Interaction of flooding and salinity stress on baldcypress (*Taxodium distichum*). Tree Physiology, 1996, 16(1-2): 307-313.

[21] Marschner H. Mineral nutrition in higher plants. Plant, Cell and Environment, 1988, 11: 147-148.

[22] Stevenson F J, Cole M A. Cycles of Soil: Carbon, Nitrogen, Phosphorus, Sulfur, Micronutrients. Second Edition. John Wiley & Sons Inc., 1999.

[23] Sela M, Telor E, Fritz E, et al. Localization and toxic effects of cadmium, copper, and uranium in azolla. Plant Physiology, 1988, 88(1): 30-36.

[24] Wu L, Antonovics J. Zinc and copper uptake by Agrostis stolonifera, tolerant to both zinc and copper. New Phytologist, 1975, 75(2): 231-237.

[25] Lidon F C, Henriques F S. Effects of copper toxicity on growth and the uptake and translocation of metals in rice plants. Journal of plant nutrition, 1993, 16(8): 1449-1464.

[26] Franklin N M, Rogers N J, Apte S C, et al. Comparative toxicity of nanoparticulate ZnO, bulk ZnO, and $ZnCl_2$ to a freshwater microalga (*Pseudokirchneriella subcapitata*): the importance of particle solubility. Environmental Science and Technology, 2007, 41(24): 8484-8490.

降雨特征及小气候对秦岭油松林降雨再分配的影响*

陈书军 陈存根 曹田健 侯 琳 李荣华 张硕新

摘要

为了研究降雨特征及小气候对秦岭天然次生油松林冠层降雨再分配的影响,连续3年观测降雨事件发生时,油松林林外降雨、穿透降雨、树干茎流以及降雨特征和小气候因子,然后进行各因子间的相关性和多元回归分析。结果表明:降雨量级是影响林冠降雨再分配的最主要因素,穿透降雨量与降雨量、降雨量和降雨历时交互项、相对湿度呈正相关,与降雨历时和温度呈负相关。树干茎流量与降雨量、降雨量和温度交互项呈正相关,与温度呈负相关。林冠截留量与降雨量、降雨历时和温度呈正相关,与降雨量和降雨历时、温度的交互项呈负相关。雨前干燥期、风向、风速、蒸发速率、光合有效辐射和净辐射与穿透降雨量、树干茎流量和林冠截留量均无相关性。

关键词: 穿透降雨; 树干茎流; 林冠截留; 降雨量; 小气候; 生态水文

森林对降雨具有调节、转换和再分配功能,林冠将降雨重新再分配为穿透降雨、树干茎流和林冠截留,是森林生态系统水分传输的第一界面层[1]。林冠截留减少了进入林地的雨量,通过穿透降雨和树干茎流的方式将降雨输入到林内,不仅改变了降雨的空间分布格局,而且影响了与之相关联的物质循环和能量转换[2]。林冠对降雨分配过程的影响,是森林涵养水源、减少地表径流重要的生态水文过程[3],在森林生态系统水文循环和水量平衡中占有极其重要的地位,对局地或整个集水区的水分循环有着重要影响[4]。森林的存在所引起的生态系统内降雨时空分布的变化,是一个复杂的、动态的变化过程,受到林分特征[5]、降雨特征[6]、林分内外局地小气候因素直接或者综合的影响[7]。国内外学者针对这一过程已开展了广泛的研究,不同森林类型和气候特征致使穿透降雨率、树干茎流率和林冠截留率存在较大差异。如温带针阔叶林穿透降雨率变化范围为70%~90%[8-10],树干茎流率一般低于5%,很少超过10%[11,12],针叶林林冠截留率一般在20%~48%[11],我国主要森林生态系统林冠截留率在11%~37%[13]。

---

* 原载于:水科学进展,2013,24(4):513-521.

油松（*Pinus tabulaeformis*）是我国温性针叶林中分布最广的森林群落之一[14]，在秦岭林区分布较广，是秦岭山地的顶级群落之一，也是典型的地带性植被群落[15]。国内学者对油松林降雨再分配已有较多的研究，但大多数研究对象均为人工林，且林龄较低，研究地林分面积分布较小，主要研究了不同区域和林龄油松林冠层降雨再分配的规律和大小。本文以秦岭近成熟林阶段的天然次生油松林作为研究对象，通过对关键生态区域重要的水源涵养林（油松林）穿透降雨、树干茎流、降雨事件发生时的降雨特征以及林内和林外气象因子的长期定位观测，主要目的是研究不同降雨事件和天气条件下降雨再分配的特征、主要影响因子及其各影响因子间的相互关系，为正确全面评价油松林生态系统的生态水文功能和水源涵养林的经营管理提供理论依据。

1 研究区概况

研究地位于陕西省宁陕县境内的秦岭森林生态系统国家野外科学观测研究站，火地沟2号集水区油松林综合观测场内。地处北亚热带北缘（33°18′N，108°20′E），秦岭南坡中山地带，海拔1550～1700m，坡向西南向，坡度20°～30°，年均气温8～10℃，年均降水量900～1200mm，降雨多集中于7～9月，降雪从10月末到次年4月初，年均蒸发量800～950mm，土壤为山地棕壤，成土母岩主要为花岗岩和片麻岩[16]。林分主要为20世纪50、60年代采伐后，天然更新的次生林，林龄大多在51～60a，郁闭度0.85。乔木层伴生树种主要有锐齿栎（*Quercus aliena* var. *acuteserrata*）、华山松（*Pinus armandi*）、漆树（*Toxicodendron verniciflluum*）、青榨槭（*Acer davidii*）等。林下灌木主要有栓翅卫矛（*Euonymus phellomas*）、刚毛忍冬（*Lonicera hispida pall*）和白檀（*Symplocos paniculata*）、披针胡颓子（*Elaeagnus lanceolata*）等。林下草本主要有中华鳞毛蕨（*Dryopteris chinensis*）、野青茅（*Deyeuxia sylvatica*）、青菅（*Carex leucochlora*）、东亚唐松草（*Thalictrum minus*）、羽裂华蟹甲草（*Sinacalia tangutica*）等。

2 研究方法

2.1 气象要素观测

（1）林内气象因子

油松林综合观测场内建有27.5m高梯度因子观测塔一座，分层观测不同高度林分内气象因子的变化情况。数据采集器为CR23X，24h观测，每60min输出一次数据。根据整个林分的实际情况，从林内地面层到林分顶层1.5m高处共分为6层。第1层至第6层，分别为地表层（1.5m）、近地表层（7.0m）、林冠下层（14.0m）、林冠中层（19.5m）、林冠顶层（25.0m）、林冠顶层1.5m处（26.5m）。各层分别观测风速（A100LM）、风向（W200P）、空气温湿度（HMP45C_L）、净辐射（CNR-1）、光合有效辐射（LQS1070、LI190SB_L）及冠层温度（IRTS-P）。

（2）林外气象因子

距油松林综合观测场 500m 外的空旷地，建有自动气象站一座（UT30）。数据采集器为 CR1000，24h 观测，每 60min 输出一次数据。主要观测因子包括降雨量（TE525MM_15、52202_25）、风速（A100LM_L25）、风向（W200P_L25）、总辐射（LI200X_25）、光合有效辐射（LI190SB_25）、日照（CSD1）、空气温湿度（HMP45C_25）、蒸发（FS-01）T 及大气压（CS105）。

2.2 穿透降雨、树干茎流观测

在林分内选择郁闭度适中的位置 5 处，距树干 0.5m 起，在林下沿等高线向外安置离地面 1m 高，直径 20cm，长 4m 的 U 型穿透降雨收集器，然后将所收集的雨水导入集水器。选择 12 株标准木，距地面 1.3m 处起，分别用聚乙烯塑料管蛇形向下缠绕树干，用玻璃胶粘牢，沿管内侧削 4cm 长，0.8cm 宽的削面 2~4 个，将树干茎流导入管内，管基部连接集水器。最后再将穿透降雨和树干茎流集水器分别与翻斗式流量计（6506H、6506G，澳大利亚）和事件记录器相连，自动记录流量和各项记录具体的产生时间。

2.3 数据分析

通过野外布设的各类自动记录仪器，同步观测降雨事件发生期间各项小气候因子和降雨的变化特征。选取 2006~2008 年生长季（5~10 月份），各组仪器均正常工作，有完整记录的 100 组降雨数据，运用 SPSS 统计分析软件对各项观测因子与穿透降雨量、树干茎流量和林冠截留量进行相关性分析，确立相关因子、主导因子与各相关因子的交互项，然后再分别进行多元线性回归分析。

3 结果与分析

3.1 观测期间降雨特征

如图 1 所示，降雨量级<10mm 的降雨发生频率超过半数，达 56%，>50mm 的频率相对较低，仅为 7%，平均降雨量为 16.2mm，变异系数为 1.14。平均降雨强度为 1.68mm/h，最小雨强为 0.12mm/h，最大雨强达 6.80mm/h，变异系数为 0.98，雨强<0.5mm/h 的降雨事件占 24%，75%降雨事件的雨强<2.0mm/h。平均降雨历时为 12h，最长历时为 80h，最短历时为 1h，变异系数为 0.92，历时 8~12h 发生频率最高，为 29%，≥24h 的发生频率最低，仅为 6%。平均雨前干燥期为 61.6h，最长为 359.5h，最短为 4h，变异系数变化较大，为 1.28，雨前干燥期为 5~12h 和≥72h 的发生频率最高，分别为 28%和 27%，其他各区段频率分布变化不大。以上观测结果表明，研究期间的降雨，主要以小雨为主，表现出雨强较小，历时较长的特征。

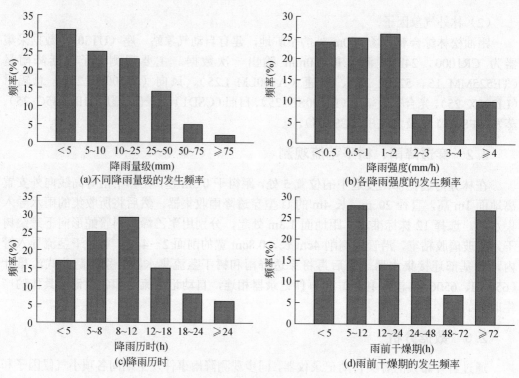

图 1　研究期间降雨特征

3.2　林冠层降雨的输入与分配

如图 2 所示，随着降雨量级增大，穿透降雨率呈递增趋势，从降雨量级<5mm 的 46.6%，上升到降雨量级≥75mm 时的 70.2%。林冠截留率的变化规律与穿透降雨率相反，降雨量级<5mm 时，截留率达最高为 53.4%，然后随降雨量级的增加而逐渐降低，当降雨量级≥75mm 时，截留率最低，仅为 21.4%。

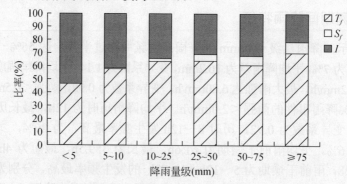

图 2　不同降雨量级油松林穿透降雨率 T_f、树干茎流率 S_f 和林冠截留率 I 的变化

降雨量级处于 5~10mm 时，产生树干茎流，树干茎流率为 0.8%，随着降雨量级增大，树干茎流率显著增高，降雨量级≥75mm 时，树干茎流率达 8.3%，几乎增长了近

10 倍[17]。在 10～25mm，25～50mm 和 50～75mm 这三个降雨量级下，穿透降雨率和林冠截留率上下略有起伏，但树干茎流率变幅较大，上升明显。分析表明，降雨事件发生时，降雨量级是影响林冠降雨再分配的最主要因素，降雨量级不同，林冠对大气降雨再分配的大小和规律会产生相应的变化。类似的研究结论，在其他不同类型林分中也有报道[18,19]。

3.3 小气候因子变化特征

植被的存在改变了下垫面的构成状况，增加了粗糙度，改变了空气动力学效应，从而影响森林的辐射、热量和水分平衡。在成片的森林地区，由于地理位置、地形特点、林木种类、林型结构等因素的综合影响，而形成一种特殊的森林小气候，与周围环境不断进行物质和能量的转换，影响并改变森林内及其周围地区的气候和环境条件，从而对林冠降雨再分配产生一定的影响[18]。

由表 1 可见，观测期间降雨发生时小气候因子变化特征为，风向（W_{d_6}）主要以东南风向为主，发生频率达 54.6%，变异系数为 0.30。风速相对较小，大部分风速<1.0m/s，最高风速为 1.61m/s，从林冠层上部向下风速逐渐减小，林冠顶层（W_{s_5}）和林冠顶层上 1.5 米处（W_{s_6}）风速最高，平均风速几乎是林内的 3.5 倍。平均蒸发率（E）为 0.14mm/h，最大值为 0.53mm/h，变异系数为 0.98。森林的存在对热量影响较大，林内、林外以及林内各层间温度均存在一定差异，林内明显高于林外，第一层（T_{a_1}）平均温度比第六层（T_{a_6}）高 0.35℃，冬季尤为显著，最高可达 0.94℃，最高和最低温度变幅林内小于林外，各层温度的变异系数变化不大。相对湿度在不同层次间表现出一定的变化规律，第 1 层（R_{h_1}）最高，第 5 层（R_{h_5}）最低，与温度的变化相似，林内高于林外，各层间变异系数变化较小。由于枝叶的阻挡、吸收和反射，光合有效辐射各层间变化幅度较大，林内第 2 层（P_{ar_2}）最低，为 14.8mol/m^2，第 6 层（P_{ar_6}）最高，达 126.3mol/m^2，两者相差 8.5 倍左右。净辐射与光合有效辐射变化趋势相同，林内第 2 层（R_{net_2}）最低，为 2.6MJ/m^2，第 6 层（R_{net_6}）显著升高，达 55.4MJ/m^2，两者相差近 21.5 倍。

表 1　观测期间降雨条件下油松林不同层次小气候因子的变化

项目	符号	单位	平均值	最大值	最小值	变异系数
第 6 层风向	W_{d_6}	°	135.60	219.60	45.76	0.30
第 2 层风速	W_{s_2}	m/s	0.09	0.72	0	1.23
第 5 层风速	W_{s_5}	m/s	0.33	1.52	0.04	0.74
第 6 层风速	W_{s_6}	m/s	0.32	1.61	0.01	0.86
蒸发速率	E	mm/h	0.14	0.53	0	0.98
第 1 层温度	T_{a_1}	℃	13.73	24.03	0.81	0.37
第 2 层温度	T_{a_2}	℃	13.48	24.21	0.33	0.38
第 3 层温度	T_{a_3}	℃	13.48	24.43	0.34	0.39
第 4 层温度	T_{a_4}	℃	13.50	24.63	0.33	0.39
第 5 层温度	T_{a_5}	℃	13.53	24.39	0.31	0.39
第 6 层温度	T_{a_6}	℃	13.38	24.26	0.08	0.39

续表

项目	符号	单位	平均值	最大值	最小值	变异系数
第1层相对湿度	R_{h_1}	%	94.7	98.7	61.8	0.06
第2层相对湿度	R_{h_2}	%	93.8	98.9	59.2	0.06
第3层相对湿度	R_{h_3}	%	93.5	99.2	57.9	0.07
第4层相对湿度	R_{h_4}	%	92.4	98.8	57.2	0.07
第5层相对湿度	R_{h_5}	%	90.6	97.7	57.3	0.07
第6层相对湿度	R_{h_6}	%	92.3	100.0	57.2	0.07
第2层光合有效辐射	P_{ar_2}	mol/m^2	14.8	154.6	0	1.57
第3层光合有效辐射	P_{ar_3}	mol/m^2	22.1	224	0	1.53
第4层光合有效辐射	P_{ar_4}	mol/m^2	27.5	233.3	0	1.35
第6层光合有效辐射	P_{ar_6}	mol/m^2	126.3	1333.3	0	1.47
第2层净辐射	R_{net_2}	MJ/m^2	2.6	65.9	-14.1	4.92
第3层净辐射	R_{net_3}	MJ/m^2	7.2	152.5	-12	3.03
第6层净辐射	R_{net_6}	MJ/m^2	55.4	596.4	-11	1.54

3.4 降雨再分配与降雨特征和小气候因子的相关性分析

降雨发生时小气候因子和降雨特征在一次降雨事件过程中不断发生变化，使得林冠截留功能波动性大，稳定性小，导致林冠对降雨再分配的能力和大小产生相应的变化。相关性分析表明（表2），穿透降雨量 T_f 分别与 P、D_u、T_a 和 R_h 显著相关，与 A_{dt}、W_{d_6}、W_s、E、P_{ar}、R_{net} 无相关性。其中与 P 相关系数最高，达 0.978，最低为 T_{a_6}，相关系数为 0.242。树干茎流量 S_f 分别与 P、T_a 显著相关，与 D_u、A_{dt}、W_{d_6}、W_s、E、R_h、P_{ar}、R_{net} 无相关性。截留量 I 分别与 P、D_u 和 T_a 显著相关，与 A_{dt}、W_{d_6}、W_s、E、R_h、P_{ar}、R_{net} 无相关性。

从观测数据分析可知，降雨量 P 对 T_f、S_f 和 I 均有显著影响，降雨历时 D_u 对 T_f 和 I 有显著影响，对 S_f 影响较小。通常雨前干燥期 A_{dt} 越长，林分越干燥，林冠截持水分能力越强。但当降雨量较大时，林分会很快被湿润，A_{dt} 对降雨再分配的影响相对会减弱[20]。本研究中 A_{dt} 虽然与 T_f 表现为负相关，与 S_f 和 I 为正相关，但对 T_f、S_f 和 I 的影响并不显著。主要原因可能是，本研究中降雨间隔期较短的降雨比例较高，降雨事件相对集中，且占总降雨量比重较大，导致 A_{dt} 的影响未能体现出来。

风可以增加林冠蒸发，减少林冠的存储能力，从而对林冠截留产生影响[21]。但本研究中风向和风速对 T_f、S_f 和 I 的影响均不显著。Staelens 等[18]研究认为，风对山毛榉林树干茎流和林冠截留量无明显影响。段旭等[22]对六盘山华北落叶松和白桦林冠截留的研究也认为，在所有涉及影响林冠截留的因素中，风速的影响最小，均与本研究结论相似。有一些研究认为，风对 T_f、S_f 和 I 会产生一定的影响[23,24]，如 Van Stan 等[25]对山毛榉和杨树树干茎流影响的研究认为，风可以增加树干茎流量。风向和风速对降雨再分配的影响，主要是由林木本身特性和局地地形所决定[2]。本研究地位于秦岭深山地带，周围崇山峻岭，森林覆盖率超过 90%，林分郁闭度高，面积比较大，特殊的地形本身就不会产生大风，即使有风，当风进入林分后，由于林冠的摩擦和阻挡，对风有很好的抑制和削减作用。相对于一些落叶或者阔叶树种，油松树体表面粗糙，树皮开裂、疏松，以及附着在树干上的一些附生物，本身持水能力较强受风的影响较小。所以，风向和风速对 T_f、S_f 和 I 的影响不显著。

表 2 油松林穿透降雨量、树干茎流量和林冠截留量与降雨特征及小气候因子的相关分析

	项目	穿透降雨量	树干茎流量	截留量
降雨特征	降雨量 P	0.978（**）	0.871（**）	0.888（**）
	降雨历时 D_u	0.462（**）	0.262（n.s.）	0.473（**）
	雨前干燥期 A_{dt}	−0.103（n.s.）	0.123（n.s.）	0.137（n.s.）
小气候因子	W_{d_6}	−0.143（n.s.）	−0.091（n.s.）	−0.114（n.s.）
	W_{s_2}	−0.158（n.s.）	0.017（n.s.）	−0.130（n.s.）
	W_{s_5}	−0.142（n.s.）	−0.081（n.s.）	−0.095（n.s.）
	W_{s_6}	−0.182（n.s.）	−0.090（n.s.）	−0.136（n.s.）
	E	0.179（n.s.）	0.143（n.s.）	0.170（n.s.）
	T_{a_1}	0.256（*）	0.349（*）	0.216（*）
	T_{a_2}	0.250（*）	0.336（*）	0.208（*）
	T_{a_3}	0.247（*）	0.332（*）	0.204（*）
	T_{a_4}	0.246（*）	0.332（*）	0.203（*）
	T_{a_5}	0.245（*）	0.335（*）	0.202（*）
	T_{a_6}	0.242（*）	0.336（*）	0.201（*）
	R_{h_1}	0.316（**）	0.190（n.s.）	0.131（n.s.）
	R_{h_2}	0.330（**）	0.200（n.s.）	0.181（n.s.）
	R_{h_3}	0.344（**）	0.237（n.s.）	0.181（n.s.）
	R_{h_4}	0.301（**）	0.223（n.s.）	0.148（n.s.）
	R_{h_5}	0.290（**）	0.243（n.s.）	0.161（n.s.）
	R_{h_6}	0.293（**）	0.197（n.s.）	0.187（n.s.）
	P_{ar_2}	−0.043（n.s.）	0.011（n.s.）	−0.030（n.s.）
	P_{ar_3}	−0.034（n.s.）	0.039（n.s.）	−0.022（n.s.）
	P_{ar_4}	−0.029（n.s.）	0.013（n.s.）	−0.031（n.s.）
	P_{ar_6}	−0.058（n.s.）	−0.029（n.s.）	−0.035（n.s.）
	R_{net_2}	0.108（n.s.）	0.001（n.s.）	0.066（n.s.）
	R_{net_3}	0.064（n.s.）	0.004（n.s.）	0.038（n.s.）
	R_{net_6}	−0.045（n.s.）	−0.009（n.s.）	−0.016（n.s.）

** 表示显著水平为 0.01，* 表示显著水平为 0.05，n.s. 表示无相关性。

温度作为林冠层水分蒸发和植物蒸腾的主要驱动影响因子之一，与蒸发高低变化规律一致，对降雨再分配会产生一定的影响[22, 26]。徐小牛等[27]对日本冲绳岛天然次生常绿阔叶林的研究认为，温度决定着湿润林冠的蒸发速率，从而对林冠截留产生影响。本研究中林内各层温度与 T_f、S_f 和 I 均存在相关性，但蒸发速率与三者之间无相关性。主要原因可能是，所观测的降雨事件中，温度最高值仅为 24.6℃，平均相对湿度却超过 92%，在这种低温高湿条件下，林冠层表面水分蒸发损失相对较低，对降雨再分配的影响相对较弱。

相对于空旷地，林内空气相对湿度变幅较大，随着降雨事件的发生，相对湿度在短时间内会显著增高，变幅在 8%~20%[24]。Crockford 和 Richardson[1]研究认为，林内空气相对湿度的大小会对林分蒸散速率产生影响，是影响林冠截留高低的因素之一。Gerrits 等[28]对山毛榉林的研究认为，相对湿度对林冠截留有一定影响。但也有研究认为，相对湿度对林冠截留影响较弱[18, 26]。本研究认为相对湿度与 T_f 呈正相关，但与 S_f 和 I 无相关性。

辐射是林冠蒸发能量的主要来源之一，它的强弱与林冠层水分蒸发量的多少直接相关，从而对林冠截留产生影响[7]。与 Herbst 等[29]的研究相似，本研究中各层的光合有效辐射和净辐射对林冠蒸发影响较弱，与 T_f、S_f 和 I 无相关性。

3.5 穿透降雨量、树干茎流量和林冠截留量与降雨特征及小气候因子的多元回归

降雨事件发生时，降雨特征和小气候因子各不同相同，每次降雨事件间相互独立。降雨再分配各组分的大小，不是由单一的因子决定，而是同时受多个因子的综合影响。穿透降雨量 T_f 的多元回归方程如下。

$$T_f = 0.137 + 0.104P + (0.004P - 0.034)D_u - (0.001P + 0.002)T_a + (0.005P + 0.001)R_h \quad (1)$$
$$(R^2 = 0.966)$$

式（1）表明，降雨越多，相对湿度越高，林内产生的穿透降雨越多[30]。降雨量大于 8.5mm 时，随着降雨历时的增长，穿透降雨量呈递增趋势。温度与穿透降雨呈负相关，温度增高，穿透降雨量减少。主要原因可能是，降雨事件发生的整个过程都存在蒸发，温度增高，水分蒸发损失会增大，穿透降雨量相应减少[31]。相对湿度高时，枝叶表面比较湿润，林冠蒸腾蒸发减弱，降落到树体表面的水分不易被截持和保留，林冠截留量相应减少，穿透降雨量增加[32]。

树干茎流量 S_f 的多元回归方程如下。

$$S_f = 0.379 + 0.009P + (0.005P - 0.075)T_a \quad (R^2 = 0.826) \quad (2)$$

式（2）表明，树干茎流量与降雨量、降雨量和温度的交互项呈正相关，与温度呈负相关。降雨量小于 15mm 时，树干茎流量随温度的升高而降低，与温度对穿透降雨的影响相仿，主要是枝干上所截持水分的蒸发，会随温度的增高而增大。降雨量大于 15mm 时，枝干吸水达到一定程度，持水能力下降，利于树干茎流的形成和向下输送，同时蒸

发损失占降雨比重相应变小，树干茎流量随温度的升高呈递增趋势。

林冠截留量 I 的多元回归方程如下。

$$I = -0.532 + 0.385P + (0.056 - 0.005P)D_u + (0.056 - 0.002P)T_a \quad (R^2 = 0.828) \quad (3)$$

式（3）表明，林冠截留量与降雨量、降雨历时和温度呈正相关，与它们和降雨量的交互项呈负相关。在降雨量相近，雨量较小的情况下，如降雨量小于11.2mm时，降雨历时越长，单位时间内雨强越小，穿过林冠进入林内的雨量相应减少，表现为林冠截留量高，穿透降雨量低的特征[7]。降雨量大于11.2mm时，降雨历时越长，枝叶和树干越来越湿润，林冠层对降雨的截持效能逐渐减弱，到达一定阈值后，历时越长，林内雨量越多，林冠截留量呈递减趋势，穿透降雨量则呈递增趋势[33]。与何常清等[23]对岷江川滇高山栎林的研究相似，降雨量大、降雨持续时间长时，林冠贮水量达饱和后，开始滴落成为穿透降雨，间接减小了林冠截留量。降雨过程中以及降雨终止后，截持在林木表面的雨水，会不断地通过蒸发的方式损失掉，是构成林冠截留的主体部分之一[34]。蒸发量与温度高低呈正相关，特别是降雨较小，历时较长时，受温度影响较大，如降雨量小于28mm时，林冠截留量随温度的升高而增大。

4 结论

1）3年野外观测表明，研究期间的降雨，主要以小雨为主，表现出雨强较小，历时较长的特征。降雨事件发生时，尽管影响穿透降雨量、树干茎流量和林冠截留量的因素较多，但降雨量级仍是决定其大小的最主要因素，降雨量级越小，林冠层的截留分配效应越明显。

2）相关性分析表明，穿透降雨量与降雨量、降雨历时、温度和相对湿度显著相关；树干茎流量与降雨量和温度显著相关；林冠截留量与降雨量、降雨历时和温度显著相关；雨前干燥期、风向、风速、蒸发速率、光合有效辐射和净辐射与穿透降雨量、树干茎流量和林冠截留量均无相关性。

3）多元回归表明，穿透降雨量与降雨量、降雨量和降雨历时交互项、相对湿度呈正相关，与降雨历时和温度呈负相关；树干茎流量与降雨量、降雨量和温度交互项呈正相关，与温度呈负相关；林冠截留量与降雨量、降雨历时和温度呈正相关，与降雨量和降雨历时、温度的交互项呈负相关。

参 考 文 献

[1] Crockford R H, Richardson D P. Partitioning of rainfall into throughfall, stemflow and interception: effect of forest type, ground cover and climate. Hydrological Processes, 2000, 14: 2903-2920.

[2] Link T E, Unsworth M, Marks D. The dynamics of rainfall interception by a seasonal temperate rainforest. Agricultural and Forest Meteorology, 2004, 124: 171-191.

[3] Price A G, Carlyle-moses D E. Measurement and modelling of growing-season canopy water fluxes in a mature mixed deciduous forest stand, northern Ontario, Canada. Agricultural and Forest Meteorology, 2003, 119: 69-85.

[4] Savenije H H G. The importance of interception and why we should delete the term evapotranspiration from our vocabulary. Hydrological Processes, 2004, 18 (8): 1507-1511.

[5] Germer S, Elsenbeer H, Moraes J M. Throughfall and temporal trends of rainfall redistribution in an open tropical rainforest, south-western Amazonia (Rondônia, Brazil). Hydrology and Earth System Sciences, 2006, 10 (3): 383-393.

[6] Cuartas L A, Tomasella J, Nobre A D, et al. Interception water-partitioning dynamics for a pristine rainforest in Central Amazonia: marked differences between normal and dry years. Agricultural and Forest Meteorology, 2007, 145: 69-83.

[7] Xiao Q F, Mcpherson E G, Ustin S L, et al. Winter rainfall interception by two mature open-grown trees in Davis, California. Hydrological Processes, 2000, 14 (4): 763-784.

[8] Berger T W, Untersteiner H, Schume H, et al. Throughfall fluxes in a secondary spruce (*Picea abies*), a beech (*Fagus sylvatica*) and a mixed spruce-beech stand. Forest Ecology and Management, 2008, 255 (3-4): 605-618.

[9] Levia D F, Frost E E. Variability of throughfall volume and solute inputs in wooded ecosystems. Progress in Physical Geography, 2006, 30 (5): 605-632.

[10] Llorens P, Domingo F. Rainfall partitioning by vegetation under Mediterra-nean conditions: a review of studies in Europe. Journal of Hydrology, 2007, 335 (1-2): 37-54.

[11] Carlyle-moses D E. Throughfall, stemflow, and canopy interception loss fluxes in a semi-arid Sierra Madre Oriental matorral community. Journal of Arid Environments, 2004, 58 (2): 180-201.

[12] Iida S, Tanaka T, Sugita M. Change of interception process due to the succession from Japanese red pine to evergreen oak. Journal of Hydrology, 2005, 315 (1-4): 154-166.

[13] Liu S R, Sun P S, Wen Y G. Comparative analysis of hydrological functions of major forest ecosystems in China. Chinese Journal of Plant Ecology, 2003, 27 (1): 16-22.

[14] 吴刚, 冯宗炜. 中国油松林群落特征及生物量的研究. 生态学报, 1994, 14 (4): 415-422.

[15] 景丽, 朱志红, 王孝安, 等. 秦岭油松人工林与次生林群落特征比较. 浙江林学院学报, 2008, 25 (6): 711-717.

[16] 张胜利, 李靖, 韩创举, 等. 南水北调中线工程水源林生态系统对水质的影响——以秦岭南坡中山地带火地塘林区为例. 水科学进展, 2006, 17 (4): 559-565.

[17] 陈书军, 陈存根, 邹伯才, 等. 秦岭天然次生油松林冠层降雨再分配特征及延滞效应. 生态学报, 2012, 32 (4): 1142-1150.

[18] Staelens J, De schrijver A, Verheyen K, et al. Rainfall partitioning into throughfall, stemflow, and interception within a single beech (*Fagus sylvatica* L.) canopy: influence of foliation, rain event characteristics, and meteorology. Hydrological Processes, 2008, 22 (1): 33-45.

[19] 孙向阳, 王根绪, 李伟, 等. 贡嘎山亚高山演替林林冠截留特征与模拟. 水科学进展, 2011, 22 (1): 23-29.

[20] 鲍文, 包维楷, 何丙辉, 等. 岷江上游油松人工林对降水的截留分配效应. 北京林业大学报, 2004, 26 (5): 10-16.

[21] Šraj M, Brilly M, Mikoš M. Rainfall interception by two deciduous Mediterranean forests of contr-

asting stature in Slovenia. Agricultural and forest meteorology, 2008, 148: 121-134.

[22] 段旭, 王彦辉, 于澎涛, 等. 六盘山分水岭沟典型森林植被对大气降雨的再分配规律及其影响因子. 水土保持学报, 2010, 24 (5): 120-125.

[23] 何常清, 薛建辉, 吴永波, 等. 岷江上游亚高山川滇高山栎林地降雨再分配. 应用生态学报, 2008, 19 (9): 1871-1876.

[24] Guevara-escobar A, González-sosa E, Véliz-chávez C, et al. Rainfall interception and distribution patterns of gross precipitation around an isolated Ficus benjamina tree in an urban area. Journal of Hydrology, 2007, 333: 532-541.

[25] Van stan J T, Siegert C M, Levia D F, et al. Effects of wind-driven rainfall on stemflow generation between codominant tree species with differing crown characteristics. Agricultural and Forest Meteorology, 2011, 151: 1277-1286.

[26] 陈引珍, 何凡, 张洪江, 等. 缙云山区影响林冠截留量因素的初步分析. 中国水土保持科学, 2005, 3 (3): 69-72.

[27] 徐小牛, 王勤, 平田永二. 亚热带常绿阔叶林的水文生态特征. 应用生态学报, 2006, 17 (9): 1570-1574.

[28] Gerrits A M J, Pfister L, Savenije H H G. Spatial and temporal variability of canopy and forest floor interception in a beech forest. Hydrological Processes, 2010, 24: 3011-3025.

[29] Herbst M, Rosier P T W, Mcneil D D, et al. Seasonal variability of interception evaporation from the canopy of a mixed deciduous forest. Agricultural and Forest Meteorology, 2008, 148: 1655-1667.

[30] 曹云, 黄志刚, 郑华, 等. 柑桔园林下穿透雨的分布特征. 水科学进展, 2007, 18 (6): 853-857.

[31] 何常清, 薛建辉, 吴永波, 等. 应用修正的 Gash 解析模型对岷江上游亚高山川滇高山栎林林冠截留的模拟. 生态学报, 2010, 30 (5): 1125-1132.

[32] Brauman K A, Freyberg D L, Daily G C. Forest structure influences on rainfall partitioning and cloud interception: A comparison of native forest sites in Kona, Hawaii. Agricultural and Forest Meteorology, 2010, 150: 265-275.

[33] Mcjannet D, Wallace J, Reddell P. Precipitation interception in Australian tropical rainforests: I. Measurement of stemflow, throughfall and cloud interception. Hydrological Processes, 2007, 21: 1692-1702.

[34] Shi Z J, Wang Y H, Xu L H, et al. Fraction of incident rainfall within the canopy of a pure stand of *Pinus armandii* with revised Gash model in the Liupan Mountains of China. Journal of Hydrology, 2010, 385: 44-50.

秦岭天然次生油松林冠层降雨再分配特征及延滞效应

陈书军　陈存根　邹伯才　张硕新　王得祥　侯　琳

摘要　为了研究秦岭典型地带性植物油松林冠层降雨再分配特征及延滞效应，选择陕西宁陕县秦岭森林生态系统国家野外科学观测研究站55龄天然次生油松林，从2006～2008年（5～10月份）对林外降水、穿透降雨和树干茎流进行定位观测。利用其中100次实测数据进行分析研究，结果表明：总降雨量为1576.4mm，穿透降雨量为982.9mm，树干茎流量为69.5mm，冠层截留量为524.0mm，分别占总降雨量的62.4%、4.4%和33.2%。降雨分配与降雨量级密切相关，降雨量级增大，穿透降雨率和茎流率呈增大趋势，截留率呈降低趋势，变化幅度分别为46.6%～68.9%、0.8%～9.2%、53.4%～22.0%。穿透降雨量、树干茎流量和林冠截留量与林外降雨量之间的关系分别为 TF=0.6548P-0.4937，R^2=0.9596；SF=-0.2796+0.0452P+0.0005P^2，R^2=0.8179；I=0.5958$P_{0.8175}$，R^2=0.8064。降雨事件发生后，穿透降雨和树干茎流出现的时间与降雨发生的时间并不同步，均表现出一定的延滞性，随着降雨量级增大，滞后时间表现出逐渐缩短的趋势［(78.5±8.8) min～(16.0±0.0) min，(111.0±33.0) min～(41.2±0.0) min］。降雨终止时，特别是当降雨量>10.0mm，穿透降雨终止时间也存在一定的延滞性［(3.2±2.6) min～(12.0±0.0) min］。但树干茎流终止时间先于降雨终止时间，降雨量级越小，树干茎流终止时间愈早［(-58.3±21.5) min～(-9.8±0.0) min］。

关键词：秦岭；油松林；穿透降雨；树干茎流；林冠截留

随着人类对淡水需求量的不断增加，关键生态区域的重点水源涵养林保护和功能研究已成为各国政府、科学家和民众十分关注的焦点问题。森林具有净化水质、涵养水源、调节径流、保持水土、减免自然灾害等众多的生态功能。秦岭横贯我国中部，其南北分

---

* 原载于：生态学报，2012, 32（4）：1142-1150.

属北亚热带和暖温带，是我国南北气候区的天然分界线，长江、黄河两大水系的自然分水岭。秦岭林区总面积 484.9 万 hm^2，林分面积 252.4 万 hm^2，天然林面积达 218.8 万 hm^2，占林分总面积的 86.7%[1]，秦岭是"南水北调中线工程"重要的水源涵养区，同时也是我国中部地区的重要生态屏障。

油松（*Pinus tabulaeformis*）是我国温性针叶林中分布最广的森林群落，也是我国北方广大地区最主要的造林树种之一[2, 3]。其生态水文功能已有众多学者进行研究，从 19 世纪 80 年代至今，分别在河北隆化[4]、陕西秦岭[5]、山西太岳山[6]、内蒙古黄土沟壑区[7]、黄土丘陵区[8]、黄土高原区[9, 10]、岷江上游[11]、辽西低山丘陵区[12]、华北土石山区[13, 14]、内蒙古半干旱石质山区[15]、华北石质山区[16]等不同地域，不同林龄油松林冠降水截留分配规律进行了相关研究，为深入了解油松林生态系统的生态水文功能和作用提供了大量可靠的数据。但研究林分多为林场营造的人工林，营林地比较干旱瘠薄，林龄普通偏小，而且分布面积不大。油松在秦岭林区分布较广，是秦岭山地的顶级群落之一，为典型的地带性群落[17]。国家自 1998 年实施封山育林政策以来，森林覆盖率逐步提高，油松群落得到较好的恢复。通过对天然次生油松林的林冠截留、穿透降雨、树干茎流量和局地小气候因子的长期定位观测，探讨在经过自然更新和恢复，生长旺盛，结构良好，处于近成熟林阶段（51～60 龄）下[3]，油松林林冠层对不同降雨事件再分配的作用规律和延滞效应。更加深入地了解油松林的水文过程和作用机制，对当前秦岭水源涵养林的保护、经营与管理提供基础理论依据。

1 研究区概况

研究地设在陕西省宁陕县境内的秦岭森林生态系统国家野外科学观测研究站火地沟 2 号集水区天然次生油松林综合观测场内。位于秦岭南坡中山地带中部，地处北亚热带北缘（$33°18'N$，$108°20'E$），海拔 1550～1700m，年均气温 8～10℃，年降水量 900～1200mm，年蒸发量 800～950mm，降雨多集中于 5～10 月，其降雨量占全年降水 85%以上。土壤为花岗岩和变质花岗岩母质的山地棕壤。坡向为西南向，坡形多变，坡度范围 20°～30°。现有森林为原生植被在 20 世纪 50、60 年代主伐后恢复起来的天然次生林，郁闭度 0.85。主要伴生树种有锐齿栎（*Quercus aliena* var. *acuteserrata*）、华山松（*Pinus armandi*）、红桦（*Betula albosinensis*）、漆树（*Toxicodendron verniciflleum*）、青榨槭（*Acer davidii*）等。常见草本有野青茅（*Deyeuxia sylvatica*）、青菅（*Carex leucochlora*）、东亚唐松草（*Thalictrum minus*）等，镶嵌分布于林隙。油松林综合观测场面积 $5000m^2$，大多数油松林龄在 51～60a，具体林分特征见表 1。

2 研究方法

2.1 林外降雨（*P*）测定

在距油松林样地 500m 的林外，安装自动气象站（UT30 Weather Station，美国），通过 CR1000 数据采集器测定和记录林外降雨量和降雨过程。

2.2 穿透降雨（TF）测定

在林分内选择郁闭度适中的位置 5 处，距树干 0.5m 起沿等高线向外安置直径 20cm，长 4m 的 U 型收集器，收集器距地面 1m 高。相当于以树干为中心，沿同心圆布设多个雨量筒，有利于收集林冠下不同位置处的穿透降雨。雨水收集器再与翻斗式流量计（6506H，澳大利亚）和 HOBO 事件记录器相连，自动记录穿透降雨量和具体产生过程。

穿透降雨量根据下列公式计算。

$$TF = \frac{1}{n}\sum_{i=1}^{n} TF_i / FA_i \qquad (1)$$

式中，TF 为研究地单位面积平均穿透降雨量（mm），n 为林内 U 型收集器重复数（$n=5$），TF_i 为每次降雨第 i 个林内收集器穿透降雨量（mm³），FA_i 为第 i 个收集器面积（mm²）。

表 1 秦岭油松林林分特征

树龄（a）	树高（m）	平均胸径（cm）	冠层厚度（m）	冠幅（东西向）（m）	冠幅（南北向）（m）	叶面积指数 LAI	天空可见度 DIFN	坡度（°）	坡向
41~50	17.3±3.5	16.4±2.3	6.2±2.3	3.8±1.4	3.9±1.1				
51~60	23.4±3.0	23.6±1.9	8.5±3.4	3.6±1.9	4.1±1.8	2.07±0.31	0.21±0.04	20~30	SW
61~70	26.5±3.5	32.3±1.8	11.7±2.4	3.6±1.8	4.1±1.4				

注：由陕西秦岭森林生态系统国家野外科学观测研究站提供。

2.3 树干茎流（SF）测定

选择林分分布均匀，林冠枝叶结构能代表平均林冠的样木 12 株，每株用直径 2cm 聚乙烯塑料管从胸径处，由上往下蛇形缠绕于树干上，用玻璃胶粘牢，沿管内侧削 4cm 长，0.8cm 宽的削面，每圈 2~4 处，使茎流可顺削面导入管内，管基部连接集水器收集茎流，再与翻斗式流量计（6506G，澳大利亚）和 HOBO 事件记录器相连，自动记录树干茎流量和具体产生过程。

树干茎流量根据下列公式计算。

$$SF = \frac{1}{n}\sum_{i=1}^{n} SF_i / FA_i \qquad (2)$$

式中，SF 为研究地单位面积平均树干茎流量（mm），n 为观测林木的个数（$n=12$），SF_i 每次降雨第 i 棵树的树干茎流量（mm³），FA_i 为第 i 棵树林冠投影面积（mm²）。

2.4 林冠截留量（I）

$$I = P - (TF + SF) \qquad (3)$$

式中，I 为林冠截留量（mm），P 为林外降雨量（mm），TF 为穿透降雨量（mm），SF

为树干茎流量（mm），忽略降雨过程中的蒸发量。

3 结果与分析

3.1 林冠层对降雨的再分配

本观测期年均降雨量（933.5±196.8）mm，其中5～10月降雨量占全年降雨88%以上。选择（2006～2008年，5～10月份）其中100次有效的完整降雨事件进行分析。由表2可知，降雨总量为1576.4mm，通过林冠层后降雨被重新分配，其中穿透降雨量982.9mm，树干茎流量69.5mm，冠层截留量524.0mm，分别占总降雨量的62.4%、4.4%和33.2%。根据降雨量大小将100场降雨过程划分为6个降雨级别，发生频率最高的降雨量级是<5mm，达31次，但占总降雨量比重最低，仅为5.6%。降雨量级为25～50mm的降雨事件发生14次，但此降雨量级的降雨量、穿透降雨量和截留量占总降雨量、总穿透降雨和总截留量比重最高，分别为29.9%，30.7%和27.8%。

3.1.1 穿透降雨与林外降雨关系

降雨量级对林冠穿透降雨量会产生直接的影响，降雨量级不同，降雨的透过率也随之发生相应的变化。基本趋势为降雨透过率随降雨量级的增大而增大。从降雨量级<5mm时的46.6%，上升到降雨量级≥100mm时的68.9%，几乎增长了近1.5倍（表2）。根据实测数据分析表明，穿透降雨和林外降雨关系呈LINEAR函数关系（图1），拟合方程如下。

$$TF = 0.6548P - 0.4937 \quad (n=100, R^2=0.9596) \tag{4}$$

式中，TF为穿透降雨量（mm），P为降雨量（mm）。

表2 秦岭油松天然次生林不同降雨量级降雨再分配

降雨量级（mm）	次数	林外降雨（mm）	穿透降雨		树干茎流		截留	
			量（mm）	比例（%）	量（mm）	比例（%）	量（mm）	比例（%）
<5	31	88.8	41.4	46.6	0.0	0.0	47.4	53.4
5～10	25	182.5	105.8	57.9	1.5	0.8	75.2	41.2
10～25	23	355.4	226.2	63.6	7.7	2.2	121.5	34.2
25～50	14	471.1	301.3	64.0	24.4	5.2	145.5	30.9
50～100	6	374	236.3	63.2	26.3	7.0	111.4	29.8
≥100	1	104.6	72.1	68.9	9.6	9.2	23.0	22.0
合计	100	1576.4	982.9	62.4	69.5	4.4	524.0	33.2

3.1.2 树干茎流与林外降雨关系

树干茎流观测表明（表2），在降雨量级为5～10mm时，开始出现茎流，此后茎流率随着降雨量级增大而增大，变化幅度为0.8%～9.2%。在100次降雨事件中产生树干茎流的事件为 45 次，树干茎流与次降雨的关系通过曲线拟合，进行分析和比较，树干茎流量与林外降雨量的关系呈 QUADRATIC 函数关系（图2），拟合方程如下。

$$SF = 0.2796 + 0.0452P + 0.0005P^2 \quad (n=45,\ R^2=0.8179) \tag{5}$$

式中，SF 为林内穿透水量（mm），P 为林外降雨量（mm）。

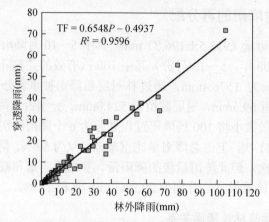

图 1　油松林穿透降雨与林外降雨的关系

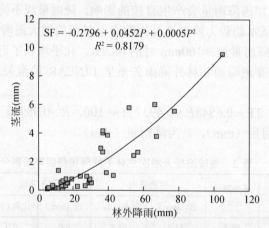

图 2　油松林树干茎流与林外降雨的关系

3.1.3　林冠截留与林外降雨关系

大气降雨通过冠层后，部分降水会被林冠层截留。林冠总截留率为 33.2%，在不同的降雨量级中，林冠截留率变化幅度为 53.4%～22.0%，林冠截留率随降雨量级的增大而减小，减缓的幅度也随降雨量级的增大而逐渐降低（表2）。林冠截留与次降雨的关系通过曲线拟合的比较，POWER 函数模拟林冠截留量和林外降雨量之间的关系效果最佳（图3），拟合方程如下。

$$I = 0.5958P^{0.8175} \quad (n=100,\ R^2=0.8064) \tag{6}$$

式中，I 为林冠截留量（mm），P 为林外降雨量（mm）。

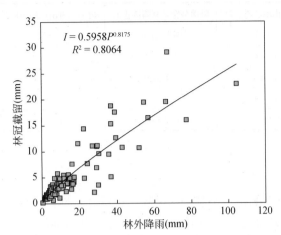

图 3　油松林林冠截留与林外降雨的关系

3.2　林冠层对降雨输入的延滞效应

当单次降雨事件发生后，穿透降雨和树干茎流出现的时间与降雨产生的时间并不同步。如表 3 所示，降雨量级为<5mm 时，穿透降雨在降雨事件发生 78.5min±8.8min 后才被观测到，降雨量级≥100mm，仅需 16.0min±0.0min，两者相差近 5 倍，但都表现出一定的延滞性。而且随着降雨量级增大，滞后时间表现出逐渐缩短的趋势。降雨停止时，穿透降雨终止时间也表现出一定的延滞性。降雨量<10.0mm 时，穿透降雨终止先于降雨终止的时间；降雨量>10.0mm 时，逐渐出现延滞性，特别是降雨量级较大时，这种效应会更明显，从3.2min±2.6min（降雨量级 10~25mm）到 12.0min±0.0min（降雨量级≥100mm）。

只有当大气降雨达到一定量后，才会产生茎流。相对于穿透降雨，茎流出现的时间更滞后，几乎是穿透降雨所需时间的 2 倍。茎流产生的滞后时间随降雨量级的增大显著缩短，从降雨量级 5~10mm 时的 111.0min±33.0min，减小到降雨量级≥100mm 时的 41.2min±0.0min。茎流终止时间表现为先于大气降雨终止时间，一旦降雨变小，水量不能保证，茎流就会很快终止。如表 3 所示，当降雨量级为≥100mm 时，茎流终止时间先于大气降雨终止时间 9.8min±0.0min，降雨量级为 5~10mm 时，为 58.3min±21.5min，降雨量级越小，茎流终止时间越早。

表 3　油松林在不同降雨量级输入条件下穿透降雨和树干茎流出现和终止的滞后时间

降雨量级（mm）	穿透降雨出现的滞后时间（min）	降雨与穿透降雨终止的时间差（min）	树干茎流出现的滞后时间（min）	降雨与树干茎流终止的时间差（min）
<5	78.5±8.8	-12.9±3.4	0.0±0.0	0.0±0.0
5~10	57.2±9.6	-5.7±3.3	111.0±33.0	-58.3±21.5

续表

降雨量级（mm）	穿透降雨出现的滞后时间（min）	降雨与穿透降雨终止的时间差（min）	树干茎流出现的滞后时间（min）	降雨与树干茎流终止的时间差（min）
10~25	53.4±4.4	3.2±2.6	100.2±13.1	-43.3±16.8
25~50	43.8±4.3	6.9±1.9	74.3±13.9	-21.6±13.2
50~100	32.5±6.1	10.3±1.7	66.5±5.3	-25.7±18.2
≥100	16.0±0.0	12.0±0.0	41.2±0.0	-9.8±0.0

4 结论与讨论

林冠对降水的再分配是一个复杂的过程，受林分特性，如树种、林冠结构、树龄、林冠厚度等[18-21]以及降雨和气象因素，如降雨量、降雨强度、降雨历时、温湿度、风、雨前干燥期和雨前枝叶湿度等的综合影响[22-25]。可对局地或整个集水区的水分循环产生重要的影响[26]。

研究期间100次有效的完整降雨事件，总降雨量为1576.4mm，通过林冠层后降雨被重新分配，其中穿透降雨量982.9mm，占总降雨量的62.4%。不同降雨量级，穿透降雨率的变化幅度为46.6%~68.9%，基本趋势为降雨透过率随降雨量级的增大而增大。我国不同研究区域油松林穿透降雨率为54.62%~82.18%（表4），本研究穿透降雨率与以内蒙古东部地区油松人工林带人工林的基本相当（62.8%）[7]，与以往研究结果比较，相对偏小，显著低于党坤良和雷瑞德[5]在同一研究区域1987~1988年的测定结果（79.77%），表明穿透降雨率随着林龄增长，冠层厚度增深、冠幅增大和叶量增多，穿透降雨率逐渐变小。穿透降雨量与林外降雨量呈明显的线性关系，与陈丽华等[14]和胡珊珊等[16]的研究结果一致。

树干茎流量和出现时间，因植物种类和降雨量而异[27]。本研究中出现45次树干茎流的降雨事件，茎流总量为69.5mm，占总降雨量的4.4%。树干茎流以"点"的形式向林地输入，虽然树干基面积相对于林地面积所占比重较低，与林内穿透降雨相比贡献较小[28]，但茎流中水溶性养分离子浓度高，且直接输入到树干基部附近[29]，并沿着根的生长方向直接进入土壤[30]，对加速植物生长和促进养分循环发挥着重要的作用[31]。降雨量级是影响茎流量的主导因子，不同降雨量级，茎流率的变化幅度为0.8%~9.2%，随着降雨量级的增大呈递增趋势，与范世香等[32]和夏体渊等[33]的研究结论相似。根据回归方程推断，降水量≥5.9mm，树干茎流开始为正值，实测数据为降雨量级在5~10mm时，开始出现茎流相仿。植物种类不同，茎流率差别很大，但大多数研究认为，林木的树干茎流量通常较少，占降雨量的比值一般≤5%，很少超过10%[34-35]。本研究茎流率为4.4%，处于此范围之内。树干茎流除受降雨量、降雨强度、降雨历时和冠层特征的影响外，同时还受树干的粗细、树体表面的粗糙度、干燥度和直立程度等影响[24]。通过对本研究和已往研究结果的分析，发现树干茎流率的变化规律性不明显，与林龄、树高等的相关性不强。

本研究林冠总截留率为 33.2%，不同降雨量级，林冠截留率的变化幅度为 53.4%～22.0%，林冠截留量随着降雨量而增加，两者呈幂函数关系，与岷江上游油松林[11]，中亚热带樟树林[36]等对降水的截留分配规律相似。针叶林林冠截留率通常在 20%～48%[35]，相同生境条件下，针叶林林冠截留率一般较阔叶林高[37]。由表4可见，我国不同研究区域油松林冠层截留率在 15.9%～36.97%，本研究截留率虽然在 20%～48%的范围内，但与我国其他区域油松林研究结论相比，截留率相对较高，明显超出同一研究区域党坤良和雷瑞德[5]报道的 31 龄油松截留率（20.19%）。主要原因可能是，本研究林分为天然次生林，林龄 51～60 龄，已达近成熟林阶段，树高、冠幅和冠层厚度均较其他研究林分大，林木生长旺盛，林分连片生长，着生地地形复杂等原因会部分削弱一些局地气象因素的影响，导致截留率相对较高。仅从森林对降雨截留有效性方面而言，随着林分越来越成熟，林冠截留能力会不断加强。但林冠所截留的降雨最终将被消耗于蒸发，实质会减少林内地表的实际雨量和土壤水分的有效补充，是雨水资源的无益损失。

降雨经林冠向下传递，进入林内，最终落入土壤的整个过程中，从接触到林冠起，林冠就会对降雨产生再分配作用，同时表现出穿透降雨和树干茎流与降雨事件发生和终止时间上的不一致性。通过 3 年的野外实测，发现降雨事件发生时降雨强度的变化总是呈现从小变大再减缓的规律。降雨事件开始发生时，降雨强度相对较小，林冠和枝叶表面比较干燥，雨滴从上向下穿透林冠过程中，如果碰到枝叶，很容易被吸附、阻挡和截持，除部分雨滴直接从林冠间隙穿过，前期大量雨滴将会被林冠阻挡或截持。林外降雨与林内降雨的产生在时间会出现明显的差异，从而产生对降雨输入的延滞效应。陈丽华等[14]对华北土石山区油松人工林研究也存在相似规律。降雨量级越小，林内雨出现时间越迟，随着降雨量级增大，林冠从干燥到湿润的过程所需时间变短，林内雨出现时间相应变短。如降雨量级<5mm 时，穿透降雨在降雨事件发生 78.5min±8.8min 后才被观测到；降雨量级≥100mm 时，仅需 16.0min±0.0min，两者相差近 5 倍。降雨量>10.0mm 时，穿透降雨终止时间逐渐表现出延滞性，降雨量级越大越明显，从 3.2min±2.6min（降雨量级 10～25mm）到 12.0min±0.00min（降雨量级≥100mm）。由于森林对降雨存在延滞效应，可适度迟缓林外大气降雨到达林内的时间，延迟地表径流的产生，从而对削减洪峰，减少水土流失、泥石流和滑坡等地质灾害产生一定的积极作用[38]。

表 4　已发表的不同研究地油松林冠截留率、穿透降雨率和树干茎流率的观测值

研究地点	年平均降水量（mm）	林龄（a）	树高（m）	穿透降雨率（%）	茎流率（%）	截留率（%）	研究时间	来源
秦岭，陕西	933.5±196.8	55	23.6±1.9	62.4	4.4	33.2	2006～2008 年	本研究
秦岭，陕西	900～1200	31	10	79.77	0.04	20.19	1987～1988 年	[5]
太岳山，山西	757.8	34	11.94	82.0	2.1	15.9	1992～1993 年	[6]
赤峰，内蒙古	360	18	4	62.8	4.9	32.3	1989～1993 年	[7]
宜川，陕西	574.4	34	11	71.6	3.3	25.1	1995～2000 年	[10]

续表

研究地点	年平均降水量（mm）	林龄（a）	树高（m）	穿透降雨率（%）	茎流率（%）	截留率（%）	研究时间	来源
茂县，四川	945.3	23	6.43	54.62	8.41	36.97	2002～2003年	[11]
大凌河流域，辽宁	450～580	28	4.7	73.55	3.37	23.08	2002～2004年	[12]
密云，北京	669	33	7.1	67.65	0.68	31.67	2004～2006年	[13]
密云县城北山区，北京	669	—	7.1	64.4	0.55	25.4	—	[14]
呼和浩特，内蒙古	400	30	4.59	65.83	1.13	33.04	2006～2006年	[15]
易县，河北	641.2	50	5～6	82.18	1.07	16.75	2005～2008年	[16]

树干茎流的滞后效应更为明显，几乎是穿透降雨所需时间的2倍。主要是由于只有当大气降雨达到一定量后，除林冠枝叶外，树皮本身吸附的水分也达到饱和后，雨水才会沿着树干向下传递，产生茎流。滞后时间随着降雨量级的增大显著缩短，不同降雨量级，滞后时间的变化幅度为111.0min±33.0min～41.2min±0.0min。Owens等[27]研究表明，降雨强度较大时，降雨一个小时后出现茎流，与本研究结果相似。同时认为降雨终止后，茎流会延迟一个小时，与本研究结论有一定差别。本研究的观测结果为，茎流终止时间表现为先于大气降雨终止时间，如降雨量级≥100mm时，茎流终止时间先于大气降雨终止时间9.8min±0.0min，降雨量级5～10mm时，为58.3min±21.5min，明显地表现出降雨量级越小，茎流终止时间愈早的趋势。可能原因是成年油松树皮较厚，平均厚度2.5cm，多孔柔软且开裂，吸水能力较强，从林冠到树干再到树基，有较长的传输距离，在此过程中不仅存在着树皮、树干的吸附，同时还会产生一定量的蒸发[39-41]。只有水量达到一定程度，雨水才可汇流向下，降雨一旦变小，水量不能保证，茎流便相应终止。

参 考 文 献

[1] 韩福利. 秦岭林区森林资源保护的意义和可持续经营利用. 陕西林业科技, 2006, (4): 69-72.

[2] 吴刚, 冯宗炜. 中国油松林群落特征及生物量的研究. 生态学报, 1994, 14 (4): 415-422.

[3] 罗天祥, 李文华, 赵士洞. 中国油松林生产力格局与模拟. 应用生态学报, 1999, 10 (3): 257-261.

[4] 董世仁, 郭景唐, 满荣洲. 华北油松人工林的透流、干流和树冠截留. 北京林业大学学报, 1987, 9 (1): 58-68.

[5] 党坤良, 雷瑞德. 秦岭火地塘林区不同林分水源涵养效能的研究. 土壤侵蚀与水土保持学报, 1995, 1 (1): 79-84.

[6] 曾杰, 郭景唐. 太岳山油松人工林生态系统降雨的第一次分配. 北京林业大学学报, 1997, 19 (3): 22-26.

[7] 赵焕胤, 朱劲伟, 王维华. 半干旱地区油松人工林带降水截留作用分析. 东北林业大学学报, 1997, 25 (6): 67-70.

[8] 魏天兴, 余新晓, 朱金兆. 山西西南部黄土区林地枯落物截持降水的研究. 北京林业大学学报,

1998，20（6）：1-6.
[9] 赵鸿雁，吴钦孝. 黄土高原人工油松林林冠截留动态过程研究. 生态学杂志，2002，21（6）：20-23.
[10] 赵鸿雁，吴钦孝，刘国彬. 黄土高原人工油松林水文生态效应. 生态学报，2003，23（2）：376-379.
[11] 鲍文，包维楷，何丙辉，等. 岷江上游油松人工林对降水的截留分配效应. 北京林业大学学报，2004，26（5）：10-16.
[12] 魏晶，吴钢. 辽西低山丘陵区人工油松林和沙棘林的水文生态效应. 生态学报，2006，26（7）：2087-2092.
[13] 肖洋，陈丽华，余新晓，等. 北京密云水库油松人工林对降水分配的影响. 水土保持学报，2007，21（3）：154-157.
[14] 陈丽华，杨新兵，鲁绍伟，等. 华北土石山区油松人工林耗水分配规律. 北京林业大学学报，2008，30（增刊 2）：182-187.
[15] 莎仁图雅，田有亮，郭连生. 大青山区油松人工林降雨分配特征研究. 干旱区资源与环境，2009，23（6）：157-160.
[16] 胡珊珊，于静洁，胡堃，等. 华北石质山区油松林对降水再分配过程的影响. 生态学报，2010，30（7）：1751-1757.
[17] 景丽，朱志红，王孝安，等. 秦岭油松人工林与次生林群落特征比较. 浙江林学院学报，2008，25（6）：711-717.
[18] Asdak C，Jarvis P G，van Gardingen P，et al. Rainfall interception loss in unlogged and logged forest areas of Central Kalimantan，Indonesia. Journal of Hydrology，1998，206（3/4）：237-244.
[19] Chappell N A，Bidin K，Tych W. Modelling rainfall and canopy controls on net-precipitation beneath selectively-logged tropical forest. Plant Ecology，2001，153（1/2）：215-229.
[20] Loescher H W，Powers J S，Oberbauer S F. Spatial variation of throughfall volume in an old-growth tropical wet forest，Costa Rica. Journal of Tropical Ecology，2002，18（3）：397-407.
[21] Germer S，Elsenbeer H，Moraes J M. Throughfall and temporal trends of rainfall redistribution in an open tropical rainforest，south-western Amazonia（Rondonia，Brazil）. Hydrology and Earth System Sciences，2006，10（3）：383-393.
[22] Rutter A J，Morton A J. A predictive model of rainfall interception in forests. III. Sensitivity of model to stand parameters and meteorological variables. Journal of Applied Ecology，1977，14(2)：567-588.
[23] Gash J H C. An analytical model of rainfall interception by forests. Quarterly Journal of the Royal Meteorological Society，1979，105（443）：43-55.
[24] Crockford R H，Richardson D P. Partitioning of rainfall into throughfall，stemflow and interception：effect of forest type，ground cover and climate. Hydrological Processes，2000，14(16/17)：2903-2920.
[25] Xiao Q F，McPherson E G，Ustin S L，et al. Winter rainfall interception by two mature open-grown trees in Davis，California. Hydrological Processes，2000，14（4）：763-784.
[26] Savenije H H G. The importance of interception and why we should delete the term evapotranspiration from our vocabulary. Hydrological Processes，2004，18（8）：1507-1511.
[27] Owens M K，Lyons R K，Alejandro C L. Rainfall partitioning within semiarid juniper communities：

effects of event size and canopy cover. Hydrological Processes, 2006, 20 (15): 3179-3189.

[28] Manfroi O J, Koichiro K, Nobuaki T, et al. The stemflow of trees in a Bornean lowland tropical forest. Hydrological Processes, 2004, 18 (13): 2455-2474.

[29] Park G G. Throughfall and stemflow in the forest nutrient cycle. Advances in Ecological Research, 1983, 13: 57-134.

[30] Martinez M E, Whitford W G. Stemflow, throughfall and channelization of stemflow by roots in three Chihuahuan desert shrubs. Journal of Arid Environments, 1996, 32 (3): 271-281.

[31] 马雪华. 在杉木林和马尾松林中雨水的养分淋溶作用. 生态学报, 1989, 9 (1): 15-20.

[32] 范世香, 裴铁番, 蒋德明, 等. 两种不同林分截留能力的比较研究. 应用生态学报, 2000, 11 (5): 671-674.

[33] 夏体渊, 吴家勇, 段昌群, 等. 滇中3种林冠层对降雨的再分配作用. 云南大学学报（自然科学版）, 2009, 31 (1): 97-102.

[34] Iida S, Tanaka T, Sugita M. Change of interception process due to the succession from Japanese red pine to evergreen oak. Journal of Hydrology, 2005, 315 (1/4): 154-166.

[35] Carlyle-Moses D E. Throughfall, stemflow, and canopy interception loss fluxes in a semi-arid sierra madre oriental matorral community. Journal of Arid Environments, 2004, 58 (2): 180-201.

[36] 闫文德, 陈书军, 田大伦, 等. 樟树人工林冠层对大气降水再分配规律的影响研究. 水土保持通报, 2005, 25 (6): 10-13.

[37] Silva I C, Rodriguez H G. Interception loss, throughfall and stemflow chemistry in pine and oak forests in northeastern Mexico. Tree Physiology, 2001, 21 (12/13): 1009-1013.

[38] Wei X, Liu S, Zhou G, et al. Hydrological processes in major types of Chinese forest. Hydrological Processes, 2005, 19 (1): 63-75.

[39] Herwitz S R, Levia D F Jr. Mid-winter stemflow drainage from bigtooth aspen (*Populus grandidentata* Michx.) in central Massachusetts. Hydrological Processes, 1997, 11 (2): 69-175.

[40] Pypker T G, Bond B J, Link T E, et al. The importance of canopy structure in controlling the interception loss of rainfall: examples from a young and an old-growth Douglas-fir forest. Agricultural and Forest Meteorology, 2005, 130 (1/2): 113-129.

[41] 薛建辉, 郝奇林, 吴永波, 等. 3种亚高山森林群落林冠截留量及穿透雨量与降雨量的关系. 南京林业大学学报（自然科学版）, 2008, 32 (3): 9-13.

半干旱黄土丘陵沟壑区降水特征值和下垫面因子影响下的水土流失规律

卫伟 陈利顶 傅伯杰 巩杰

摘要

水土流失是多种因素交互作用的结果,尤其在半干旱黄土丘陵沟壑区更具有典型性。选择定西市安家沟小流域(35°35′N,104°39′E)为典型区域,在其阴坡中部布设了 15 个径流小区,包括 5 种土地利用类型(农田、牧草地、灌丛、乔木林、自然草地)和 3 个坡度(10°、15°、20°),进行径流、侵蚀和降雨前期土壤水分的测定。利用偏相关和方差分析等统计方法,研究了降水特征值及不同下垫面因子对水土流失的影响规律。结果表明,在黄土丘陵沟壑区:①径流量主要受降水量、最大 30min 雨强、以及降水量与最大 30min 雨强之积的影响。影响侵蚀量的决定性降水因子为降水量和最大 30min 雨强的乘积 PI30 与最大 30min 雨强,而与降水量、平均雨强的相关性较差。②坡度在 10°~20°范围内的变化对径流量没有显著影响,而对侵蚀量有显著影响,在该范围内随着坡度的升高侵蚀量增加(尤以 15°~20°范围内增加较快)。③土地利用类型是影响径流侵蚀的一个极其关键的因素。在坡度和土地利用类型双重作用下,以土地利用为主导因素,坡度作用次之。即在一定范围内,人为扰动强烈且坡度高的农田和人工草地最易遭受土壤侵蚀。④径流与侵蚀主要与表层 0~20cm 的土壤水分呈正相关关系,而与下部其他层次土壤水分无明显相关关系。

关键词:水土流失;降水;坡度;土地利用类型;前期土壤水分

水土流失继续对干旱半干旱地区的生态环境和人民生产生活构成严重制约和威胁。目前关于影响地表径流与土壤侵蚀规律的研究已经很多,概括起来主要包括 3 个方面,即下垫面性质(如坡度[1]、坡形[2,3]、坡长[4,5]、植被[6,7]、土壤结构[8])、气候因素(如降水量[9]、降雨动能[10,11]、雨滴击溅[12])、人为干扰(如农业实践[13,14])等因子对径流产沙的影响。这些研究针对某一特定因子对径流侵蚀的影响做了大量详细

* 原载于:生态学报,2006,26(11):3847-3853.

的探讨。然而，影响水土流失的因素以及水土流失发生发展的过程极为复杂，各个因子之间互相作用，互相叠加或抵消，共同导致了径流和土壤侵蚀。尽管也有不少学者探讨了降雨及下垫面等综合作用下的地表径流与土壤侵蚀规律[15,16]，但多是基于短期零散的观测数据、模拟实验以及模型预测的研究，长期连续的定点监测研究则相对较少。而后者能够有效避免片面结论的产生以及由于模型模拟失误所造成的预测失实问题[17,18]，是研究长期生态演变过程所必需的。

本文基于典型黄土丘陵沟壑小流域内多年连续的实测数据，运用统计分析的方法，分析了降水因子（降水量、降水历时、平均雨强、最大30min雨强、降水量与最大30min雨强的乘积）、下垫面性质（坡度、土地利用类型、前期土壤水分）等指标及其综合影响对径流侵蚀的作用，试图明确回答以下两个问题：①半干旱黄土丘陵沟壑区多因子组合对径流量和侵蚀量影响的规律性；②在此基础上对比分析不同土地利用方式的土壤侵蚀特征，从而提出何种土地利用即植被配置最能有效地遏制或降低干旱半干旱区的土壤侵蚀。

1 研究区概况

研究区位于甘肃省陇中地区的定西市安家沟小流域（35°35′N，104°39′E），属于典型的半干旱黄土丘陵沟壑区。流域内多年年平均气温 6.3℃，≥5℃年积温 2933.5℃，≥10℃年积温 2239.1℃，极端最高气温 34.3℃，极端最低气温-27.1℃。年均降水量 427.0mm，其中60%以上集中在7~9月份，且多暴雨。年蒸发量1510mm，空气相对湿度65.8%，太阳辐射591.1J/cm^2，年日照时数2409h，无霜期141d，干燥度1.15，属于中温带半干旱气候。

该地区的土壤主要是黄土母质基础上发育起来的灰钙土和盐渍土，平均厚度为40~60cm。土壤黏粒介于33.12%~42.17%，有机质含量介于0.37%~1.34%，2m土层内的土壤容重介于1.09~1.36g/cm^3，平均1.2g/cm^3，孔隙率在50%~55%。土壤具有垂直节理，土质疏松，湿陷性强，极易发生土壤侵蚀。侵蚀模数可高达1.0万t/（km^2·a），这几年经过连续治理，情况虽有所好转，但侵蚀依然严重。

流域内自然植被覆盖度低。自然覆盖度阳坡一般为25%~35%，阴坡及部分梁顶为50%~60%。植被带属于森林草原带干草原区。乔木种主要有油松（*Pinus tabulaeformis* Carr.）、侧柏（*Platycladus orientalis* L.）、山杏（*Prunus armeniaca* L.）等，灌木有沙棘（*Hippophae rhamnoides* L.）、柠条（*Caragana* ssp.），草本植物为紫花苜蓿（*Medicago sativa* L.）、红豆草（*Onobrychis viciifolia* Scop.）、针茅（*Stipa bungeana* Trin.）。主要的农作物有马铃薯（*Solanum tuberosum* L.）、春小麦（*Triticum aestivum* L.）、玉米（*Zea mays* L.）及豌豆（*Caragana kansuensis* Pojark.）等。

2 研究方法

2.1 试验小区的布设

试验小区布设在安家沟小流域阴坡中部，位于定西市水土保持研究所的野外气象站附近，海拔 2115m。共设有 15 个径流小区，小区的垂直投影面积为 5m×10m（农田、人工草地、自然草地）和 10m×10m（灌丛和乔木林）。包括 5 种土地利用类型：农田、人工草地、自然草地、灌丛和乔木林，主要作物分别是春小麦（*Triticum aestivum* L.cv Leguan）、紫花苜蓿（*Medicago sativa* L.）、针茅（*Stipa bungeana* Trin.）、沙棘（*Hippophae rhamnoides* L.）和油松（*Pinus tabuliformis* Carr.）；各类型小区分别按照 10°、15°、20° 3 个等级的坡度布设（表1）。小区于 1985 年建成，农田内为春小麦单作。沙棘和油松的初始定植密度分别为 1m×1m 和 20m×40m。针茅群落盖度为 85%～100%，草本层高度 20～40cm。

表1 不同组合类型的下垫面处理

土地利用类型	坡度		
	10°	15°	20°
农田	R_{1-1} 或 E_{1-1}（P_4）	R_{1-2} 或 E_{1-2}（P_1）	R_{1-3} 或 E_{1-3}（P_{11}）
牧草地	R_{2-1} 或 E_{2-1}（P_2）	R_{2-2} 或 E_{2-2}（P_5）	R_{2-3} 或 E_{2-3}（P_{12}）
灌丛	R_{3-1} 或 E_{3-1}（P_3）	R_{3-2} 或 E_{3-2}（P_6）	R_{3-3} 或 E_{3-3}（P_9）
乔木林	R_{4-1} 或 E_{4-1}（P_8）	R_{4-2} 或 E_{4-2}（P_7）	R_{4-3} 或 E_{4-3}（P_{10}）
自然草地	R_{5-1} 或 E_{5-1}（P_{13}）	R_{5-2} 或 E_{5-2}（P_{15}）	R_{5-3} 或 E_{5-3}（P_{14}）

注：R、E、P 分别代表径流、侵蚀和径流小区。

2.2 数据的测定与处理

利用 SM_1 型雨量计和 SJ_1 型虹吸式雨量计测算降水数据；同时监测每次降水是否产生径流和侵蚀，记录对应的径流量和侵蚀量；根据天气预报对每次降雨来临前的前期土壤水分进行测量，利用土钻取 1m 深土样，共分 5 个层次，即 0～20cm、20～40cm、40～60cm、60～80cm、80～100cm。每个层次两个重复，装入铝盒内。采用电子天平称其初始重量，然后采用烘干法（105℃烘箱内，8h）烘干，二次称重后计算土壤含水量。

选择整理 1986～1998 年的 13a 连续的实测配套数据进行分析处理，利用 SPSS13.0 中的相关分析、偏相关分析以及方差分析等统计方法进行降水特征值和下垫面性质因子与径流量、侵蚀量的关系探讨。

3 结果分析

3.1 降雨因子（特征值）与水土流失

在黄土高原半干旱地区，降水是唯一的水分输入渠道，同时也是该地区径流侵蚀发生发展的原动力[10]。分析降水因子与径流侵蚀之间的关系，必须从降水量、降水历时、平均雨强等相关指标入手，但由于这些因子之间存在显著的自相关关系，利用一般的相关分析难以排除这种干扰，从而对结果造成影响。而利用偏相关分析的手段，不仅可以有效剔除干扰因子，还可以准确判断对因变量贡献更大的因子。因此本研究采用偏相关方法分别对安家沟小流域的多年降水特征值与对应不同处理下的径流量和侵蚀量进行了分析，结果见表2和表3。

表2 不同处理下的径流量与各降雨因子偏相关分析

处理类型	降雨量 P	降水历时 D	平均雨强 A.I	最大30min雨强 I_{30}	雨量×雨强 PI_{30}
R_{1-1}	0.257**	−0.097	−0.177	0.296**	0.376***
R_{2-1}	0.317***	−0.072	−0.056	0.266***	0.242**
R_{3-1}	0.287**	−0.048	−0.007	0.004	0.275**
R_{4-1}	0.305***	−0.177	−0.088	0.271**	0.192*
R_{5-1}	0.414***	−0.152	−0.161	0.332**	−0.100
R_{1-2}	0.283**	−0.125	−0.202*	0.401***	0.154*
R_{2-2}	0.327***	−0.128	−0.151	0.440***	0.147*
R_{3-2}	0.272**	−0.066	−0.045	0.046	0.150
R_{4-2}	0.229*	−0.131	−0.045	0.182*	0.270**
R_{5-2}	0.492***	−0.158	−0.151	0.292***	0.039
R_{1-3}	0.344***	−0.197	−0.222*	0.485***	0.024
R_{2-3}	0.300	−0.187*	−0.110	0.311***	0.039*
R_{3-3}	0.311	−0.068	−0.059	0.068	0.133
R_{4-3}	0.318	−0.204	0.121	0.291**	0.143
R_{5-3}	0.414	−0.150	0.142	0.372***	0.084

* $0.01 < P < 0.05$；** $0.001 < P < 0.01$；*** $P < 0.001$。

由表2可知，在控制其他降水特征值不变的条件下，不同处理下的径流量和降水量、最大30min雨强呈显著正相关关系；其次为PI_{30}（降水量和最大30min雨强的积）；和平均雨强的相关性较差，只有个别处理与之呈现出较为显著的相关性。因此可以认为，主要影响径流量的降水特征值为降水量、最大30min雨强以及二者之积。

表3 不同处理下的侵蚀量与各降雨因子偏相关分析

处理类型	降雨量 P	降水历时 D	平均雨强 A.I	最大30min雨强 I_{30}	雨量×雨强 PI_{30}
E_{1-1}	−0.128	−0.069	−0.102	0.014*	0.366**
E_{2-1}	−0.111	−0.035	−0.026	−0.013	0.306**
E_{3-1}	0.151	−0.093	−0.032	0.058	−0.041
E_{4-1}	−0.192*	0.030	0.012	−0.155	0.420**
E_{5-1}	0.094	−0.124	−0.155	0.221*	−0.074
E_{1-2}	−0.145	−0.038	−0.031	−0.065	0.386**
E_{2-2}	−0.063	−0.079	−0.055	0.075*	0.196*
E_{3-2}	0.092	−0.059	−0.149	0.194*	−0.063
E_{4-2}	−0.192*	0.033	−0.014	−0.111	0.396**

续表

处理类型	降雨量 P	降水历时 D	平均雨强 A.I	最大 30min 雨强 I_{30}	雨量×雨强 PI_{30}
E_{5-2}	0.102	−0.120	−0.180	0.261**	−0.099
E_{1-3}	−0.160	−0.041	−0.052	−0.026	0.378**
E_{2-3}	−0.164	−0.091	−0.264**	0.337**	−0.195*
E_{3-3}	−0.175	−0.012	0.001	−0.098	0.372**
E_{4-3}	−0.086	−0.042	−0.103	0.023*	0.261**
E_{5-3}	0.133	−0.146	−0.173	0.283**	−0.082

* $0.01<P<0.05$；** $0.001<P<0.01$。

表 3 的侵蚀量与降水因子偏相关分析表明，首先，在控制对应其他变量的条件下，主要影响侵蚀量的因子为降水量和最大 30min 雨强的乘积，其次为最大 30min 雨强，与降水量、平均雨强的相关性较差。因此，PI_{30} 和 I_{30} 是降水因子中影响侵蚀量的两个决定性因子。对比表 2 和表 3，发现一个有趣的规律：降水量的多少直接影响着径流量，但从降水量的多少来判断是否产生侵蚀或产生多少侵蚀是不可靠的。

3.2 下垫面性质与水土流失

3.2.1 坡度、土地利用类型与水土流失

对 15 个不同处理类型，3 个等级的坡度，以及 5 种土地利用类型进行归类，分别利用方差分析（ANOVO）分析、LSD 多重比较（最小显著性差异，$P<0.05$）对这 3 组影响因子和径流量与侵蚀量的关系进行方差差异性分析，结果如下。

从图 1 中可以看出，在坡度和土地利用类型共同作用下，不同下垫面之间的径流量差异显著；而相同土地利用类型下，不同坡度之间径流量无明显差异（$F=0.84$，$P>0.05$）；不同土地利用类型之间的径流量差异极为显著。可知土地利用类型是影响径流量的一个关键因素。

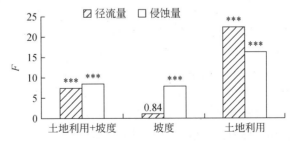

图 1 坡度、土地利用类型等因子与水土流失的 F 值检验

***$P<0.001$。

至于坡度对径流量的作用，由于本研究布设的径流小区的坡度等级较少，且范围仅局限于 10°～20°，因此并不能简单地认为坡度的变化对径流没有影响，但至少在上述

变化范围内对径流量的影响极小（$P>0.05$）。

从图1中同时看出不同坡度和土地利用类型处理下的侵蚀量差异极为显著（$P<0.001$），说明这两个因素对侵蚀量的影响都比较大。依据统计学基本原理，对变异的度量，方差分析中唯一有用的就是离差平方和。方差分析方法就是从总离差平方和中分解出可追溯到指定来源的部分离差平方。因而离差平方和越大，可认为对总体的贡献率越高。单因素方差分析结果显示，土地利用的离差平方和（其值为351 019.2）远大于坡度的离差平方和（其值为86 265.5）。因而，认为土地利用类型的作用要大于坡度对侵蚀量的影响。

进一步分析不同坡度因子对侵蚀量的贡献大小，由于均数的相对值比较能够说明各个自变量对因变量的效应大小，利用方差分析中的LSD多重比较的方法，求得均差如图2。

由图2知，20°～15°、20°～10°、15°～10°的均差分别达到12.24、16.64、4.41。尽管15°～10°的P值差异性不够显著，但均差仍达到4.41。说明在10°～20°坡度范围内，侵蚀量随着坡度的增加而增加，在10°～15°范围内增加较缓慢，15°～20°范围内增加较快。

利用相同的方法进行不同土地利用类型对径流量和侵蚀量的方差分析，比较其均差，结果分别见表4和表5。

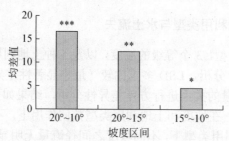

图2 不同坡度因子作用下的侵蚀量均差及显著性比较

* $0.01<P<0.05$；** $0.001<P<0.01$；*** $P<0.001$。

由表4可知，5种土地利用类型中，小麦和其他4种土地利用类型之间的均差均为正，可知5种土地利用类型中，农田最容易产生径流；牧草地仅次于农田，也比较容易产生径流；乔木林比自然草地更易产生侵蚀；径流量最小的是灌丛。农地和牧草地受到人为的扰动最剧烈最频繁；多年的油松林下形成了大量的针叶枯落物，对于改善土壤结构，提高蓄水能力发挥了作用；特别是多年生的沙棘在黄土高原地区阴坡长势极好，其良好的覆盖度、特有的根蘖繁殖能力、发达的根系以及枯枝落叶层使得其对径流产生了极大的降低和削弱作用。这方面的研究已有很多，如准格尔旗德胜西乡黑毛兔沟种植沙棘7a后，植被覆盖度达61%，侵蚀模数由4万t/（$km^2 \cdot a$）[19]。国内研究表明，根系固土作用的大小，与根系生物量和分布密切相关。沙棘根系分布较浅，多呈水平分布，水平根幅度一般在2m×2m，最大可达6m×8m，在土壤表层形成网状的根系层，对保持水土具有很大的作用[20]。

表 4　不同土地利用类型作用下的径流量均差比较

	农田	牧草地	灌丛	乔木林	自然草地
农田	—	0.17	1.18***	0.44**	0.86***
牧草地		—	1.02***	0.29	0.69***
灌丛			—	−0.75***	−0.32*
乔木林				—	0.43**
自然草地					—

* $0.01 < P < 0.05$；** $0.001 < P < 0.01$；*** $P < 0.001$。

表 5　不同土地利用类型作用下的侵蚀量均差比较

	农田	牧草地	灌丛	乔木林	自然草地
农田	—	33.50***	39.96***	34.02***	31.88***
牧草地		—	6.46	0.52	1.62
灌丛			—	−8.08	−2.14
乔木林				—	5.94
自然草地					—

*** $P<0.001$。

由表5知，小麦和另外4种土地利用类型的均差明显，其产生的侵蚀量最大；苜蓿的侵蚀量次之。这和径流量的分析相一致。其余几种土地利用类型之间均差的 P 值不显著（$P>0.05$），说明其对侵蚀量的影响差异不太明显，但从均差的比较可以看出，仍以沙棘的抗侵蚀作用最好，其次为荒草地和油松。

综合表4和表5，发现受干扰最严重的农田和人工草地产生了较严重的水土流失；而灌丛和荒草地的保水保土效果最佳；另外栽植多年的针叶林（如油松林）具有较强的水土保持能力，其原因可能是由于地表凋落物的累积和土壤性状的改善，从而较好地控制了径流侵蚀。

3.2.2　前期土壤水分状况与水土流失

由于降雨之前的土壤含水量状况影响土壤保持水土的能力、下渗速率、土壤的通透性以及土体的物理结构，对地表径流和土壤侵蚀会产生一定的作用[21]。因而研究分析不同层次土壤水分状况和水土流失的关系，也应该成为探讨水土流失成因的一个重要方面。由于土壤水分的不同层次之间存在自相关关系，为避免其对结果造成干扰，对不同层次的土壤水分和径流量与侵蚀量进行偏相关分析，结果见图3。

由图3可知，径流量和侵蚀量和0～20cm表层的土壤水分呈现较为显著的相关关系，径流量还与20～40cm土壤水分有微弱的相关性，但由于作用很小，故可以不予考虑。因此，表层土壤水分含量的高低，会对径流和侵蚀发生发展产生一定的影响，而下层的土壤水分状况则基本上不影响径流侵蚀。这可能与黄土高原多暴雨，径流多为瞬时超渗产流有直接关系。

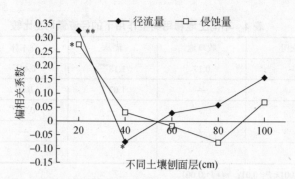

图3 不同层次前期土壤水分与水土流失量的偏相关分析
* $0.01 < P < 0.05$；** $0.001 < P < 0.01$。

4 讨论与结论

4.1 讨论

1）本文对影响地表径流与土壤侵蚀的输入因子（降水）和承受因子（下垫面）进行了重点探讨但事实上，影响径流和侵蚀过程的因素是错综复杂的，各个因素之间相互影响、叠加或消减。比如，在研究土壤侵蚀时，地表径流本身的作用也不容忽视。径流是降雨和下垫面作用的产物，而其本身也是产生侵蚀的一个重要因子。利用斯皮尔曼相关分析得到径流量和侵蚀量的相关系数为 0.337（$P<0.001$），相关关系显著。表明产生径流是发生水力侵蚀的先决条件，即径流是产生水力侵蚀的必要非充分条件。因此，在考虑影响侵蚀的因素时（如回归模拟分析等），应该将径流量当作一个变量列入。

2）关于降雨历时与径流侵蚀的关系在控制其他因子不变的情况下，降雨历时和径流侵蚀均无明显相关性（表2和表3）。但在降雨强度不变的情况下（如模拟降雨），降雨量和降雨历时呈显著正相关[22]，降雨历时和径流侵蚀的关系变得较为复杂[23]。一种情况是降雨强度弱，只有地表径流产生，降雨历时和径流量呈显著正相关，而与侵蚀量无关；第二种情况是降雨强度大，发生超渗产流，径流和侵蚀都有发生，降雨历时和径流量及侵蚀量呈显著正相关。但在自然界，天然降雨强度时刻处于随机变化中[24]，因此偏相关分析不太适合模拟降雨。对于降雨历时和径流侵蚀的关系探讨，下结论时需要慎重。

3）关于最大降雨强度截至目前，关于黄土丘陵区最大雨强方面的研究已经很多，如最大 5min、10min、15min、30min、60min 降雨强度等[25, 26]。具体哪一个降雨强度指标与径流侵蚀的关系最为显著，有待进一步比较。本文选择较为常用的最大 30min 降雨强度作为指标，旨在说明最大降雨强度对径流侵蚀具有显著的影响。

4.2 结论

分析表明，在半干旱黄土丘陵沟壑区，不同土地利用类型坡地径流量和侵蚀量与降雨特征值（雨量、雨强、历时等）和下垫面因子（坡度、土地利用、前期土壤水分状况等）关系密切。

1）径流量主要受降水量、最大 30min 降雨强度以及二者之积的影响。首先，影响侵蚀量的因子为降水量和最大 30min 降雨强度的乘积 PI_{30}，其次为最大 30min 降雨强度，与降水量、平均雨强的相关性较差。降水量的多少直接影响着径流量，但根据降水量的多少来判断是否产生侵蚀或产生多少侵蚀则是非常不可靠的。

2）坡度在 10°～20°范围内的变化对径流量没有显著影响，而对侵蚀量有显著影响，在该范围内随着坡度的增加，侵蚀量增加（尤以 15°～20°范围内增加较快）。

3）土地利用类型是影响径流侵蚀的一个极其关键的因素，不同土地利用类型之间的径流与侵蚀存在显著差异。5 种土地利用类型中，农田（小麦）和人工草地（苜蓿）最易产生径流和侵蚀；以在黄土高原区长势旺盛的灌木丛（沙棘）和人为干扰少的荒草地降低和遏制径流侵蚀的效果最佳；多年生常绿乔木林（油松）的作用次之。

4）在坡度和土地利用类型双重作用下，以土地利用为主导因素，坡度次之。即在研究所涉及的几种土地利用类型中，低坡度下的灌木及荒草地抵御土壤侵蚀的效果最佳，而坡度高的农田和人工草地最易遭受土壤侵蚀。

5）降雨前的土壤水分状况也是影响径流和侵蚀的一个重要因子。径流量和侵蚀量主要与表层 0～20cm 的土壤水分呈现正相关关系，而与下部其他层次土壤水分无明显的相关关系。

参 考 文 献

[1] Dennis M F，Bryan R B. The relationship of soil loss by interrill to slope gradient. Catena，2000，38（3）：211-222.

[2] Heerdegen R G，Beran M A. Quantifying source areas through land surface curvature and shape. Journal of Hydrology，1982，57（3）：359-373.

[3] Wheater H S，Shaw T L，Rutherford J C. Storm runoff from small lowland catchments in southwest England. Journal of Hydrology，1982，55（1-4）：321-337.

[4] Li S，Cai Q，Wang Z. Effect of slope length on runoff and soil erosion. Journal of Arid Land Resources and Environment，1998，12（1）：29-35.

[5] 孔亚平，张科利，唐克丽. 坡长对侵蚀产沙过程影响的模拟研究. 水土保持学报，2001，15（2）：17-24.

[6] Chirino E，Bonet A，Bellot J，et al. Effects of 30-year-old Aleppo pine plantations on runoff, soil erosion, and plant diversity in a semi-arid landscape in south eastern Spain. Catena，2006，65（1）：19-29.

[7] Arsenault É，Bonn F. Evaluation of soil erosion protective cover by crop residues using vegetation indices and spectral mixture analysis of multispectral and hyperspectral data. Catena，2005，62（2）：157-172.

[8] Ludwig B，Boiffin J，Chadœuf J，et al. Hydrological structure and erosion damage caused by concentrated flow in cultivated catchments. Catena，1995，25（95）：227-252.

[9] Nearing M A，Jetten V，Baffaut C，et al. Modeling response of soil erosion and runoff to changes in precipitation and cover. Catena，2005，61（2-3）：131-154.

[10] Hammad A H A, Børresen T, Haugen L E. Effects of rain characteristics and terracing on runoff and erosion under the Mediterranean. Soil and Tillage Research, 2006, 87 (1): 39-47.

[11] Hamed Y, Albergel J, Pépin Y, et al. Comparison between rainfall simulator erosion and observed reservoir sedimentation in an erosion-sensitive semiarid catchment. Catena, 2003, 50 (1): 1-16.

[12] Erpul G, Gabriels D, Janssens D. Assessing the drop size distribution of simulated rainfall in a wind tunnel. Soil and Tillage Research, 1998, 45 (3-4): 455-463.

[13] Zhang G S, Chan K Y, Oates A, et al. Relationship between soil structure and runoff/soil loss after 24 years of conservation tillage. Soil and Tillage Research, 2007, 92 (1-2): 122-128.

[14] Chan K Y. Impact of tillage practices and burrows of a native Australian anecic earthworm on soil hydrology. Applied Soil Ecology, 2004, 27 (1): 89-96.

[15] 卢金发, 刘爱霞. 黄河中游降雨特性对泥沙粒径的影响. 地理科学, 2002, 22 (5): 552-556.

[16] 张丽萍, 倪含斌, 吴希媛. 黄土高原水蚀风蚀交错区不同下垫面土壤水蚀特征实验研究. 水土保持研究, 2005, 12 (5): 126-127.

[17] Nearing M A, Jetten V, Baffaut C, et al. Modeling response of soil erosion and runoff to changes in precipitation and cover. Catena, 2005, 61 (2): 131-154.

[18] Aizen V, Aizen E, Glazirin G, et al. Simulation of daily runoff in Central Asian alpine watersheds. Journal of Hydrology, 2000, 238 (1): 15-34.

[19] 吴淑芳, 刘建凯. 黄土高原沙棘林抗侵蚀性能研究综述. 国际沙棘研究与开发, 2005, 3 (2): 43-47.

[20] 张勇. 浅论黄土高原水土流失治理与沙棘生态建设. 沙棘, 2005, 18 (3): 33-35.

[21] Castillo V M, Gómez-Plaza A, MartíNez-Mena M. The role of antecedent soil water content in the runoff response of semiarid catchments: a simulation approach. Journal of Hydrology, 2003, 284 (1): 114-130.

[22] 陈洪松, 邵明安, 张兴昌, 等. 野外模拟降雨条件下坡面降雨入渗、产流试验研究. 水土保持学报, 2005, 19 (2): 5-8.

[23] 甘枝茂. 黄土高原地貌与土壤侵蚀研究. 西安: 陕西人民出版社, 1990.

[24] 常福宣, 丁晶, 姚健. 降雨随历时变化标度性质的探讨. 长江流域资源与环境, 2002, 11 (1): 79-83.

[25] 谢云, 刘宝元, 章文波. 侵蚀性降雨标准研究. 水土保持学报, 2000, 14 (4): 6-11.

[26] 田凤霞, 王占礼, 牛振华, 等. 黄土坡面土壤侵蚀过程试验研究. 干旱地区农业研究, 2005, 18 (6): 141-146.

黄土丘陵沟壑区极端降雨事件及其对径流泥沙的影响*

卫 伟 陈利顶 傅伯杰 巩 杰 黄志霖

摘要 半干旱黄土丘陵沟壑区是我国水土流失的重灾区。在全球气候变化的大背景下，极端降雨事件时有发生，加重了区域水土流失防治的难度。因此，科学界定极端降雨事件、进而探讨其发生规律及其对径流侵蚀的影响尤为重要。通过整理定西市安家沟流域（35°35′N，104°39′E）17年的降水和径流侵蚀数据进行统计分析。以降雨量和最大30min雨强为指标，采用世界气象组织的标准划分了极端降雨事件。结果发现：①研究区内极端次降雨事件的雨量和雨强的临界值分别为40.11mm和0.55mm/min，次降雨量的多年平均值为18.87mm。17年间共发生12次极端事件，5月、7月、8月份的发生概率分别为16.67%、50%和33.33%。因此最佳防治时间段为7、8月份。②聚类分析表明极端降雨事件可分为三类：降雨量和雨强都大于临界值，占25%；降雨量大于临界值，而雨强小于临界值，占41.67%；雨强大于临界值，降雨量大于多年平均值而小于临界值，占33.33%。③在极端降雨事件作用下，径流系数和侵蚀模数要比对应的多年平均值高。总体而言，降雨量和雨强都很高的极端事件的破坏性最强，但高历时低雨强的极端事件所产生的破坏也不容低估。④沙棘林在生长演替的过程中显著增强了抵御土壤侵蚀的能力，对极端降雨事件有很好的防治作用。抵御极端降雨最弱的是坡耕地，主要是由于受到坡度大、植被覆盖率低以及人为干扰等因素的影响。

关键词：黄土丘陵沟壑区；极端降雨；径流泥沙；影响

半干旱黄土丘陵沟壑区是我国水土流失的重灾区[1]。大量研究表明，该区严峻的水土流失态势是自然和人为因素共同影响的结果[2,3]。在众多影响因素中，降雨作为水土流失的原动力[4,5]，极大地影响甚至决定着土壤侵蚀的强度和运行规律。尤其是近些年

---

* 原载于：干旱区地理，2007，30（6）：896-901。

来,在全球气候急剧变化的大背景下,极端降雨事件时有发生,它的危害和破坏性比一般的降雨事件更为严重。因此,深入研究半干旱黄土丘陵沟壑区极端降雨事件的发生规律和分布特征,以及在其主导下的不同土地利用/土地覆被的径流与侵蚀特征,对于进一步深入理解和探讨水土流失的发生机理、采取更为有力的土壤侵蚀防治策略、开展卓有成效的植被恢复都有所裨益和启示。然而涉及降雨因素的土壤侵蚀效应,目前国内主要集中在降雨侵蚀力、雨滴击溅、植被对降雨的截流作用等方面的研究上[6-8],而对于如何划分和界定极端降雨事件、进而深入探讨其发生规律及对径流和侵蚀的影响等研究则相对匮乏。

基于1986~2004年(5~9月份,其中2000~2001年数据缺失)连续17年野外实测径流小区的降水和径流泥沙数据,在统计软件SYSTAT1.0和SPSS13.0的支持下,对17年共计151次产流性降雨事件进行分析,在科学划分极端降雨事件的基础上,致力于探讨并试图解决以下几个问题:①分析研究区内极端降雨事件发生的基本特征及其季节分布;②对比分析极端降雨事件所产生的水土流失与一般降雨事件所产生的水土流失,借以探讨二者不同的破坏程度;③分析植被演替过程对极端降雨事件破坏性的遏制力度,以便从更加细微的角度为水土流失防治提供理论和决策依据。

1 研究方法

1.1 研究区简介

研究区位于黄河流域的中部、甘肃省定西市的安家沟小流域(35°35′N,104°39′E),海拔高度1900~2250m,年平均降水量为427mm,年潜在蒸发量1510mm,无霜期141天。空气相对湿度为65.8%,干燥度1.15,气候区划属于中温带半干旱区,水土保持区划属于黄土丘陵沟壑区第Ⅵ副区。该流域沟壑密度为3.14km/km^2,沟深30~50in[①],地貌属典型梁峁状黄土丘陵沟壑区。土壤为黄绵土和沟道盐渍土,粉壤土质地,垂直土层深度2m内的土壤容重1.1~1.4g/cm^3,土壤平均孔隙度为55%植被属于典型的森林草原带干草原区,自然植被群落以禾本科、豆科、菊科草本植物为主,间有少量零星灌木分布,主要品种为沙棘(*Hippophae rhamnoides* L.)、柠条(*Caragana korshinskii* Pojark.)等。人工草本主要为紫花苜蓿(*Medicago sativa* L.)和红豆草(*Onobrychis viciifolia* Scop.),人工乔木主要有侧柏(*Platycladus orientalis* L.)、山杏(*Prunus armeniaca* L.)、油松(*Pinus tabulaeformis* Carr.)等。针茅(*Stipa bungeana* Trin.)作为典型的天然草本物种在研究区域内广泛分布。

1.2 径流小区的布设

在安家沟小流域的阴坡中部共布设15个径流小区,配置五种土地利用类型,分别为:旱作坡耕地(春小麦)、人工牧草地(紫花苜蓿)、灌木林地(沙棘)、乔木林地(油松)和荒草地(针茅);每种土地利用类型分别设有10°、15°、20°三个等级的坡度。乔木林地和灌木林地的小区面积为10m×10m,其他三种为5m×10m。旱作坡耕地和人工

---

① 1in=2.54cm。

牧草地按照当地习惯进行耕作管理。沙棘和油松的初植密度为 1m×1m 和 3m×2m，没有采取任何抚育措施。

1.3 降雨及径流泥沙测定

利用研究区内标准气象观测站内的 SM_1 型雨量计和 SJ_1 型虹吸式雨量计监测降雨量和降雨的全过程，记录降雨持续时间、各时段的降雨量，并根据需要测算瞬时降雨强度；径流产生后，量算径流池内水的总体积，计算径流量；搅拌均匀后用 250ml 容量瓶重复取水和泥沙混合样品，沉淀后取出泥沙烘干，推算对应的侵蚀量。观测时间为 1986 年至 2004 年（2000～2001 年缺失数据）的 5～9 月份。

1.4 极端降雨事件的划分标准

极端降雨事件一般是指与历史同期相比出现较少的小概率降雨事件，具有危害性高、突发性强等特征。本研究采用世界气象组织（World Meteorological Organization，WMO）关于极端气候事件的定义标准进行极端降雨事件的划分。WMO 规定：如果某个（些）气候要素的时、日、月、年值达到 25 年以上一遇，或者与其对应的多年平均值（一般应达 30 年左右）的"差"超过其二倍的均方差时，这个（些）气候要素值就属于"异常"气候值。出现"异常"气候值的事件就是"极端气候事件"。而降雨是表征气候属性的一个最基本要素，因此将符合上述标准的降雨事件归属于"极端降雨事件"。同时，选择合适的降雨特征值对于科学划分和评价极端降雨事件至关重要。尽管大量研究已证明最大 30 分降雨强度与土壤侵蚀的拟合效果最好[9]，但其他因子如降雨量、降雨历时等也很重要[10]，因此需要综合考虑。依据这样一个基本的原则，我们选择降雨量、最大 30 分降雨强度两个降雨特征值为基本衡量指标，结合研究地区 16 年连续的降雨数据，求算多年降雨事件的平均值和其二倍的均方差，最终确定了极端降雨事件的划分标准如下。

1）在满足降雨量大于多年平均值的前提下，最大 30 分降雨强度超过 WMO 的相关规定（即其值与多年平均值的差值大于其二倍的均方差）；

2）如果最大 30 分降雨强度达不到 WMO 规定，降雨量必须达到或超过 WMO 的相关规定（即其值与多年平均值的差值大于其二倍的均方差）。

满足以上两个条件中任意一个的降雨事件均属于一次极端降雨事件。

1.5 数据的分析与处理

所有数据在统计软件 SPSS13.0 和 SYSTAT1.0 下处理；主要数据处理方法为聚类分析、描述性统计分析等。

2 结果与分析

2.1 极端降雨事件的特征及其分布

根据研究区多年的降雨资料，结合世界气象组织的上述规定，推算出研究区内极端次降雨事件的降雨量和最大 30 分降雨强度的临界值分别为 40.11mm 和 0.55mm/min，次降雨量（产流降雨事件）多年平均值为 18.87mm。依据设定的标准，对 151 次降雨事件进行综合分析，发现其间共有 12 次极端降雨事件（表 1）。

表 1　极端降雨事件的时间分布及其特征

序列	时间（日/月/年）	降雨量（mm）	降雨历时（min）	最大雨强（mm/min）
1	19/5/1986	49.1	885	0.65
2	1/7/1986	47.0	1260	0.33
3	3/8/1988	22.0	210	0.61
4	26/7/1989	43.4	365	0.78
5	4/8/1992	24.2	140	0.65
6	8/7/1993	40.1	1420	0.11
7	28/7/1996	41.9	940	0.15
8	6/8/1997	42.2	690	0.33
9	13/7/1999	28.6	425	0.67
10	23/5/2002	26.5	28	0.95
11	1/8/2003	46.9	1435	0.12
12	25/7/2004	66.6	440	0.63

由表 1 可知，极端降雨事件的发生带有很强的随机性和偶然性。总体而言，近些年来几乎每年都会有一次极端降雨事件发生。与此同时，不同月份发生的次数差异很大：7 月份 6 次，发生概率为 50%；8 月份为 4 次，发生概率为 33.33%；5 月份为 2 次，发生概率为 16.67%，而 17 年间每年的 6 月份和 9 月份均没有极端降雨事件出现。即：大约 82% 的极端降雨事件都集中分布于 7 月份和 8 月份，5 月份较少，而其他月份没有该类型的降雨事件发生。因而防治极端降雨事件及其危害的最佳时间段为每年的 7 月和 8 月份，其次为 5 月份。基于降雨量和最大 30 分降雨强度两个特征值，利用层次聚类分析中的 Cosine 方法对以上 12 次极端降雨事件进行聚类，结果如图 1 所示。

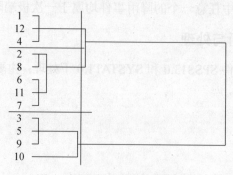

图 1　极端降雨事件的聚类分析

由图 1 可知，极端降雨事件可以划分为三种情况：一是降雨量和最大 30 分降雨强度都大于临界值（降雨事件 1、4、12），这类极端降雨事件的特征是雨强高而降雨历时适中，从而导致大的降雨量，占 25%；二是降雨量大于临界值，而最大 30 分降雨强度没有达到临界值（降雨事件 2、6、7、8、11），这类事件一般是特征是历时较长而导致大的降雨量，占 41.67%；三是雨强大于临界值，而降雨量大于多年平均值而小于临界值（降雨事件 3、5、9、10），这类事件一般无论历时、雨量和雨强都小于第一类极端事件，占 33.33%。

2.2 极端降雨事件下的径流侵蚀规律

2.2.1 径流规律分析

对极端降雨事件下各土地利用类型的径流量与 17 年间所有降雨事件的多年平均径流系数进行比较，如图 2 所示。

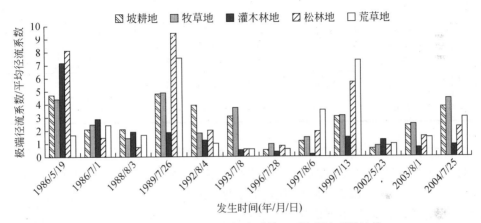

图 2　极端降雨事件下的径流系数/平均径流系数比较

由图 2 可以看出，总体而言，在极端降雨事件作用下，各土地利用类型的径流系数要比平均状态下的径流系数高，也就是说，极端降雨事件作用下，地表径流更容易发生，其危害性更强。而在一般降雨事件下所发生径流的量及其危害程度都比较低。极端降雨作用下，坡耕地的径流量最高时可以高于平均值 4.80 倍；牧草地最高为平均值的 4.88 倍；灌草地最高为平均值的 7.13 倍；油松林地最高为均值的 9.50 倍；荒草地最高值为均值的 7.55 倍。

按照上面划分的三类极端降雨事件标准，分别对比其径流系数，发现降雨量和最大雨强都大于临界值的降雨（事件 1、4、12）所产生的径流最为严重；最大雨强较大而雨量低于临界值的降雨（事件 3、5、9、10）次之；但当降雨历时足够长，即便最大雨强较弱，同样可以产生较为严重的径流，以降雨事件 6 最为典型，其破坏性不容低估。

2.2.2 侵蚀规律分析

对极端降雨事件下各土地利用类型的侵蚀量与各自对应的 17 年间的所有降雨事件的多年平均侵蚀模数进行比较，如表 2 所示。

表2 极端降雨事件下的侵蚀模数/平均侵蚀模数比较

序列	发生时间(日/月/年)	坡耕地	牧草地	灌木林地	松林地	荒草地
1	19/5/1986	47.58	9.53	345.18	76.12	41.72
2	1/7/1986	4.68	26.00	19.73	15.66	31.42
3	3/8/1988	20.18	16.79	28.90	9.93	16.46
4	26/7/1989	368.53	353.60	84.00	222.46	90.45
5	4/8/1992	1.55	1.65	0.00	1.29	0.00
6	8/7/1993	7.79	27.42	0.00	2.80	12.3
7	28/7/1996	0	0	0.00	0.00	0.00
8	6/8/1997	7.37	23.62	0.00	3.04	38.23
9	13/7/1999	36.09	61.41	4.75	39.32	29.76
10	23/5/2002	6.02	4.31	3.16	1.83	6.34
11	1/8/2003	65.30	54.88	1.21	1.21	15.34
12	25/7/2004	86.34	67.94	3.06	6.37	29.56

由表2可知，在极端降雨事件作用下，土壤侵蚀模数的变异幅度远大于径流系数的变异幅度。如坡耕地侵蚀模数最高时可高出平均值368.53倍。牧草地最高为均值的353.6倍，灌木林地最高为均值的345.18倍，油松林地最高为均值的222.46倍，荒草地最高为均值的90.45倍。并且出现最高值的时间不一致，表明降雨只是影响土壤侵蚀的一个因素，土地利用类型也发挥着重要作用。同时，分析1989年7月26日出现的这次极端降水，发现其最大降雨强度是历次降雨中最大的，从而验证了最大30分降雨强度是影响土壤侵蚀的关键因素。

对比分析第6次（1993年7月8日）和第7次（1996年7月28日）降雨事件，发现在相近的雨强和雨量作用下，土壤侵蚀发生的程度迥异，这可能和不同年份的下垫面状况发生变化有关。总体来看，降雨量和最大雨强都很高的极端事件所产生的侵蚀较之雨强和雨量都小的极端事件要严重许多，如降雨事件1和4，这和他们产生的径流效应的规律很接近。

特别指出的是，灌木林（沙棘）在不断生长演替的过程中，对极端降雨事件可能诱发的水土流失逐渐发挥良好的抵御作用，其径流侵蚀效应随时间的推移有明显下降趋势。这与沙棘的生长特性关系密切。国内研究表明，根系固土作用的大小，与根系生物量和分布密切相关。沙棘根系分布较浅，多呈水平分布，水平根幅度一般在2m×2m，最大可达6m×8m，在土壤表层形成网状的根系层，对保持水土具有很大的作用[11]。同时，沙棘在黄土高原地区阴坡长势极好，其良好的覆盖度（4年以后冠幅盖度可达95%以上）、特有的根蘖繁殖能力以及枯枝落叶层使得其对径流产生了极大的降低和削弱作用。

3 讨论与结论

1)研究区内极端次降雨事件的雨量和雨强的临界值分别为 40.11mm 和 0.55mm/min，次降雨量的多年平均值为 18.87mm。17 年间共发生 12 次极端事件，5 月、7 月、8 月份的发生概率分别为 16.67%、50%和 33.33%。因此最佳防治时间段为 7、8 月份。聚类分析表明极端降雨事件可分为三类：降雨量和雨强都大于临界值的占 25%；降雨量大于临界值，而雨强小于临界值，占 41.67%；雨强大于临界值，降雨量大于多年平均值而小于临界值，占 33.33%。

2）一般而言，在极端降雨事件作用下，各土地利用类型的径流系数和侵蚀模数要比对应的多年平均值高。总体而言，降雨量和最大雨强都很高的极端事件破坏性最强，但高历时低雨强的极端事件所产生的水土流失也不容低估。

3）沙棘林在生长演替的过程中显著增强了抵御土壤侵蚀的能力，对极端降雨事件有很好的防治作用。抵御极端降雨最弱的是坡耕地，主要是由于受到坡度大、植被覆盖率低以及人为干扰等因素的影响。可以采取坡改梯田、消减坡长、退耕还林（草）以及地表覆盖等其他辅助措施来强化抵御土壤侵蚀的能力[12, 13]。

4）关于极端降雨事件的界定问题。极端降雨事件的界定目前尚没有一个通用的科学标准，不同的学者采用不同的定义标准。国内多将极端降雨等同于暴雨。如在气象学中是以日降雨量达到 50mm 为界线，将达到或超过这个数值的降雨列为暴雨。高超等人在研究极端降雨事件对农业非点源污染物迁移的影响时，就采用日降雨量>50mm 作为划分极端降雨事件的尺度[14]。而于秀晶等将 1961～1990 年逐年日降水量序列（由少到多）的第 90 个百分位值的 30 年平均值作为划分吉林省极端降雨事件的阈值，日降雨量超过该值即为一次极端降雨事件[15]。但是由于这些标准没有照顾到区域降水的差异性，因而并不适合于黄土高原地区。而方正山、刘尔铭分别拟定了黄土高原地区的暴雨标准，但没有和土壤侵蚀相联系，是单纯的降雨特征参数[16]。本研究以影响水土流失的两个关键降雨特征值——最大 30 分雨强和降雨量为综合指标，按照世界气象组织规定的国际标准进行划分，有效克服了由于单一降雨特征值划分降雨事件的所导致的片面性甚至错误结论，以及无法将研究土壤侵蚀和划分极端事件有机结合等瓶颈问题。同时该方法还能有效解决由于区域降水的空间差异问题，可以用于不同区域的极端降雨事件划分及相关研究。

5）关于影响径流侵蚀的其他因素。事实上，径流和侵蚀发生的机理及其动态是非常复杂的过程。除了受降雨和土地利用影响外，其他很多因素如坡度特征、降雨发生时的前期土壤含水量等都对水土流失产生不同程度的影响[17, 18]。如针对 1996 年 7 月 28 日这次极端降雨事件，其地表径流系数就很小，土壤侵蚀模数为 0。分析其原因，发现除了降雨强度相对较小这一原因外，在这次降雨事件之前相当长的一段时间内基本没有降雨发生，其前期土壤水分含量很低，土体十分干燥，这是导致径流和侵蚀弱的另外一个重要原因。因此，在实际操作中，一定要全面考虑一切可能诱发水土流失的因素，给予有效的综合防治。

参 考 文 献

[1] Shi H, Shao M. Soil and water loss from the Loess Plateau in China. Jounud of Arid Environments, 2000, 45: 9-20.

[2] Wei W, Chen L, Fu B, et al. The effect of land uses and rainfall regimes on runoff and soil erosion in the loess hilly area, China. Journal of Hydrology, 2007, 335: 247-258.

[3] Fu B. Soft erosion and its control in the loess plateau of China. Soil Use and Management, 1989, 5: 76-82.

[4] 卢金发, 刘爱霞. 黄河中游降雨特性对泥沙粒径的影响. 地理科学进展, 2002, 22 (5): 552-556.

[5] 卫伟, 陈利顶, 傅伯杰, 等. 半干旱黄土丘陵沟壑区降水特征值和下垫面因子影响下的水土流失规律. 生态学报, 2006, 26 (11): 277-284.

[6] Hamed Y, Albergel J, Pépin Y, et al. Comparison between rainfall simulator erosion and observed reservoir sedimentation in an erosion-sensitive semiarid catchment. Catena, 2003, 50 (1): 1-16.

[7] Erpul G, Gabriels D, Janssens D. Assessing the drop size distribution of simulated rainfall in a wind tunnel. Soil and Tillage Research, 1998, 45 (3-4): 455-463.

[8] 马三保, 郑妍, 马彦喜. 黄土丘陵区水土流失特征与还林还草措施研究. 水土保持研究, 2002, 9 (3): 55-57.

[9] 甘枝茂. 黄土高原地貌与土壤侵蚀研究. 西安: 陕西人民出版社, 1990.

[10] 张士锋, 刘昌明, 夏军, 等. 降雨径流过程驱动因子的室内模拟实验研究. 中国科学, 2004, 34 (3): 280-289.

[11] 张勇. 浅论黄土高原水土流失治理与沙棘生态建设. 沙棘, 2005, 18 (3): 33-35.

[12] 卫伟, 彭鸿, 李大寨. 黄土高原丘陵沟壑区生态环境现状及对策——以延安市杜甫川流域为例. 西北林学院学报, 2004, 19 (3): 179-182.

[13] 沈国舫. 写在"西部大开发中的生态环境建设问题"笔谈之前. 林业科学, 2000, 36 (5): 2-4.

[14] 陈莉. 黄土丘陵区软埝梯田建设与维护监测研究. 水土保持研究, 2004, 11 (2): 41-44.

[15] Bedaiwy M N, Rolston D E. Soil surface densification under simulated high intensity rainfall. Soil Technology, 1993, 6 (4): 365-376.

[16] 高超, 朱继业, 朱建国, 等. 极端降水事件对农业非点源污染物迁移的影响. 地理学报, 2005, 60 (6): 991-997.

[17] 于秀晶, 王凤刚. 吉林省极端降水的变化特征及其与环流异常的关系. 吉林气象, 2005, (4): 5-8.

[18] Castillo V M, Gómez-Plaza A, MartíNez-Mena M. The role of antecedent soil water content in the runoff response of semiarid catchments: a simulation approach. Journal of Hydrology, 2003, 284 (1): 114-130.

黄土小流域水沙输移过程对土地利用/覆被变化的响应*

卫 伟　陈利顶　温 智　吴东平　陈 瑾

摘要　以甘肃定西安家沟小流域为典型研究区，基于 TM、ALOS 遥感影像解译和地面长期水文数据，深入分析了 1997~2010 年流域土地利用变化特征及其产流产沙效应。结果显示：①14 年间，流域林灌草面积分别增加 160.23%、176.33%和 80.75%；坡耕地、居民地、裸地和梯田面积分别减少 25.57%、0.16%、48.45%和 21.52%。以 2005 年为时间节点，发现前期灌草增加较多、裸地减少明显，后期则是乔木增加比例和坡耕地减少比例更为显著，彰显出不同历史阶段植被恢复的策略变化。②流域出口多年平均径流量和输沙量分别由前期的 18 249m^3 和 6 383kg 锐减至后期的 2 292m^3 和 2 267kg，流域土地利用/覆被有效增加是其主要驱动。③春冬季节，由于降雨稀少、径流泥沙的本底值很低，前后两个阶段的水沙输移量差异较小，土地利用/覆被变化的影响相对尚不显著。但在夏秋季节，随着降雨事件增多，土地利用/覆被变化减水减沙的效应趋于显性化。

关键词：土地利用变化；植被覆盖；产流输沙；黄土高原；流域尺度

全球环境变革和人类活动加剧的大背景下，土地利用/覆被变化成为地球系统科学的重要研究内容[1, 2]。在流域尺度上，土地利用变化严重影响水文环境、水量平衡、地表理化过程、生态系统动态及其服务，而这种影响尤以干旱半干旱脆弱生态区最为显著[3, 4]。

地处中国西北的黄土高原，属于典型的半干旱脆弱生态区。长期以来，由于不合理的土地利用方式和恶劣敏感的生态环境本底，水土流失的严重性堪称世界之最，成为制约当地经济发展和生态环境改善的瓶颈与顽疾[5]。为了遏制水土流失、促进人地关系和谐发展，国家相继启动了退耕还林、封山育林和流域综合治理等一系列生态建

---

* 原载于：生态环境学报，2012，21（8）：1398-1402.

设工程，从而进一步改变了该地区的土地利用/覆被格局，深刻影响区域水文循环和泥沙输移过程[6]。截至目前，围绕黄土高原不同尺度土地利用/覆被变化对径流输沙影响这一话题，国内外学者进行了大量研究和有益探索，并取得重要进展[7]。但是，由于历史和资料原因，半干旱区长期连续的生态水文监测研究依旧相对薄弱，从一个较长时间序列探讨土地利用变化对径流泥沙动态影响尚显不足，其深层次机理有待进一步诠释[8]。鉴于此，本研究以甘肃定西典型黄土丘陵小流域为核心区，基于影像解译和水文观测数据，定量分析了土地利用/土地覆被变化特征及其对流域产流输沙的影响机理，旨在深入揭示二者间的互动关系，为黄土高原及类似地区土地利用规划、水土资源保持与管理提供科学依据。

1 研究区域

安家沟小流域（35°35′ N，104°39′ E）位于甘肃省定西市东郊，属典型的半干旱黄土丘陵沟壑区。海拔 1900~2000m，流域面积 8.91km²。多年平均气温 6.3℃，≥10℃年积温 2239℃，极端最高气温 34.3℃，极端最低气温-27.1℃。年均降水量 427.0mm，其中 60%以上集中在 7~9 月份，且多暴雨。年蒸发量 1510mm。空气平均相对湿度 65.8%，太阳辐射 591.1J/m²，年日照时数 2409h，无霜期 141d。土壤以黄绵土为主，平均厚度 40~60m，垂直节理，土质疏松，湿陷性强，极易发生水土流失。

植被带为森林草原带干草原区。全流域已发现野生植物 23 科、79 种，栽培植物 23 科、64 种。乔木种主要有油松（*Pinus tabuliformis* Cart.）、侧柏（*Platycladus orientalis* L.）、山杏（*Prunus armeniaca* L.）等，灌木有沙棘（*Hippophae rhamnoides* L.）、柠条（*Caragana* ssp.），草本植物为紫花苜蓿（*Medicago sativa* L.）、红豆草（*Onobrychis vichfolia* Scop.）、针茅（*Stipabungeana* Trin.）。主要的农作物有马铃薯（*Solanum tuberosum* L.）、春小麦（*Triticum aestivum* L. CV Leguan）、玉米（*Zea mays* L.）、胡麻（*Linum usitatissimum* L.）以及豌豆（*Caragana kansuensis* Pojark.）等。

2 研究方法

2.1 图像处理

根据 1997 年、2005 年航片和 2010 年 ALOS 影像数据，以 1:10 000 地形图为依据，在 ERDAS 和 ARCGIS 软件支持下，对上述 3 期遥感数据进行处理，统计生成 1997 年、2005 年和 2010 年研究区土地利用/覆被数据，并通过空间叠加分析，得到该地区土地利用/覆被变化的动态信息，生成相关专题图。采用中国科学院资源环境数据库中的土地利用分类系统，并结合安家沟小流域的实际特点，将流域内的土地利用类型划分为 8 类：梯田、坡耕地、有林地、灌木林地、草地、居住地、裸地和水域。

2.2 数据采集

2.2.1 径流量

径流量测控采用流域卡口站监测法，安家沟为梯形测流槽。降雨量较小时用接流筒按体积法施测，洪水时用率定水位流量关系曲线和浮标法测速计算流量，两种方法同步进行，对照检查。浮标系数在平水时采用 0.85，大洪水时采用中泓一点法施测，浮标系数采用 0.65。径流流量计算公式如下。

$$Q = V \times H \times L \times \alpha$$

式中，Q 为流量（m³/s），V 为流速（m³/s），H 为水深（m），L 为岸边距（m），α 为浮标系数。

2.2.2 含沙量

泥沙观测每天取样次数与测流次数基本相同，为控制含沙量变化可适当增加取样次数。平水期每日观测时距相等，洪水期视水情设定测量次数，测距为几分钟到数小时，采用置换法获取含沙量，逐日平均流量及逐日平均含沙量采用算数平均法计算。

取 r_s=2.65、r_w=1.0，则

$$W_s = 1.6 \times (W_{ws} - W_2)$$

$$S = \frac{W_s}{V} \times 1000$$

式中，W_{ws} 为取样瓶及洪水质量（g）；W_2 为与样品同体积的清水质量及瓶质量（g）；W_s 为净沙质量（g）；r_s、r_w 分别为沙和水的容质量；V 为样品体积（ml），S 为含沙量（kg/m³）。

2.3 数据处理

所有相关、回归分析及显著性检验在统计软件 SPSS 中进行。

3 结果与分析

3.1 安家沟流域土地利用/覆被变化

根据遥感解译结果，并结合野外调查校准，获得安家沟小流域的土地利用变化特征（表1）。该流域 1997～2010 年土地利用变化具有以下两个特点：①有林地、灌木林地、草地面积大幅增加，居住地和水域少量增加；②梯田、坡耕地和裸地面积大幅减少。其中，1997～2005 年有林地、灌木林地、草地比例分别增加 15.04%、120.28%、64.54%，梯田、坡耕地、裸地分别减少了 8.46%、4.37%、37.95%。此后，由于流域社会经济的发展和退耕还林工作力度加大，2005～2010 年有林地、灌木林地、草地比例又分别增加 126.20%、25.45%、9.85%，梯田、坡耕地、裸地分别减少 14.26%、22.17%、16.93%。

表1 安家沟小流域土地利用面积及其变化

土地类型	面积（hm²）			变化百分比（%）		
	1997年	2005年	2010年	1997~2005年	2005~2010年	1997~2010年
梯田	442.34	404.9	347.17	-8.46	-14.26	-21.52
坡耕地	99.58	95.23	74.12	-4.37	-22.17	-25.57
有林地	38.09	43.82	99.12	+15.04	+126.20	+160.23
灌木林地	64.51	142.1	178.26	+120.28	+25.45	+176.33
草地	38.86	63.94	70.24	+64.54	+9.85	+80.75
居民地	44.79	44.83	44.72	+0.09	-0.25	-0.16
裸地	180.83	112.2	93.21	-37.95	-16.93	-48.45
水域	0	1.98	2.16	—	9.09	—

进一步分析各土地利用类型的空间转移过程，结果发现，各土地利用类型互化强度较大，梯田、坡耕地、裸地向有林地、灌木林地的转换成为土地利用变化的主导过程；另外，部分裸地被开垦为耕地，耕地退耕还草，而草地又向有林地转化，最终使得研究区内坡耕地的覆盖面积显著减少，有林地和流域植被覆盖度增加，土地利用结构得到优化。同时，发现在前期灌草增加的比例相对较多，而裸地减少明显；后期则是乔木增加比例和坡耕地减少比例更为显著，这和不同时期流域植被恢复依赖的主要先锋植物种有密切关系，一定程度上彰显出黄土高原地区不同历史阶段植被恢复的策略变化，但这种策略的科学性到底如何？尚有待更长序列的时间验证。

3.2 土地利用/覆被变化对年水沙输移的影响

基于航片判读和遥感影像解译分析，结合地面长期水文监测数据，以2005年为时间节点将研究时段一分为二，深入探讨安家沟1997~2010年土地利用/覆被变化对流域多年平均水沙输移程度的影响（图1）。结果显示，与前期（1997~2005年）相比，后期（2005~2010年）流域径流输沙量显著降低。多年平均径流量由前期的18 249m³锐减至2 292m³，减幅高达88%；而多年平均输沙量则由7 283kg降低至1 967kg，降低幅度为73%。进一步分析和研究发现，尽管水土流失和径流泥沙的具体迁移过程会受降雨、地形、植被、人为干扰等多种因素的调控和支配，但导致安家沟小流域水沙输移量发生显著变化的主要原因依旧是流域土地利用格局优化和植被覆被度的提高（图2）。

由图2可知，安家沟小流域内的植被覆盖度呈逐年增加之态势，其值由1997年的15%稳步上升至2005年的31%和2010年的39%；而与之对应，年水沙输移量则呈明显下降趋势。植被覆盖度的上升显示出生态环境治理的成效，主要归功于多年以来的退耕还林、封山育林和流域综合治理工程的共同实施。尽管流域出口径流量和输沙量在不同年份之间会有一定程度的上下波动，但其走向和变化趋势依然明显，即随植被覆盖度增加而显著下降。而事实上，前人的大量研究也支持和佐证了本研究的相关结论，即通过有效的土地利用调整和优化，完全可以实现提高半干旱脆弱生态区植被覆被度、控制水土流失和改善区域生态环境状况之目标[9, 10]。

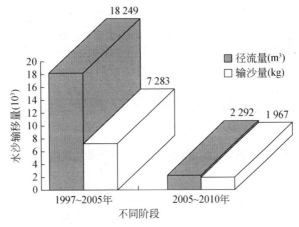

图 1 不同阶段流域出口多年平均水沙输移量变化

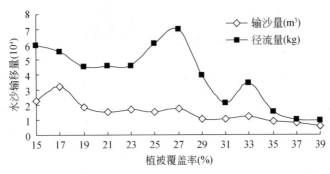

图 2 植被覆盖度和水沙输移量的动态关系

3.3 土地利用/覆被变化对月产流输沙的影响

流域径流产沙过程不仅表现出显著的年际波动规律，同时也呈现季节分明的月变化。而事实上，年内各月份水文变化过程是切实影响当地生态和人居环境的关键因素。因此，为了更好地诠释这种动态规律，结合监测数据探讨了不同历史阶段各月份径流量和输沙量与土地利用/覆被变化之间的可能关系，结果见图 3 和图 4。

由图 3 和图 4 可知，与前期（1997~2005 年）相比，后期（2005~2010 年）各月份的平均径流量和输沙量都呈降低趋势。但前后 2 个时期在不同季节均存在较大差异。在春冬季节，由于降雨稀少、径流泥沙的本底值很低，2 个阶段的水沙输移量差异较小，土地利用/覆被变化的影响相对尚不显著。但在夏秋季节，随着降雨事件增多，土地利用/覆被变化减水减沙的效应趋于显性化。以 7~8 月份的平均径流量为例，前期这两个月的平均值分别为 5 882m³ 和 7 049m³，而后期其值锐跌至 300m³ 和 454m³。平均输沙量也有类似变化规律，分别由前期的 2 298kg 和 2 313kg 降低至 459kg 和 495kg，降低幅度极大。而根据当地气象资料，并未发现 2 个时期的降雨特征值有显著变化。因此，结合图 3 的相关结果，可以肯定的是，土地利用/覆被变化是前后两个时期相同季节差异显著的核心驱动力。

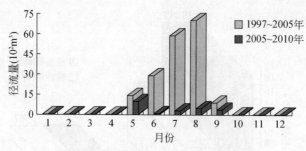

图3 不同阶段流域出口径流量月变化

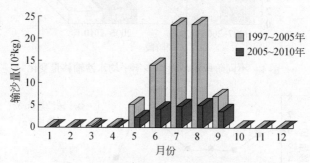

图4 不同阶段流域出口输沙量月变化

4 主要结论

1）研究期间，安家沟流域林、灌、草面积增加显著；坡耕地和裸地面积显著减少；由于退耕还林和劳务输出，居民地没有显著扩张，部分梯田退耕成林灌草而致使其总面积下降。以2005年为时间节点，发现前期灌草增加相对较多，后期乔木增加更显著，一定程度上体现出不同阶段植被恢复的策略变化。

2）从多年平均状况来看，相对于前期，后期流域出口多年平均水沙输移量下降幅度极大。后期各月份水沙量也均低于前期，土地利用结构优化和植被覆盖度增加是其主要驱动。

3）从年内变化来看，流域产流产沙主要集中在5～9月份，与降水的季节分布一致。春冬季节，由于降雨稀少、径流泥沙的本底值很低，前后两个阶段的水沙输移量差异较小，土地利用/覆被变化的影响相对尚不显著。但在夏秋季节，随着降雨事件增多，土地利用/覆被变化减水减沙的效应趋于显性化。

参 考 文 献

[1] Mahmood R, Leeper R, Quintanar A I. Sensitivity of planetary boundary layer atmosphere to historical and future changes of land use/land cover, vegetation fraction, and soil moisture in Western Kentucky, USA. Global and Planetary Change, 2011, 78: 36-53.

[2] Booth D B, Karr J R, Schauman S, et al. Reviving urban streams: land use, hydrology, biology,

and human behavior. Journal of the American Water Resources Association, 2004, 40(5): 1351-1364.

[3] Zhou D C, Zhao S Q, Zhu C. The grain for green project induced land cover chang in the Loess Plateau: a case study with Ansai County, Shanxi Province, China. Ecological Indicators, 2012, 23: 88-94.

[4] Bi H X, Liu B, Wu J, et al. Effects of precipitation and land use on runoff during the past 50 years in a typical watershed in Loess Plateau, China. International Journal of Sediment Research, 2009, 24: 352-364.

[5] Chen L D, Wei W, Fu B J, et al. Soil and water conservation on the loess plateau in China: review and prospective. Progress in Physical Geography, 2007, 31: 389-403.

[6] 张晓明, 余新晓, 武思宏, 等. 黄土丘陵沟壑区典型流域土地利用/土地覆被变化水文动态响应. 生态学报, 2007, 27 (2): 414-423.

[7] 唐丽霞, 张志强, 王新杰, 等. 晋西黄土高原丘陵沟壑区清水河流域径流对土地利用与气候变化的响应. 植物生态学报, 2010, 34 (7): 800-810.

[8] Wei W, Chen L D, Fu B J, et al. The effect of land uses and rainfall regimes on runoff and soil erosion in the semi-arid loess hilly area, China. Journal of Hydrology, 2007, 335: 247-258.

[9] 王晓燕. 黄土高原植被破坏与重建过程中土壤侵蚀强度变化. 生态环境学报, 2009, 18 (3): 1083-1087.

[10] Boulain N, Cappelaere B, Séguis L, et al. Water balance and vegetation change in the Sahel: a case study at the watershed scale with an eco-hydrological model. Journal of Arid Environments, 2009, 73: 1125-1135.

Application of Gash analytical model and parameterized Fan model to estimate canopy interception of a Chinese red pine forest*

Chen Shujun　Chen Cungen　Chris B. Zou　Elaine Stebler
Zhang Shuoxin　Hou Lin　Wang Dexiang

Abstract

Loss of precipitation by canopy interception constitutes a substantial portion of the water budget in a forested ecosystem, and accurate models to simulate canopy interception are critical for effective management of forest water resources. We modeled the canopy interception of an evergreen Chinese red pine (Pinus tabulaeformis) forest using the Gash analytical model and the parameterized empirical Fan model and compared the modeled results with directly measured data. Based on 100 rainfall events between 2006 and 2008, the estimated canopy interception ratio was 35.9% from the Gash model and 53.6% from the Fan model, compared to 33.2% from the direct measurement. The differences between measured and modeled values from the Gash model ranged from −0.3% to +7.1% for different rainfall amounts and from +1.9% to +3.2% for different years. The Fan model satisfactorily simulated interception for large rainfall events(>50mm)with differences from −3.4% to +1.3%, but substantially overestimated interception loss for smaller rainfall events (+21.2% to +37.2%). The Gash analytical model adequately simulated the canopy interception of Chinese red pine forest. The parameterized Fan model compared favorably to the Gash model in simplicity but not in precision. The Fan model required only incidental precipitation data to run after parameterization, but substantial improvement

---

* 原载于: Journal of Forest Research, 2013, 18 (4): 335-344.

in modeling precision is needed before it can be used for this forest.

Keywords: Ecohydrological effects; *Pinus tabulaeformis*; Qinling Mountains; Stemflow; Throughfall

1 Introduction

The amount of precipitation intercepted and subsequently evaporated from a forest canopy can be substantial [1,2]. Improving prediction of potential interception water loss of different species, stand densities, and canopy spatial arrangements is of considerable value in the management of forests for water resources [3-5]. Interception water loss of forest canopy is determined by a suite of factors characterizing precipitation input, i.e. precipitation intensity and duration, the interception surface property such as gap fraction, canopy roughness, phenology, and most importantly, the meteorological conditions driving the evaporation process [6-9]. The stochasticity of meteorological conditions and transience of vegetation attributes make the precise projection of incident interception loss a formidable challenge. During the course of the last century, major models have been developed for predicting canopy interception for different vegetation types and under different climatic regimes. Horton's early canopy interception model was built based on the maximum canopy holding capacity and the evapotranspiration during the precipitation [10]. Later, Rutter developed and then improved a physically based model driven by evaporation and using a running water balance approach [6,11]. Gash [7] simplified the Rutter model into a storm-based analytical model, requiring only mean rainfall and evaporation rates, the temporal pattern of rainfall, and certain forest canopy parameters to run. The Gash model stresses the importance of meteorological conditions and separates individual rain events into phases of wetting, saturation, and drying. Storms are considered discrete events if there is sufficient time in between for the drying of the canopy and stems. Gash modified his original model to change from a unit area to a unit canopy area which greatly improved the use of the model for sparsely covered forests [12]. Mulder [13], Zeng et al. [14], Van Dijk and Bruijnzeel [2], and Murakami [15] developed various modifications of the Gash model; these are commonly referred to as Gash-type models. These Gash and Gashtype models have been applied to various canopies and climates throughout the world [16]. Muzylo et al. [16] provide a quite comprehensive review of interception modeling prior to 2008 and cover 15 major models in detail. Recently, the relative importance of meteorological conditions in determining interception loss [8,17], and the interactive effect of climatic condition and vegetative properties [9], were recognized. These studies suggest the need for a thorough examination of any model that does not specifically consider meteorological conditions.

In China, as is true in Japan and many other Asian countries, forests are predominantly distributed in the upper reaches of the river systems or on the ridges of mountains with limited access. Routine and reliable climatic data are usually lacking for those regions and the application of the Gash model or any other data-intensive modeling framework would be a challenge. In addition, model improvement is urgently needed for forests in regions with intensive storms and high rainfall rates such as occur in southeastern Asian [16].

In China, the Fan model [18,19] is the only non-empirical model developed for forests with intensive storm events. The Fan model was developed for a China fir plantation (*Cunninghamia lanceolata*) in southcentral China and has a hybrid model framework between an empirical model and purely physical-based model. We call it a 'parameterized empirical model' here for simplicity. The uniqueness of this hybrid model framework is that the relationship between rainfall interception and rainfall intensity is assumed to follow a parabolic function, in which the parameter β needs to be defined in advance for different forests. The coefficient (β) was determined by vegetation properties and can be estimated by determining canopy interception ratios of a few large rainfall events. The model can be directly driven by incidental precipitation once the forest specific coefficient value β is parametrized. Such a model holds promise for comparison of the effects of different forest disturbances and vegetation management practices on interception water loss.

The Qinling Mountains, one of the major mountain ranges in China, are oriented east–west with a total length of over 1600km, their width ranging from 10 to 300km north to south. They are the major headwater source for one of the three proposed routes of the South-to-North Water Transfer Project. The routes are designed to have a collective capacity of transferring 44.8billion m³/year of water to alleviate water shortage in the arid north [20]. Chinese red pine (*Pinus tabulaeformis*) is one of the dominant tree species of the evergreen conifer forest at elevations between 1400 and 1700m above sea level. The Qinling Mountain range was brought into the State Natural Forest Protection Project in 1999 to be managed solely for soil and water conservation. As a result, new initiatives were undertaken by the Chinese government to accelerate the process of afforestation and improve ecological conditions. Management of a forest for its soil conservation and water provision depends largely on understanding the ecohydrological processes associated with the natural recovery, or alternatively the management practices of such forest. This requires development of new models or validation of existing models that will be useful for evaluating the hydrological effects of different management practices, such as water loss to canopy interception.

The objectives of this study were threefold. First, we wanted to estimate the parameters and coefficients necessary to run two models, the Gash analytical model and the Fan model, to simulate canopy interception of a mature Chinese red pine forest. Second, we wanted to compare our observed data with the simulated results from both models. Third, by evaluating the performance of these two very different models, we intended to identify a proper modeling tool that forest and water resource managers could use for future applications within this temperate forest type. The Gash model was chosen because it has been validated for many different forest types and we anticipated that it could also be successfully applied to this forest type. The revised Gash model for sparse canopies[12] was not used because our measured throughfall and canopy storage capacity values were calculated per unit of ground area rather than per unit of canopy cover. The little known Fan model was chosen in order to determine if a simple parameterized, vegetation-structure only based model framework could effectively simulate the primarily climate-driven process of water flux from canopy interception. If we can show that the Fan model is adequate, then the Fan model would provide resource managers with a tool that could be easily applied to the remote red pine forest regions of China where continuous measurement of meteorological data is a challenge.

2 Materials and methods

2.1 Study area

This research was carried out in Experimental Watershed 2 at the Qinling National Forest Ecosystem Research Station at Huoditang, Ningshan County, China (33°18′N, 108°20′E). Watershed elevations vary from 1550 to 1700m and the research site was approximately 1km^2. The mean annual temperature is 8–10°C. The annual precipitation is between 900 and 1200mm, of which over 85% fall during the growing season (May and October). The annual potential evapotranspiration is 800–950mm. The soil is mountainous brown soil generated from granite parent materials with a depth ranging from 30 to 50cm and has high organic matter content[21]. The current forest is a secondary forest naturally regenerated after the clear-cuts of the 1950s and 1960s. Now, the dominant canopy tree species is Chinese red pine with other tree species such as *Quercus aliena* var. *acuteserrata*, *Pinus armandii*, *Betula albo-sinensis*, *Toxicodendron vernicifluum*, and *Acer davidii* occasionally present in the canopy layer. Understory cover is low and mainly present in forest canopy gaps. The age of the majority of Chinese red pine trees is between 51 and 60 years. The canopy cover is 85% and the leaf area index is 2.07±0.31 based on site measurement using a canopy analyzer (LAI-2000; Li-Cor, Lincoln, NB, USA).

A 27.5-m tower was built on the research site in 2005. The meteorological data were collected at five heights from the ground, namely, 1.5m (near ground layer), 14m (below canopy), 19.5m (inside canopy), 25m (canopy surface), and 26.5m (above canopy). Wind speed and direction (Model 03001; R.M. Young, Traverse City, MI, USA), air temperature (CS107; Campbell Scientific, Logan, UT, USA), air humidity (HMP45C; Vaisala, Woburn, MA, USA), canopy temperature (IRTS-P; Campbell Scientific) and net radiation (CNR-1; Kipp and Zonen, Bohemia, NY, USA) were measured at each height. A fully equipped meteorological station (UT30; Campbell Scientific) was built in an open area approximately 500m to the south of the study site; it automatically recorded gross rainfall

(TE-525I; Texas Electronics, Dallas, TX, USA), wind direction and speed (MetOne 034B; Grants Pass, OR, USA), air temperature (CS107), relative humidity (HMP45C) and barometric pressure (CS105; Campbell Scientific).

2.2 Direct quantification of canopy interception

Throughfall and stemflow data were collected on the research site from June 2006 to October 2008. Throughfall (TF) was measured using U-shaped collectors made by cutting PVC pipe (4m in length and 0.2 m in diameter) into two pieces. Each piece was sharpened at the outer edge so that each collector would have a precise collection area of 0.8m^2. Five average-sized Chinese red pine trees [as determined by diameter at breast height (DBH) and canopy diameter] were randomly chosen in a 200m × 200m area. Each tree was assigned one throughfall collector. The collector was set at a height of 1.0m above the ground and oriented along the topographical contour from the tree trunk toward the canopy edge, slightly sloping toward the edge of the canopy for free drainage. Each collector was connected to an automatic tipping bucket flow gauge (6506H; Unidata; Australia) with a capacity of 125ml per tip.

Throughfall was calculated as:

$$TF = \frac{1}{n}\sum_{i=1}^{n}(TF_i / A)$$

where TF is in mm, n is the number of throughfall collectors ($n = 5$), TF_i is the throughfall for the ith collector (mm^3), and A is the collection area of each collector (8×10^5mm^2 in this study).

Stemflow (SF) was collected by wrapping a piece of polyethylene plastic tubing of 2cm diameter around the tree trunk at about 1.5m above ground level and spiraling around the trunk 6 times. To ensure a good seal between the tubing and the tree trunk, tree bark was carefully removed using a small saw and sharp knife to ensure smooth contact. The upper surface of the tubing was shredded to create 30-40 openings of approximately 0.8cm wide and 2.5cm long to facilitate entry of stemflow. From our on-site observation, these openings were sufficient to capture stemflow input in this forest stand. Twelve Chinese red pine trees, comparable in diameter and canopy structure to those used for throughfall measurements, were chosen for stemflow measurements. Three adjacent trees were grouped to share one automatic tipping bucket flow gauge (6506G; Unidata) with a capacity of 75ml per tip. The output of each tipping bucket was recorded by a HOBO datalogger (HOBO Event; Onset Computer, Bourne, MA, USA) with a time stamp for each tipping event. Stemflow was calculated as:

$$SF = \frac{1}{n}\sum_{i=1}^{n} SF_i / FA_i$$

where SF is in mm, n is the number of stemflow collectors ($n = 4$), SF_i is stemflow measured from ith collector (mm^3), and FA_i is the canopy area directly measured for the three trees associated with the ith collector (mm^2) [22].

Snow at our site accounts for only a small part of the annual precipitation and was not considered in this study. From the gross rainfall measured in the open area, if 4h elapsed without a tip then the next tip was considered as the start of a new rainfall event. Rainfall events without reliable throughfall and stemflow measurement due to equipment failure were excluded. There were a total of 100 rainfall events with complete throughfall and stemflow recording during June 2006 to October 2008; refer to Chen et al. [22] for more details.

3 Model descriptions

3.1 Gash model

The Gash model [7] represents rainfall input as a series of discrete storms with each individual storm being divided into three phases—canopy wetting, saturation, and drying. The Gash model uses the following equation to calculate total interception loss.

$$\sum_{j=1}^{n+m} I_j = n(1-p-p_t)P'_G + (\overline{E}/\overline{R})\sum_{j=1}^{m}(P_{Gj} - P'_G) + (1-p-p_t)\sum_{j=1}^{m}P_{Gj} + qS_t + p_t\sum_{j=1}^{m+n-q}P_{Gj} \quad (1)$$

This equation sums the various components of interception loss, $I = I_c + I_w + I_s + I_a + I_t$, where I is the total interception loss. $I_c = (1-p-p_t)\sum_{j=1}^{m}P_{Gj}$, for m storms insufficient to saturate the canopy ($P_G < P'_G$) and $I_w = n(1-p-p_t)P'_G - ns$, for n storms sufficient to saturate the canopy, wetting the canopy ($P_G \geq P'_G$). P_G is the gross precipitation incident on the canopy, P'_G is gross precipitation necessary to saturate the canopy, p is the free throughfall coefficient, which represents the proportion of gross precipitation that passes through the canopy without touching it, p_t is the stemflow fraction coefficient, and S is canopy storage capacity. $I_s = (\overline{E}/\overline{R})\sum_{j=1}^{m}(P_{Gj} - P'_G)$ which is the loss by evaporation from time of saturation until rainfall ceases. $\overline{E}$ is average evaporation rate and $\overline{R}$ is average rainfall intensity. $I_a = nS$, evaporation after rainfall ceases. $I_t = qS_t + p_t\sum_{j=1}^{m+n-q}P_{Gj}$, evaporation from trunks for q storms which saturate the trunks ($P_G > S_t/p_t$) where S_t is trunk storage capacity.

The precipitation required to saturate the canopy (P'_G) is a function of p and $\overline{E}$:

$$P'_G = -\overline{R}S/\overline{E} \ln\left[1 - (\overline{E}/\overline{R})(1-p-p_t)^{-1}\right] \quad (2)$$

Using the Penman–Monteith equation for saturate canopy condition, the mean evaporation rate was calculated for the forest canopy during rainfall, with canopy resistance being set to zero [23] and aerodynamic resistance was calculated as described by Gash [7].

The Gash model relates the canopy interception to the characteristics of rainfall, i.e. canopy structure, air temperature, humidity, saturated water pressure, net radiation, wind speed, and air dynamics. The model run is based on an incidental rainfall event and assumes that the canopy is completely dry before each storm.

Application of the model relies on field observation to calculate the average rainfall intensity and average evaporation rate during the rainfall, and assumes that there is no canopy dripping before canopy saturation (or throughfall). It further assumes that tree trunk evaporation happens only after rainfall stops. In addition, the total rainfall amount, throughfall, and stemflow need to be measured to define the thoughfall coefficient (p), stemflow coefficient (p_t), trunk water holding capacity (S_t) and canopy storage (S).

3.2 Fan Model

The Fan model [18,19] was developed based on the "saturation excess runoff" concept in watershed hydrology [23,24]. Conceptually, rain drops will first wet the leave or branches and then produce net rainfall only after all leaves and branches are saturated. For a hypothetical condition under which the canopy is dry and rainfall intensity, wind speed, and temperature are all constant, the amount of rainfall needed to completely saturate the canopy is the interception capacity and is determined by the canopy properties and is proportional to the canopy depth.

This model operates under the following assumptions: ①rainfall first saturates the canopy then produces throughfall; ②interception capacity for a certain canopy area is proportional to the canopy depth (or thickness) at that area; ③horizontally, canopy saturation moves from the area of thin canopy towards the area of thick canopy until the area with the thickest canopy is saturated; and ④ evaporation during this process is negligible, and at a certain point, the entire canopy is saturated and the interception (W) will not increase. Maximum interception capacity will be equal to the precipitation amount to reach this point. Although the Fan model is described in detail in Fan et al. [19], much of it is repeated here since the model has not been widely used by others.

Assuming the total canopy area is F and the interception capacity at a random point of the canopy is h, then the canopy area with interception capacity equal or smaller than h is f, and the percent canopy area with an interception capacity equal or smaller than h is . Assuming the largest interception capacity of the entire canopy is H (the precipitation required to saturate the thickest canopy), then the canopy area with interception capacity smaller than H is F. The interception capacity at a random point of canopy is: $0 \leqslant h \leqslant H$.

If $h = 0$, then $f/F = 0$; When h increases, f/F increases. When $h = H$, $f/F = 1$.

Based on "the saturation excess runoff" concept [24,25], the relationship between unsaturated area and saturated area can be described using the parabolic function.

$$\frac{f}{F} = 1 - (1 - \frac{h}{H})^\beta \quad (\beta > 0) \tag{3}$$

where β is a canopy specific coefficient reflecting the stand characteristics such as uniformity of canopy depth, stand species composition, and age structure. β determines the curvature of the relationship between f/F and h. When $\beta = 1$, it is a linear relationship.

P is the incoming gross rainfall. For any rainfall events, where P is smaller than H, the interception capacity for the entire stand is:

$$\begin{aligned} I &= \int_0^P (1 - \frac{f}{F}) dh \\ &= \int_0^P (1 - \frac{h}{H})^\beta dh \\ &= \frac{H}{\beta + 1} \left[1 - (1 - \frac{P}{H})^{\beta + 1} \right] \end{aligned} \tag{4}$$

Assuming

$$\frac{H}{\beta + 1} = W \tag{5}$$

then

$$I = W \left[1 - (1 - \frac{P}{H})^{\beta + 1} \right] \tag{6}$$

$$\text{When } P = H, \quad I = W \tag{7}$$

Therefore, W is the interception measured when the thickest canopy reaches saturation.

When $P > H$, the entire canopy is saturated and the canopy will not produce extra interception, thus $I = W$. The effect of leaf and branch growth on canopy interception can be reflected by the β, H, and W for a given stand.

For a sparse forest stand with some open canopy, the canopy interception for the closed canopy portion can be estimated by a modification of equation (6) or (7). If forest cover is α, then forest openness is $1 - \alpha$. Since there is no interception in the open area, the canopy interception can be described as:

$$\text{when } P < H, \quad I = \alpha W \left[1 - (1 - \frac{P}{H})^{\beta + 1} \right] \tag{8}$$

$$\text{and when } P \geqslant H, \quad I = \alpha W \tag{9}$$

4 Model parameterization

4.1 Gash model

A suite of parameters primarily defining the canopy structure need to be pre-determined in order to run the Gash model. These parameters include throughfall coefficient (p), stemflow fraction (p_t), trunk storage capacity (S_t) and canopy storage capacity (S).

The throughfall coefficient Throughfall coefficient (p): was determined by quantifying the gap fraction using a canopy analyzer on four overcast days during July and November in 2009. The average of p during that period was 0.21 ± 0.04.

Stemflow fraction and trunk storage capacity Stemflow fraction (p_t) and trunk storage capacity (S_t), Based on Gash and Morton [26], the stemflow fraction (p_t) and the trunk storage capacity (S_t) can be estimated as the slope and the negative interception from a linear regression of stemflow vs. rainfall. The Pearson's correlation was used to find a correlation between stemflow and precipitation (SPSS 15.0) using observed data from rain events during 2006–2008 [22]. Stemflow was strongly correlated with precipitation ($r = 0.539$, $P = 0.01$) and the linear regression between stemflow and gross precipitation is expressed as:

$$\text{Stemflow} = 0.0855 P_G - 0.8303$$
$$(R^2 = 0.7958, P<0.001, n=45) \tag{10}$$

Therefore, p_t is 0.0855, and S_t is 0.8303.

Canopy storage Accurate estimation of canopy storage (S) is critical for the Gash model [27,28]. The value of S is determined by the canopy structure [29] and modified by rainfall characteristics and meteorological factors [30]. By plotting the observed throughfall data against precipitation data following the approaches used by Leyton [31] and Link et al. [30], the canopy storage value for Chinese red pine was calculated as 1.02mm (Fig. 1). The derived values for the Gash model parameters are summarized in Table 1.

4.2 Fan model

Three parameters that need to be defined for the Fan model are the canopy specific coefficient β, maximum interception capacity (H), and W which numerically equals the maximum canopy interception. The Fan model is based on the concept that with increasing precipitation, canopy saturation will proceed from the thinnest spot to the thickest spot until the canopy is completely saturated. W and H are determined first, then β can be calculated. To define W and H, observation data are required. The analytical approach is to produce the outer boundary of the measured interception loss for rainfall amount and to identify the point where the canopy interception reaches a maximum and levels out. Fig. 2 is a plotted graph between rainfall and interception measured between 2006 and 2008 [22].

When $P \leqslant 22.0$mm, canopy interception increases linearly with increasing precipitation. When precipitation is between 22.00 and 51.5mm, there is a more gradual increase in interception loss. When $P = 51.5$mm, the water loss to canopy interception reaches a maximum at 19.4mm and remains largely unchanged at $P > 51.5$mm, therefore $I = 19.4$mm and $H = 51.5$mm. Canopy cover, α, is 85%. Then, $W = I/\alpha = 19.4/0.85 = 22.8$mm. Using Equation 7, $\beta = 1.3$.

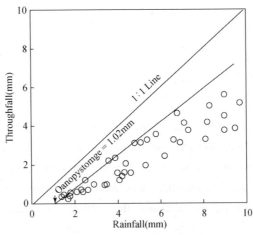

Fig.1　Estimation of canopy storage (S) using the Leyton method

Table 1　Derived model parameters for Gash 1979 and Fan models

Model	Parameter	Value
Gash model	p	0.21
	p_t	0.0855
	S_t	0.8303mm
	S	1.02mm
Fan model	a	0.85
	I	19.4mm
	W	22.8mm
	H	51.5mm

Rewriting Eqs. 8 and 9, then,

$$I = 22.8 \times 0.85 \times \left[1 - \left(1 - \frac{p}{51.5}\right)^{1.3+1}\right] \quad (P<51.5) \quad (\text{mm}) \quad (11)$$

and $I = \alpha w = 19.4$ ($P \geqslant 51.5$) (mm)　(12)

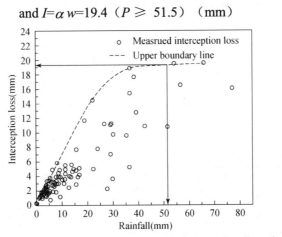

Fig. 2　Model parameterization for Fan model based on the upper boundary line of measured interception loss and rainfall from Chinese red pine forest

5 Results

5.1 Rainfall and canopy interception characteristics of Chinese red pine forest

The annual precipitation between 2006 and 2008 was 933.52±196.82mm, with over 88% occurring between May and October. From the 100 precipitation events with complete throughfall and stemflow recordings used for the simulation, the gross precipitation, throughfall, stemflow, and interception loss were 1576.4mm, 982.9mm, 69.5mm, and 524.0mm, respectively. On an annual basis, throughfall, stemflow, and interception ratios were 62.4%, 4.4%, and 33.2%, respectively [22]. Of the 100 rainfall events, 31 events were between 0mm and 4.9mm, 25 between 5.0mm and 9.9mm, 23 between 1.0mm and 24.9mm, 14 between 25mm and 49.9mm, 6 between 50mm and 99.9mm, and 1 event was >100.0mm.

5.2 Gash model output

The simulated results from Gash model for the 100 rainfall events are listed in Table 2. The total interception loss as simulated by Gash model was 566.5mm, which is 42.5mm more than measured interception loss. The interception ratio was 35.9%, 2.7% higher than the measured value.

The Gash model slightly over-estimated the interception ratios for small rainfall events. It overestimated the interception loss from rainfall events of 0-4.9mm by about 7.1%, but the difference between the estimated and observed data for 25.0-99.00mm rainfall amounts was exceptionally close, with <1% difference (Fig. 3). Overestimation by the Gash model for the small rainfall events could result from the fact that the rainfall events of 0-4.9mm in this study were long duration, with an average rainfall intensity of only 0.58mm/h, and majority of rainfall was intercepted.

Comparison of modeled interception values and observed values for each of the 3 years indicated that the Gash model slightly but consistently overestimated the rate of interception for all 3years (Fig. 4). Gash model stimulated the annual rainfall interception within the range 1.9%-3.25%, which is fairly accurate and adequate for model estimation.

Table 2 Measured values and estimated components of canopy interception loss modeled by Gash 1979 analytical model for 100 rainfall events between 2006 and 2008 in Chinese red pine forest

Component	Modeled value(mm)	%of estimated interception	Observed(mm)from 2006-2008	% of measured precipitation
I_c	6.4	1.1		
I_w	24.6	4.4		
I_s	376.2	66.4		
I_a	93.8	16.6		

Continued

Component	Modeled value(mm)	%of estimated interception	Observed(mm)from 2006-2008	% of measured precipitation
I_t	65.3	11.5		
Total interception loss ($I_c+I_w+I_s+I_a+I_t$)	566.5		524.0	33.2
Measured precipitation			1576.4	100
Measured throughfall			982.9	62.4
Measured stemflow			69.5	4.4

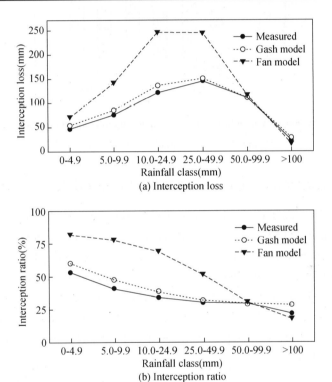

Fig.3 Comparison of the Gash and Fan modeled interception loss and interception ratio to the measured values of the Chinese red pine forest for different rainfall classes

5.3 Fan model output

The total interception loss as simulated by the Fan model for the 100 rainfall events was 844.4mm, 320mm more than observed value. The interception rate of the entire period was 53.6% of the gross precipitation, which was 20.3% higher than the measured value. Further analysis of the interception loss by different rainfall amounts indicated that the Fan model dramatically overestimated the interception loss when rainfall amount was smaller than 50mm, but it simulated the interception with rainfall amounts larger than 50mm better than

Gash model (Fig. 3). For the three different years, the Fan model substantially but constantly overestimated the canopy interception by 20.7%, 15.4%, and 24.1% for 2006, 2007, and 2008, respectively (Fig. 4).

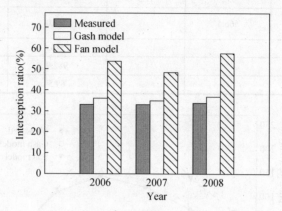

Fig.4　Comparison of annual interception ratios of Chinese red pine forest modeled by the Gash analytical model and Fan model for different years

5.4　Comparison of outputs from the two models

The Fan model satisfactorily simulated interception for rainfall events larger than 50mm with differences from −3.4% to +1.3%, but substantially overestimated interception loss for rainfall events smaller than 50mm (+21.2% to +37.2%; Fig. 3). Based on the 100 rainfall events between 2006 and 2008, the estimated canopy interception ratio was 35.9% from the Gash model and 53.6% from the Fan model, compared to 33.2% from the direct measurement (Fig. 4). The differences between measured and modeled values from the Gash model ranged from −0.3% to +7.1% for different rainfall amounts and from +1.9% to +3.2% for different years (Fig. 4).

6　Discussion and conclusion

The Gash model was developed and tested on temperate Scots pine (*Pinus sylvestris* L.)[7], and was reported to satisfactorily estimate rainfall interception for similar pine forests[16]. We anticipated that the Gash model would also be able to effectively model the rainfall interception for the Chinese red pine forest in the temperate zone and our data show that it did so. Fan et al.[19] found that their Fan model was highly effective in estimating the observed interception loss across all rainfall classes for the China fir forest in subtropical south China. However, our study found that the Fan model failed to satisfactorily model the canopy interception of Chinese red pine in the temperate zone for precipitation events of <50mm.

The reason why the Fan model was able to effectively simulate the canopy interception rate in subtropical south China [19] but not temperate Chinese red pine may be related to its modeling concept and structure. The fundamental assumption of the Fan model is that no throughfall is produced before all leaves and branches become saturated. This assumption may not always hold for several reasons. Gap fractions are always present even for areas generally classified as canopy. The canopy cover for this forest is defined as 85% from a forest survey, but the gap fraction is 21% was measured by the LAI-2000 (LiCor); therefore, use of gap fraction rather than canopy cover may better represent the free-fall space for a specific stand. Gaps allow some rain drops to pass through the canopy without hitting any part of the tree. For the canopy fraction associated with gaps, gaps decrease the depth of crown and enhance dripping from leaves and branches even before the leaves or trunks are saturated. However, the errors associated with the above-mentioned issues are reduced once a rainfall event is large enough or lasts long enough to eventually saturate most or all leaves and branches. This not only explains why the Fan model has a much better performance for rainfall events larger than 50mm but it may also partially explain why it has a better performance in the subtropical forest where rainfall storms are short in duration and high in volume [19]. For rainfall events observed in the Chinese red pine forest in this study, 93% of the rainfall events were <50mm and were not sufficient to saturate the thickest canopy for this forest. It is reasonable to assume that climatic factors such as solar radiation, wind, and temperature may have a larger influence on canopy storage capacity when rainfall duration is long and intensity is low. By not incorporating climatic factors, the Fan model would logically do poorly for small rainfall events.

The Fan model was designed to simulate canopy interception under circumstances for which meteorological data are not readily available. Therefore, further analysis of the error terms of the simulated results from this model and the Gash model may reveal the relative importance of climatic condition on the interception process. Under the conceptual framework, the Fan model de-emphasizes the importance of evaporation occurring during the rainfall event, I_s. Although I_s is a major component of the interception water loss, it varies among forest types and locations and its contribution to the interception loss ranges from 33% in temperate rainforest [30], 34% in temperate conifer forest [7], and 39.48% in warm temperate conifer forest [28] to as high as 54.7% for subtropical evergreen broad-leaved forest [32] and 69.7% for subalpine oak forest [33]. Theoretically, the Fan model will have larger estimate errors for conditions where I_s is high. Referring to the Gash model values for the Chinese red pine forest (Table 1), I_s accounts for 66.4% of the overall estimated interception loss, followed by evaporation after rainfall ceases (I_a), which accounts for another 16.6%. These are among the highest values for documented studies and therefore the Fan model would perform relatively poorly by not having a parameter to represent this process.

In addition, the Fan model suffers from the uncertainty of the critical coefficient β. β is provided to reflect the canopy biophysical characteristics; therefore, it varies once such canopy structure changes. Even though the canopy structure tends to be relatively unchanged in comparison to environmental factors, it does change under some conditions. For deciduous species, the β will change with changing seasons. Parameterization of a transient β value for the Fan model is challenging[20].

The Fan model is proposed as an alternative approach for areas that lack climatic information, not for its potential to improve the estimate precision. Although it over-estimated the canopy interception by a large margin in this study and therefore is unlikely to be of any meaningful use in the Chinese red pine forest without further calibration, it does perform reasonably well for larger rainfall events and its errors are from consistently overestimating the interception for small rainfall events, due largely to its modeling structure and assumption. Therefore, improvement to the model by including a second coefficient to take consideration of rainfall size-class or intensity may hold promise to produce a non-meteorological-driven interception model with relatively satisfactory estimating accuracy. Further considerations should be given to evaluating the model performance by directly including information related to canopy gap fraction and thickness (e.g., canopy fraction or/and LAIs). Such models could be particularly useful for the purpose of estimating the canopy interception associated with different management practices like forest thinning, trimming, and die-off in similar climatic regions. Although the Fan model does need observational data for the β parameter, this data requirement is different from needing a complete dataset to produce an empirical model. Observed data from only a few large rainfall events will provide the saturation interception value needed. The use of canopy coverage α and potential inclusion of canopy fraction or/and LAIs are viewed as strengths of the Fan model since this will easily link to different forest disturbances and management practices.

In conclusion, the Gash model satisfactorily simulated the canopy interception of the temperate Chinese red pine forest and is ideal for application for simulating rainfall interception for this forest in this region. The Fan model required only incidental precipitation data to run after parameterization and was able to satisfactorily simulate interception for large rainfall events. However, it substantially overestimated the interception of rainfall events <50mm and therefore the overall rainfall interception of Chinese red pine forest. Improvement in model parameterization to enhance the projection accuracy is recommended before this model is used for any new forest type.

Acknowledgments

This study is supported by The State Forestry Administration of China Public Service Research Funding-Research and Demonstration Project for Carbon Sequestration and Water Resource Management for Forests in Qinling Mountains (201004036), Ministry of Education of China Young Faculty Research Program (20070712026), Youth Research Foundation of

Northwest A & F University (QN2011161), and Long Term Research Program of Qinling National Forest Ecosystem Research Station (2006–2010). The Huoditang Research and Demonstration Tree Farm of Northwest A&F University provided the research site and the Qinling National Forest Ecosystem Research Station provided all background information and field equipment. Support was also provided by USGS/NWRI Grants (G09AP00146) and the Oklahoma Agricultural Experimental Station. The authors would like to thank the two anonymous reviewers for their valuable comments and suggestions to improve the paper.

References

[1] Calder I R. Evaporation in the Uplands. New York: Wiley, 1990.

[2] Van Dijk A I J M, Bruijnzeel L A. Modelling rainfall interception by vegetation of variable density using an adapted analytical model. Part 1. Model description. Journal of Hydrology, 2001, 247: 230-238.

[3] Owens M K, Lyons R K, Alejandro C L. Rainfall partitioning within semiarid juniper communities: effects of event size and canopy cover. Hydrological Processes, 2006, 20: 3179-3189.

[4] Zou C B, Ffolliott P F, Wine M. Streamflow responses to vegetation manipulations along a gradient of precipitation in the Colorado River Basin. Forest Ecology and Management, 2010, 259: 1268-1276.

[5] Dung X B, Miyata S, Gomi T. Effect of forest thinning on overland flow generation on hillslopes covered by Japanese cypress. Ecohydrology, 2011, 4: 367-378.

[6] Rutter A, Kershaw K, Robins P, Morton A. A predictive model of rainfall interception in forest. I. Derivation of the model from observation in a plantation of Corsican pine. Agricultural Meteorology, 1971, 9: 367-384.

[7] Gash J H C. Analytical model of rainfall interception by forests. Quarterly Journal of the Royal Meteorological Society, 1979, 105: 43-55.

[8] Komatsu H, Tanaka N, Kume T. Do coniferous forests evaporate more water than broad-leaved forests in Japan? Journal of Hydrology, 2007, 336: 361-375.

[9] Toba T, Ohta T. Factors affecting rainfall interception determined by a forest simulator and numerical model. Hydrological Processes, 2008, 22 (14) :2634-2643.

[10] Horton R. Rainfall interception. Monthly Weather Review, 1919, 47: 603-623.

[11] Rutter A J, Morton A J, Robins P C. A predictive model of rainfall interception in forests. II. Generalization of the model and comparison with observations in some coniferous and hardwood stands. Journal of Applied Ecology, 1975, 12 (1): 367-380.

[12] Gash J H C, Lloyd C, Lachaud G. Estimating sparse forest rainfall interception with an analytical model. Journal of Hydrology, 1995, 170: 79-86.

[13] Mulder J. Simulating interception loss using standard meteorological Data//Hutchison B, Hicks B. The Forest-Atmosphere Interaction. Dordrecht: Reidel Publishing, 1985: 77-196.

[14] Zeng N, Shuttleworth J W, Gash J H C. Influence of temporal variability of rainfall on interception loss. Part 1. Point analysis. Journal of Hydrology, 2000, 228: 228-241.

[15] Murakami S. Application of three canopy interception models to a young stand of Japanese cypress and

interpretation in terms of interception mechanism. Journal of Hydrology, 2007, 342: 305-319.

[16] Muzylo A, Llorens P, Valente F, et al. A review of rainfall interception modeling. Journal of Hydrology, 2009, 370: 191-206.

[17] Komatsu H, Shinohara Y, Kume T, et al. Relationship between annual rainfall and interception ratio for forests across Japan. Forest Ecology and Management, 2008, 256 (5): 1189-1197.

[18] Fan S X, Jiang D M, Li X H, et al. Studies on throughfall model in forest area. Acta Ecologica Sinca, 2003, 23: 1403-1407.

[19] Fan S X, Cheng Y C, Wang Z F, et al. New model of forest canopy interception to rainfall on watershed stored-full runoff theory. Journal of Nanjing Forestry University(Natural Sciences Edition), 2007, 31: 93-96.

[20] Zhang X, Xia J. Coupling the hydrological and ecological process to implement the sustainable water resources management in Hanjiang River basin. Science in China Series E: Technological Sciences, 2009, 52 (11): 3240-3248.

[21] Hou L, Lei R D, Liu J J, et al. Soil respiration in P. tabulaeformis forest during dormant period at Huoditang forest zone in the Qinling Mountains, China. Acta Ecologica Sinica, 2008, 28 (9): 4070-4077.

[22] Chen S J, Chen C G, Zou B C, et al. Time lag effects and rainfall redistribution traits of the canopy of natural secondary *Pinus tabulaeformis* on precipitation in the Qinling Mountains, China. Acta Ecologica Sinca, 2012, 32: 1142-1150.

[23] Monteith J L. Evaporation and environment. Symposia of the Society for Experimental Biology, 1965, 19: 205-234.

[24] Lin S Y. Hydrological Modeling. Beijing: China Water Power, 2001.

[25] Rui X F. Hydrological Principles. Beijing: China Water Power, 2004.

[26] Gash J H C, Morton A J. An application of the Rutter mModel to the estimation of the interceptionloss from Thetford Forest. Journal of Hydrology, 1978, 38: 49-58.

[27] Limousin L M, Rambal S, Ourcival J M, et al. Modelling rainfall interception in a Mediterranean Quercus ilex ecosystem: lesson from a throughfall exclusion experiment. Journal of Hydrology, 2008, 357: 57-66.

[28] Shi Z J, Wang Y H, Xu L H, et al. Fraction of incident rainfall within the canopy of a pure stand of *Pinus armandii* with revised Gash model in the Liupan Mountains of China. Journal of Hydrology, 2010, 385: 44-50.

[29] Dang H Z, Zhou Z F, Zhao Y S. Study on forest interception of *Picea crassifolia*. Journal of Soil Water Conservation, 2005, 19: 60-64.

[30] Link T E, Unsworth M, Marks D. The dynamics of rainfall interception by a seasonal temperate rainforest. Agricultural and Forest Meteorology, 2004, 124: 171-191.

[31] Leyton L, Reynolds E R C, Thompson F B. Rainfall interception in forest and moorland//Sopper W E, Lull H W. International Symposium on Forest Hydrology. Oxford: Pergamon Press, 1967: 63-178.

[32] Zhang G, Zeng G M, Jiang Y M, et al. Modeling and measurement of twolayer-canopy interception

loss in a subtropical evergreen forest of central-south China. Hydrology and Earth System Sciences, 2006, 10: 65-77.

[33] He C Q, Xue J H, Wu Y B, et al. Application of a revised Gash analytical model to simulate subalpine Quercus aquifolioides forest canopy interception in the upper reaches of Minjiang River. Acta Ecologica Sinica, 2010, 30 (5): 1125-1132.

Land use change: trends, drivers and consequences on surface hydrological processes

Wei Wei   Chen Cungen

Abstract

Hydrological fluctuation, sediment yield and nutrient transportation caused by rain water is a major type of land degradation and environmental deterioration in terrestrial ecosystems, threatening the harmonious development between human and nature worldwide. Land use is one of the most crucial contrived factors affecting these surface key processes. Therefore, accelerated land use conversion in today's world may possibly increase the uncertainties and complexities of surface overland/concentrated flow, water loss and soil erosion dynamics, making accurate prediction, evaluation and successful control of erosion hazard more difficult. In general, population expansion, climate change and relevant land use conversion policy are considered as three major drivers of land use change, although many other specific factors may also play constructive roles. Consequently, land use change can induce many alterations of landscape patterns and surface biophysical processes. Hence, in order to control severe water loss and soil erosion in the key susceptive regions, scientific planning and careful implementation of land use readjustment are absolutely needed. In this review paper therefore, the causes, driving forces and consequences of land use change, were stressed and elaborated. Then, the related countermeasures were also highlighted. It is thought to be of significance for land optimization and erosion control, which is expected to provide references for researchers and policymakers.

Keywords: land use change; drivers; consequences; hydrological processes; soil erosion

1 Introduction

Land use, defined as the exploitation, development, management and conservation of natural land resources dominated by human behaviors, is the most key form of interaction between human society and the planet [1,2]. Historical records show that great changes have taken place in land use pattern and land quality worldwide since many decades ago. According to the outputs of modeling predictions and land use change scenarios, this trend will continue and even become more accelerated in the following tens of years [1]. Unfortunately, many of such changes in a direction and with a rate are considered to be unsustainable and harmful for maintaining a healthy ecosystem [3]. This kind of shifts has been proved to play a vital role in reshaping earth surface features, which thus may markedly affect many important surface processes such as rainfall induced runoff, nutrient loss and soil erosion in the key sensitive areas worldwide. These hydrological and biogeochemical processes, however, play significant roles in affecting the planet's health and always can cause huge disasters. Systematic review and a better knowledge of the changing trends, drivers and major consequences of land use change on the terrestrial ecosystems, is thought to be of great significance for land use readjustment and optimization, flood prediction, hazard mitigation pollution reduction and soil-loss control, and has become a crucial issue for evaluating, planning, and managing the sustainability of the nature and society [4-6]. In this article, therefore, the trends, causes, driving forces and major consequences of land use change are elaborated and reviewed. Moreover, the challenges and countermeasures of land use change on terrestrial ecosystems and related hydrological-erosion processes are also highlighted. It is expected to provide valuable information for policymakers and academic discussions in the relevant scientific fields.

2 Land use change: an intensified fact from past to the future

As mentioned and implied before, land use practice is the most key embodiment of human behaviors and wisdoms on the field of earth science exploring and reclaiming practice through different measures [7], and has been a continuous feature since the very beginning of human civilization [1,8]. Due to this reason, some scholars even conceptualize land use activity as a "colonization of terrestrial ecosystems" [9]. The initial and ultimate purpose of land use and its conversion are generally the same: the acquisition of natural resources for immediate human needs, often at the expense of degrading environmental conditions in most cases [1]. Huge areas of natural forests, shrubs, grasses and wetlands have disappeared under strong natural forces (e.g. natural fire disturbance, disease diffusion, long-term historical climate change and tectonic movement, etc.) and consecutive improper anthropogenic activities (e.g.

war, deforestation, urbanization and infrastructure construction), and many of them were transformed into arable farmlands across almost the whole planet during the past hundreds of years [1,10]. For example, in the Chinese Loess plateau, statistics show that nearly half of earth surface area had been converted into arable croplands and pastures until 1990s. This situation however, was reverted and improved to some extent due to the implementation of the national "Grain for Green" Project after 2000 [11]. New data show that the GfG-project had been put in practice in an area of 87 000km^2 in which about 400-600million trees(timber or orchard)were planted by the end of 2005 [12]. According to the relevant regulation launched by the Chinese central government, those steep cultivated farmlands over 25° should be returned to forests or grasses in this region. Other studies also report that tremendous land-use change has occurred in many regions of the world [4,13]. Moreover, due to the enhancement of human abilities supported by advanced high technologies (e.g. Internet, RS, GIS and GPS) and modern machines, such shifts in land use are obviously to be accelerated all over the world in the forthcoming years [9].

3　Land use change: a process affected by multiple drivers

The types, patterns, intensities, processes and dynamics of land use practices, on the other hand, are influenced by manifold bio-physical and socio-economic driving forces, resulting in a considerably complex and multi-scale human-land interactive systems [13-15]. Specifically, population fluctuation (accrual or emigration) and individual/personal characteristics (e.g. age, gender, health, profession type and education level, etc.), climatic variation (e.g. changes in natural rainfall, water balance, solar radiation, temperature and evapotranspiration) and environmental evolvement (e.g. changes in surface cover, development of specific landform and large-scale geo-morphological evolution), policy readjustment related natural resources and land use activities (deforestation, reforestation, agricultural regime, forest protection and other related policies), market economy mechanism (price, supply and demand), culture and conception, spatiotemporal scale variations(different areas, scale size and time step) and even changes in land management practices all play important roles in begetting various bio-geophysical consequences of land use [1,9]. Recently, studies in SE Mediterranean Spain also declare that agricultural abandonment, forest balefire disturbance and tourist development also are partly responsible for the various outcomes of land use change [16].

Generally however, several key drivers which have been confirmed to contribute to prominent land use/land cover change are sought out and stressed as follows. Firstly, the expansion of human population is indubitably blamed as one of the most important drivers causing dramatic land-use change. For example, deforestation and steep cultivation in the loess hilly area of NW China are argued to be resulted from population expansion and human induced poverty, which subsequently decreases land surface cover and causes large-scale water loss and environmental degradation[10]. Meanwhile, rapid, large-scale and high-intensity infrastructural construction and urbanization are also blamed as the major indicators and consequences of population expansion demands [13,17,18]. Also in China, a net conversion of 1.2million ha forestland to cropland was estimated to have occurred from the late 1980s to 2000[19]. Secondly, land use practice is strongly affected by the changing climate and exterior ecological factors. For example, due to rainfall reduction and rising temperature, natural mesic grasslands in the central United States have gradually been invaded by shrub species (e.g. *C. drummondii*)and woody plants(e.g. *Juniperus* spp. and *Gleditsia triacanthos*), which alter the former pure grass-landscape step by step and turned it into a more changeable and complex community [20]. Studies also indicate that landslides and debris flow become more frequent and severer under constantly aggravating rainstorms in many mountainous areas of the world, which cause more fragile lands being unfit for human occupation[21]. Thirdly, large-scale campaigns induced by the governmental decision-makings, political systems and constituted land use policies directly influence and even determine the intensity and orientation of land use change[2]. For example, in Slovakia, small mosaics of arable land and pastures were transformed into large parcels with the size ranging from tens to several hundreds of hectares driven by a planned communist economy from 1948 to 1988. Since 1989, however, large parcels of united arable lands were again split up completely due to the transformation of political system in this country [22]. In China, for another instance, significant changes in land use pattern and quality also have taken place during the past two decades as the country embarked on market reforms and active interaction with the forces of accelerated globalization [18]. An investigation in Ethiopia found that hydro-power dam construction launched by the local and central government of this country once caused dramatic land use change in some special watersheds in the 1970s [23].

4 Land use change: begetting challenging consequences

In general, the basic purpose of land use activities is mainly for human exploiting the surface cover through different styles of land management practices [24]. The consequences and challenges of land transformation from current and long-term run, therefore, are multifunctional and cannot be simply considered as land cover change only [1]. Many other concomitant outcomes such as changes in regional landscape and topography, macro-scale morphology and physiology of plant communities, soil physiochemical features, micro-scale fauna behavior (e.g. earthworms) and microorganism activities (e.g. lichen and microbiological crusts) all should be taken into consideration [1,25]. For example, through terrace construction and conservation cultivation, many areas of sloping landforms have been changed into arable flat-terraced lands step by step in some typical hilly regions of the Loess Plateau during the past decades[10], which is helpful for reducing severe water loss, soil erosion and consequently improving land productivity to some extent. Furthermore, land use conversion also cause high-intensity and broad magnitude fragmentations of habitats and losses of global biodiversity. Generally, some scholars summarize that land use change is one of the most prime determinants of global change with major impacts on terrestrial ecosystems, global biogeochemistry and human vulnerability [24].

Moreover, changing land use practices from a relative long-term run are worried to play significant roles in accelerating higher dynamics and variability of key climatic variables [26]. Reductions in evapotranspiration and water recycling arising from land-use change may initiate a new feedback mechanism that results in decreased rainfall amounts [27]. From this standpoint of view, changes in land use pattern and quality also are blamed as critical factors governing the future availability of fresh water resources [28]. Moreover, long-term shifts in land use are even argued to affect global carbon sequestration and its releasing rate, tropospheric ozone concentration, and eventually may possibly alter the intensities and directions of climate change such as regional temperature and rainfall regimes [22]. So far, statistical records show that more about 35% of human-induced air CO_2 emissions are attributed to land use practices since the very beginning of the Industrial Revolution [1]. In fact however, land use conversion and climate change can be cause and effect for each other, and their interactions are projected to result in large shifts in vegetation community and biodiversity as well as other major ecological processes in the terrestrial ecosystems [29-31].

Specially, just focusing on the aftereffects regarding land use change, two adverse consequences can be determined. On the one hand, changes in land use through reforestation/afforestation can increase land surface cover effectively[15]. The latter, however, is more closely correlated with vegetation canopy features, and may lead to further modifications and improvement of physicochemical characteristics of surface and subsurface soil conditions [32]. Dense land cover and improved soil features, however, are confirmed to be helpful for conserving soil and water as well as nutrients. On the other hand, improper human disturbances and land use activities such as steep cultivation, overgrazing and deforestation has degraded natural ecosystems and decreased land use/land cover (LULC) in large scales and to a great extent worldwide, mainly due to the increasingly population pressure, urbanization and industrialization [27]. LULC, on the contrary, play principle roles in many physical cycles and biological processes in the surface terrestrial lands. Accordingly, unwise transitions of land use types, structures and quantities over time and space play key roles in causing severe soil erosion, changing surface water balance and inducing huge magnitude of water quality degradation.

5 Land use change: mechanism on hydrological-erosion processes

The surface layer of terrestrial ecosystem is the basic receiver and carrier of surface key bio-geophysical processes, such as overland flow, soil loss, flooding, land-sliding, non-point pollution and nutrient transportation as well as river system eutrophication. Land use however, is considered as one of the most key indicators changing earth surface features [33]. Therefore, land use change may have immediate and long-lasting impacts on surface water flow, soil erosion and pollution-transportation dynamics in the key fragile regions (e.g. the semiarid Mediterranean area and the Loess Plateau in China, etc.) where are prone to suffer from the mentioned problems. In fact, both the diminishment and acceleration of water erosion and soluble nutrient loss will appear, which mainly determined by the directions and consequences of land use conversion. For example, based on the methods of aerial photo/satellite imagery interception and land-use scenario analysis, many studies have confirmed that soil losses varied with land use types [33]. Forest encroachments in many areas of the world were found to play crucial roles in causing severe splash erosion, mass movement, drought and flood [7,10,34]. A hypothesis of "forest-hydrology-poverty nexus" alleges that deforestation significantly increases the potentials and risks of flooding, soil loss and other destructive hydrological consequences in Central America [35]. Military land use activities were also reported to contribute to the spatial variability and temporal dynamics of soil erosion [36]. More importantly, one of the key effects of land use conversion is to alter the balance of natural rainfall and evapotranspiration and the resultant runoff-sediment generation [1]. Dense coverage of vegetation induced by proper implementation of land use is proven to have strong buffering capacities on runoff generation and sediment production [26,30,37].

Therefore, land use conversion has been recognized as a major force influencing soil erosion dynamics and hydrological processes over a range of spatiotemporal scales throughout the world. For examples, extensive natural forest and short bush were converted to pastures in New Zealand after European settlement in the late 1800s, rendering the slopes more susceptible to landslide activities [25]. The enhancement of farmland from 30% to 49% in the Tocatins River watershed in Brazil led to a 24% increase of water discharge [38]. After analyzing the influence of historical land use evolution on watershed runoff by integrated modeling system GIBSI in Canada, Quilbé [39] argued that land use is a key factor that has to be taken into account when predicting potential future hydrological responses of a watershed. Due to deforestation and steep cultivation as well as overgrazing, water erosion in the key loess hilly area of China is notorious around the world, causing huge areas of land degradation and poverty [10,34,40].

Theoretically, several specific reasons may contribute to deciphering this complex phenomenon.

1) Changes in land use types, intensities, structures and covers are partly responsible for the alteration of the surface soil-landform properties and protection of the capacities of vegetation canopy, aboveground and underground biomass as well as plant litters. Differences of land use/vegetation types are proved to cause various surface cover rates and litter properties(e.g. figure, thickness and lipid content), which finally induce various erosion rates. Generally, the denser the vegetation cover, the lower the surface soil erodibility, and the weaker water erosion risks and potentials. For example, scrubland (such as sea buckthorn), rather than cropland, pine woodland and grassland, is more powerful to control runoff and soil loss in the semiarid northwestern China [33,34]. Meanwhile, land use change with the loss of surface canopy cover, is one of the major causes of surface runoff and soil erosion. Studies have indicated that the protective effect of land use/cover controls surface runoff and soil erosion effectively [30]. Rapid and widespread forest decline induced by clear-cutting, deforestation or disastrous fires can decrease rainfall interception rates via plant canopy, and reduce water reservoir served by surface litter and soil layers.

2) Changes in land use will alter the local micro-landforms and geomorphic conditions significantly. Long-term and high-intensity land use conversion with the result of landform reshaping, however, plays crucial role in the fluctuations of surface physiochemical and geo-hydrological processes. For examples, the risks regarding soil erosion and water loss under sloping farmland are far higher than those in the terraced arable land [34,41]. Conservation practices such as contouring, strip cropping, or terracing, in turn, can reduce the losses markedly [42]. Stream aggravation and soil erosion in the Macedonia was confirmed to be tied unequivocally to secular land use change [25]. Historical erosion and polluted-material transportation in southern Germany was recorded to be triggered by increased land use dynamics, and climatic fluctuations were of minor importance [25].

3) Land use change can affect soil hydraulic properties such as surface roughness, infiltrability and field-saturated hydraulic conductivity, which in turn influence the threshold and velocity of runoff generation [43]. In general however, the relationship between land use/vegetation cover and water erosion is described as either linear or exponential [32]. Different vegetative types and covers may induce various antecedent soil moisture conditions (ASMC), which contribute a lot to the distinctness of runoff generation, nutrient loss and sediment threshold [33]. For example, runoff and erosion rates in the bare micro-catchments were monitored to be the highest, while became the lowest in those catchments completely covered by vegetation, and ASMC was finally proved to be a key factor in such related consequences.

4) Changes in land use can influence the behaviors and activities of soil macro-fauna and microorganisms. Recently, an interesting field experiment was implemented in the forested areas in Northern Vietnam, and results showed that the activities and behaviors of earthworm were very sensitive to changes in land use types and vegetation covers, which in turn affect the shapes and depths of above-ground casts significantly. The surface casts however, was

proved to have high correlations with surface water infiltration and can thus change the threshold of runoff generation and soil detachment rates significantly.

5) As mentioned before, long-term and large-scale shifts in land use type, structure, pattern and quality will possibly affect surface albedo and terrain spectral reflectivity, alter the inter-balance of water-heat-energy permutation and finally the existing spatiotemporal rainfall features and regimes in the real world[1,7]. Natural rainfall features however, is the most direct driver of water loss, soil erosion, sediment yield and nutrient transportation[33]. The coupling role of changed rainfall and land use, therefore, will certainly play a more complex role in these key surface processes.

6 Concluding remarks

In general, however, how to deal with severe hydrological and erosional processes induced by dramatic land use change in the key susceptible areas remains a big challenge. As stressed before, it is mainly because that the practices of land use conversion are affected by numerous factors. Therefore, it seems to be difficult to distinguish the role of human-dominated land use activities from other factors on the surface hydrological, erosional and other major biogeochemical processes. Meanwhile, the influence of land use on such processes is also hard to quantify, especially over long-term periods and across various spatiotemporal scales where complex mutual interactions occur[39]. How to control its negative consequences, and how to establish a scientific and comprehensive planning regarding future land use change is crucial for achieving harmonious and sustainable human-nature relations. Facing all these challenges, three major aspects are elaborated here, which should be paid more attention in the future.

Firstly, detailed and doable land use planning should be constituted. The potential damages and risks of soil erosion, land sliding and pollution transportation processes caused by land use change should be evaluated comprehensively before a land use project can be implemented. Improper land use activities such as large-scale and high-intensity deforestation, overgrazing, steep cultivation, mining, urban expansion and infrastructure construction, which were blamed to cause considerable water loss, soil erosion and other types of land degradation[1,8,10,40], should be forbidden or implemented more carefully. Detailed and comprehensive land-use designing and implementation, however, is badly lacking and cannot take responsibility for water erosion control, land optimization and integrated ecosystem management at current status.

Secondly, in order to control the damages and negative consequences of water erosion and nutrient loss more effectively, basic research regarding plant species selection (e.g. new cultivars), gene breeding technology, water conservation technique and management should be enhanced. So far, insufficient technical support is still the major barrier for successful plantation and vegetation restoration, especially in some arid, semiarid and remotely mountainous regions.

Thirdly, a better knowledge and understanding of the coupling role of changes in climatic variables and land use conversion on these surface processes is badly needed. Particularly, long-term monitoring and evaluation of large-scale dramatic land use change impacts on regional rainfall characteristics as well as temperature regimes should be accelerated. To achieve this purpose, macro-landscape change based on image monitoring should be integrated with micro-ecological processes based on field investigation.

Acknowledgement

This research was supported by the Knowledge Innovation Project of the Chinese Academy of Sciences(Grant No. kzcx2-yw-421), the National Natural Science Foundation of China (40801041), the National Basic Research Program of China (2009CB421104) and the National Advanced Project of the Eleventh Five-year Plan of China (2006BAC01A06).

References

[1] Foley J A, Ruth D, Asner G P, et al. Global consequences of land use. Science, 2005, 309: 570-574.

[2] Ostwald M, Chen D L. Land-use change: impact of climate variations and policies among small-scale farmers in the Loess Plateau, China. Land Use Policy, 2006, 23: 361-371.

[3] Schneeberger N, Bürgi M, Hersperger A H, et al. Driving forces and rates of landscape change as a promising combination landscape change research—an application on the northern fringe of the Swiss Alps. Land Use Policy, 2007, 24: 349-361.

[4] DeFries R, Eshleman K N. Land-use change and hydrologic processes: a major focus for the future. Hydrological Processes, 2004, 18 (11): 2183–2186.

[5] Wang G X, Liu J Q, Kubota J P, et al. Effects of land-use changes on hydrological processes in the middle basin of the Heihe River, northwest China. Hydrological Processes, 2007, 21(10): 1370-1382.

[6] Chen Y, Xu Y, Yin Y. Impacts of land use change scenarios on storm-runoff generation in Xitiaoxi basin, China. Quaternary International, 2009, 208 (1-2): 121-128.

[7] Miles L, Kapos V. Reducing greenhouse gas emissions from deforestation and forest degradation: global land-use implications. Science, 2008, 320: 1454-1455.

[8] Rattan L. The urgency of conserving soil and water to address 21st century issues including global warming. Journal of Soil and Water Conservation, 2008, 63: 140-141.

[9] Krausmann F, Haberl H, Schulz N B, et al. Land-use change and socio-economic metabolism in Austria—Part 1: driving forces of land use change: 1950-1995. Land Use Policy, 2003, 20: 1-20.

[10] Chen L D, Wei W, Fu B J. et al. Soil and water conservation on the loess plateau in China: review and perspective. Progress in Physical Geography, 2007, 31 (4): 389-403.

[11] McVicar T R, Li L T, Niel T G V, et al. Developing a decision support toll for China's re-vegetation

program: simulating regional impacts of afforestation on average annual streamflow in the Loess Plateau. Forest Ecology and Management, 2007, 251: 65-81.

[12] Zhou H, Rompaey A V, Wang J. Detecting the impact of the "Grain for Green" program on the mean annual vegetation cover in the Shaanxi province, China using SPOT-VGT NDVI data. Land Use Policy, 2009, 26 (4): 954-960.

[13] Long H L, Tang G P, Li X B, et al. Social-economic driving forces of land-use change in Kunshan, the Yangtze River Delta economic area of China. Journal of Environmental Management, 2007, 83: 351-364.

[14] Lambin E F, Rounsevell M D A, Geist H J. Are agricultural land-use models able to predict changes in land-use intensity? Agriculture, Ecosystem and Environment, 2000, 82: 321-331.

[15] Claessens L, Schoorl J M, Verburg P H, et al. Modeling interactions and feedback machanisms between land use change and landscape processes. Agriculture, Ecosystems and Environment, 2009, 129: 157-170.

[16] Elias S, Adolfo C C, Eva A R. Land use change and land degradation in Southeastern Mediterranean Spain. Environmental Management, 2007, 40: 80-94.

[17] Sadeghi S H R, Jalili K H, Nikkami D. Land use optimization in watershed scale. Land Use Policy, 2009, 26: 186-193.

[18] Lin G C S, Ho S P S. China's land resources and land-use change: insights from the 1996 land survey. Land Use Policy, 2003, 20: 87-107.

[19] Bennett M T. China's sloping land conversion program: institutional innovation or business as usual? Ecological Economics, 2008, 65: 699-711.

[20] Briggs J M, Knapp A K, Blair J M, et al. An ecosystem in transition: causes and consequences of the conversion of mesic grassland to shrubland. Bioscience, 2005, 55: 243-254.

[21] Cheng J D, Lin L L, Lu H S. Influences of forests on water flows from headwater watersheds in Taiwan. Forest Ecology and Management, 2002, 165: 11-28.

[22] Cebecauer T, Hofierka J. The consequences of land-cover changes on soil erosion distribution in Slovakia. Geomorphology, 2008, 98: 187-198.

[23] Tefera B, Sterk G. Hydropower-induced land use change in Fincha'a watershed, Western Ethiopia: analysis and impacts. Mountain Research and Development, 2008, 28: 72-80.

[24] Verburg P H, de Steeg J, Veldkamp A, et al. From land cover change to land function dynamics: a major challenge to improve land characterization. Journal of Environmental Management, 2009, 90: 1327-1335.

[25] Brierley G, Stankoviansky M. Editorial: geomorphic responses to land use change. Catena, 2003, 51: 173-179.

[26] Feddema J J, Oleson K W, Bonan G B, et al. The importance of land-cover change in simulating future climates. Science, 2005, 310: 1674-1678.

[27] Groen M M, Savenije H H G. Do land use induced changes of evaporation affect rainfall in southeastern Africa? Physics and Chemistry of the Earth, 1995, 20: 507-513.

[28] Charles J V, Dork S. Anthropogenic disturbance of the terrestrial water cycle. Bioscience, 2000, 50 (9): 753-765.

[29] Hansen A J, Neilson R P, Dale V H, et al. Global change in forests: responses of species, communities and biomes. Bioscience, 2001, 51: 765-779.

[30] Freddy R, Ballais J L, Marre A, et al. Role of vegetation in protection against surface hydric erosion. Comptes Rendus Geoscience, 2004, 336: 991-998.

[31] IPCC. 2007. Freshwater resources and their management// Parry M L, Canziani O F, Palutikof J P, et al. Climate Change 2007: Impacts, Adaptation and Vulnerability: Contribution of Working Group II to the Fourth Assessment Report of the Intergovernmental Panel on Climate Change. Cambridge: Cambridge University Press, 2007: 174-210.

[32] Xu J X. Precipitation-vegetation coupling and its influence on erosion on the Loess Plateau, China. Catena, 2005, 64: 103-116.

[33] Wei W, Chen L D, Fu B J, et al. The effect of land uses and rainfall regimes on runoff and soil erosion in the semi-arid loess hilly area, China. Journal of Hydrology, 2007, 335: 247-258.

[34] Wei W, Chen L D, Fu B J, et al. Mechanism of soil and water loss under rainfall and earth surface characteristics in a semiarid loess hilly area. Acta Ecologica Sinica, 2006, 26: 3847-3853.

[35] Nelson A, Chomitz K M. The Forest-hydrology-poverty nexus in Central America: an heuristic analysis. Environment, Development and Sustainability, 2007, 9: 369-385.

[36] Wang G, Gertner G, Anderson A B, et al. Spatial variability and temporal dynamics analysis of soil erosion due to military land use activities: uncertainty and implications for land management. Land Degradation and Development, 2007, 18: 519-542.

[37] Wei W, Chen L D, Fu B J, et al. Responses of water erosion to rainfall extremes and vegetation types in a loess semiarid hilly area, NW China. Hydrological Processes, 2009, 23: 1780-1791.

[38] Costa M H, Botta A, Cardille J A. Effects of large-scale changes in land cover on the discharge of the Tocantins River, Southeastern Amazonia. Journal of Hydrology, 2003, 283 (1-4): 206-217.

[39] Quilbé R, Rousseau A N, Moquet J S, et al. Hydrological responses of a watershed to historical land use evolution and future land use scenarios under climate change conditions. Hydrology and Earth System Sciences, 2008, 12: 101-110.

[40] Fu B J. Soil erosion and its control in the Loess Plateau of China. Soil Use and Management, 1989, 5: 76-82.

[41] Thomaz E L. The influence of traditional steep land agricultural practices on runoff and soil loss. Agriculture, Ecosystem and Environment, 2009, 130: 23-30.

[42] Vrieling A. Satellite remote sensing for water erosion assessment: a review. Catena, 2006, 65: 2-18.

[43] Zimmermann B, Elsenbeer H, De Moraes J M. The influence of land-use changes on soil hydraulic properties: implications for runoff generation. Forest Ecology and Management, 2006, 222: 29-38.

[44] Ricliébault F, Gomez B, Page M, et al. Land use change, sediment production and channel response in upland regions. River Research and Applications, 2005, 21: 739-756.

Effects of surficial condition and rainfall intensity on runoff in a loess hilly area, China

Wei Wei   Jia Fuyan   Yang Lei   Chen Liding

Zhang Handan   Yu Yang

Abstract

Knowledge of the so-called "source-sink" pattern of surface runoff is important for soil conservation, water resources management and vegetation restoration in the dry-land ecosystems. Micro-runoff plot and rainfall simulation are effective tools in quick understanding the relations between land surface and runoff dynamics. This study made full use of these tools to examine the effect of various factors (plant species, surface cover, vegetation distribution) on runoff generation in the semiarid loess hilly area of China. Two major simulated rainfall intensities (52mm/h and 28mm/h) were designed and conducted, which can represent heavy rainstorms and moderate rainfalls in the local region, respectively. Results showed that the responses of runoff generation and dynamics were far more sensitive to high-intensity rainfalls. Rainfall events with only 1.8 times an increase in intensity and 16% decrease in duration caused a sharp increase in total discharge (13.96 times), runoff depth (16.33 times), mean flow velocity (12.17 times), peak flow velocity (9.34 times) and runoff coefficient (9.23 times), respectively. The time to runoff generation however, was shortened by 70%, which raised the alarm to caution against the risks of hydrological disasters induced by potential rainfall variation in the context of climatic change. More importantly, different plant species and surface cover play various roles in runoff generation and processes. Due to the difference in plant morphology and effective surface cover, runoff delay, total discharge retention and peak-flow reduction with shrubs (seabuckthorn) were more effective than those with secondary natural grass, followed by biological crust and bare soil. Notably, the specific positions of shrub species along the slope affects the time to runoff, specific flow process and total volume significantly. Shrubs in the lower positions acted as more powerful buffers in preventing runoff generation and surface water loss. Such findings can provide important references for runoff control, water conservation and ecosystem restoration regarding plant selection and vegetative collocation in practice in the arid and semiarid environments.

Keywords: Runoff; Rainfall simulation; Water loss; Plant species; Vegetation position; Surface cover

1 Introduction

Water scarcity is the greatest problem in semiarid and arid regions such as the Loess Plateau, the Mediterranean and other similar areas around the world [1-3]. In the context of human accelerated global warming, many dry-land areas are suffering from warmer and drier climates [4], which further worsen the status of water-carbon contradiction, increase the difficulties of plant growth and threaten the sustainability of ecosystem restoration [5]. More alarming is an increasing trend in destructive rainstorms with higher intensity and more severe erosivity, which may possibly continue on large scale, consequently increasing the potential risks of water loss in many fragile and mountainous regions [6-8]. Conserving scarce water resources in situ for better plant utilization thus becomes extremely important in such thirsty regions [9]. Runoff and water flow along the hillslope conditions, on the other hand, has long been blamed for aggravating water shortage stress, causing soil nutrient loss and hampering the process of vegetation restoration [10]. Consequently, a deep understanding of the mechanism regarding runoff performance and finding more valuable solutions to control runoff loss are significant for vegetation restoration and ecological rehabilitation, especially against the background of climatic change.

In general, runoff processes are characterized by high spatiotemporal variability in arid and semiarid ecosystems, resulting from the interaction among different environmental factors at specific scales [10,11]. There is no-doubt that the dynamics of rainfall features such as intensity and depth contribute to runoff generation and hydrological variation [12-15]. The deep rainfall-runoff relation however, is also largely regulated by such surface conditions as plant species, soil crusts, surface cover, antecedent soil moisture, vegetation buffer strips and specific positions [10,12,16,17], which makes the interactions among them uncertain over time and space. Consequently, the related factors and runoff generation processes were quite complex and difficult to quantify. For example, studies have confirmed that vegetation cover is a key factor influencing runoff generation [18], but how the source and sink of runoff transfers with plant morphology and its spatial location remains unclear [19], particularly when experiencing stochastic rainfall pulses. No sufficient information was provided for selecting suitable plant species in terms of plant morphology and spatial distribution for runoff reduction and vegetation restoration in water-limited environments [20,21]. Moreover, in such arid and semiarid zones, local hill-slopes are always characterized by spatially discontinuous vegetation, which reflects the limited supply of soil water and nutrients. The specific plant species and vegetation position can influence the location of runoff "source-sink" areas markedly [22,23]. So far, few studies have actually linked spatial vegetation position to geo-hydrological processes [11,24]. Clarifying how plant morphology and vegetation distribution reflect and respond to rainfall pulses is thus of great value. Although abundant studies regarding rainfall-runoff relations in different climatic zones were conducted across multiple scales, it still lacks a comprehensive understanding about the mechanism of runoff dynamics within different environments [17,25,26].

Unfortunately, due to water limitation and infrequent rainfalls in dry-lands, getting enough field data is always restricted by the fact that there are few naturally occurring runoff events, which hinders the progress in basic research regarding this topic. Rainfall simulation experiments, however, can overcome this drawback and provide huge amounts of data for model calibration and mechanism exploration within short periods, only because such experiments are easily conducted in the fields [15,27]. Studies across different regions have pointed out that rainfall simulation by portable simulator at a fine scale is a powerful tool in surface hydrological studies [28,29]. Furthermore, rainfall simulators can create different scenarios regarding variations in rainfall variables, helping to ascertain runoff response to rainfall and soil surface conditions. Since the end of 1930s, more than 100 rainfall simulators with less than 5m^2 plot areas as the land surface were developed [30]. So far, such fine-scale studies are more focused on the semiarid Mediterranean region in Europe and other areas such as the dry-hot valleys in southwest China [19,21,31]. Systematic research regarding the role of plant species and vegetation distribution in runoff at micro-scales was rarely conducted in the Loess Plateau of China.

The semiarid Loess Plateau in China, which covers 0.64million km^2 of the land territory, has long been criticized as one of the most degraded regions around the world due to its severe droughts and fragile ecosystem [1,2,9]. Recent studies declared that water resources may become more inadequate in many areas of the plateau due to social development and global warming [2,32]. In general, precipitation has been detected to decline while temperature and evapotranspiration have increased, although such variations remain across time and space. Meanwhile, heavy rainstorms with higher intensities in this region may possibly increase by about 8%-35% [33], which can further raise the sensitivity of land surface to rainfall pulses and thus cause higher risks of runoff loss. Although we know that plant species and vegetation patterns may play a key role in water dynamics, it is still unclear which species and what kinds of spatial patterns are more effective in runoff control. As a consequence, conserving limited water resources through enhancing infiltration in situ and reducing overland-flow at slopes becomes more challenging. Facing these issues, fine-scale studies, rather than coarse scale, can focus on specific hydrological processes and may help to answer this question [17,22,34].

In this study, 16 micro-plots were established in the growing season of 2010, in Dingxi, a semiarid loess hilly area of China. Rainfall simulation experiments were implemented for analyzing different land surface conditions (plant species, surface cover and vegetation distribution) and rainfall characteristics on runoff generation and water discharge dynamics. Specifically, three major sub-objectives were expected to be achieved: ① to analyze the response of runoff generation, surface flow rates and total runoff reduction to two major simulated rainfalls, ② to determine how different runoff indicators respond to plant species and surface coverage, and ③ to analyze the role of plant position and spatial vegetation distribution on runoff dynamics.

2　Materials and method

2.1　Study area

Our rainfall-simulation experiments were designed and conducted in the Anjiapo catchment (35°33′N-35°35′N, 104°38′E-104°41′E) in Dingxi county of Gansu province, in the western part of the Chinese Loess Plateau. In this rain-fed and semiarid catchment, a field meteorological station was established in 1985, which belongs to the Dingxi Institute of Soil and Water Conservation. According to the water deficit index (WDI) and the aridity index (ARI), this region is located in a semiarid climatic zone and is dominated by the warm-humid summers and cold-dry winters. The mean annual precipitation (based on local recorded data from 1956 to 2010) is 421mm/year, of which about 78% total rain falls during the growing season, i.e., from May to September [13]. The mean annual pan evaporation however, can reach about 1515mm. According to over 50 years (1954-2004) of monitored data in the field station, the precipitation experienced a decreasing trend while the temperature continued to increase during the past decades, which means that the local climate is becoming drier and warmer. Such a situation may further aggregate water pressure and do harm to vegetation restoration in this region.

Local soil is developed from loess material, with a mean soil depth ranging from 40 to 60m. The deepest soil layer in some areas, however, can reach and even exceed more than 100m. According to the soil classification system, the soil in this area is dominated by calcic Cambisol[35] with a clay content of 33%-42%, organic matter of 4-13g/kg, and a bulk density from 1.09 to 1.36g/cm³ within a 2m soil depth [1]. No available groundwater can be used for vegetation growth and restoration, mainly due to deep loess soil and severe drought. Limited annual rainfall is thus the only usable water resource for plants. Deep percolation can be neglected in most cases [2].

In general, the local climate is more suitable for shrub and grass species to grow, although some tree species (e.g. Chinese pine, Chinese arborvitae, poplar and willow) were planted widely in the watershed during the past several decades [1,9,13]. In recent years, due to the implementation of "grain-for-green" (the conversion of sloping cropland to forestland or grassland), "natural forest protection," labor service exports and other important projects targeting enriching local farmers and protecting the environment, steeping farmlands were largely forbidden and returned to forests and shrubs, or abandoned for natural succession. Soil biological crusts, being as key components of the surface soil in dry-land ecosystems, are also highly developed at large scales. Such microorganisms (e.g. mosses and lichens in most cases) are called "biological carpet" by local farmers, which can play important roles in soil carbon accumulation and erosion control in the arid and semiarid geographical zones [17,36]. The majority of the remainders of vegetation are secondary shrubs, grasses and some other artificial vegetation types. The dominant introduced tree species are: Chinese pine (*Pinus tabuliformis* Carr.), Chinese arborvitae (*Platycladus orientalis* L.) and apricot (*Prunus armeniaca* L.). These tree species however, are more suitable for growing in lower positions of the hill-slope with relatively higher soil moisture and nutrients. Shrub species are mainly seabuckthorn (*Hippophae rhamnoides* L.) and pea shrub (*Caragana kansuensis* Pojark.), whereas artificial and natural grass species are mainly alfalfa (*Medicago sativa* L.), sainfoin (*Onobrychis vichfolia* Scop.), common leymus (*Leymus secalinus* T.), solidleaf bluegrass (*Poastereophylla* K.) and bunge needlegrass (*Stipa bungeana* Trin.).

2.2 Field experimental design

In the early spring of 2010, sixteen experimental mini-plots in the Anjiapo catchment were established in a gentle northwest-facing hillslope near the field station of the Dingxi Institute of Soil and Water Conservation (Table 1). Before the plots were constructed, the selected site was covered by secondary natural grass and sparse seabuckthorn (*Hippophae rhamnoides* L.) shrubs which grew naturally without any human management. The grass species has grown for more than 20 years, while seabuckthorn began to invade into the hillslope since 2008, possibly due to the settlements of human or bird carried seeds. Steel material was selected and used as the basic material for each micro-plot construction. The thickness of the steel was 5mm with a height of 50cm. A 20cm depth was embedded into the soil, leaving the remaining 30cm height above ground as the border for preventing runoff loss. A discharge ditch was created at the top of each plot for controlling runoff and sediment from the upper slope. At the base of each micro-plot, a marked V-flume and a plastic drum were installed at the outlet of each plot for runoff collection. Furthermore, in order to minimize human disturbance to the inside areas of the plots, the "least perturbations method" plot-setting procedure was created and conducted during the whole installation process (Fig. 1). This method ensured that rainfall-simulation experiments could be done immediately after the plots were established. In total, four major land surface cover types were involved, which include shrub (seabuckthorn, *Hippophae rhamnoides* L.), grass (bunge needlegrass) and biological soil crust (BSCs) as well as bare soil (Fig. 2). Meanwhile, three types of shrub distributions were also designed for further detection and analysis, i.e., shrub in the upper side of the plot (SH-U), shrub in the middle side of the plot (SH-M) and shrub in the lower part of the plot (SH-L). Such treatments were used for comparing the influence of the spatial positioning of plant species on runoff generation and surface water distribution. For grass, bare soil and BSCs plots, the covered area for each runoff plot was 1.59m^2. For shrub plots, each projected plot-area was 2.55m^2 (Table 1 and Fig. 2).

Table 1 Mean characteristic of plant species and surface cover in different micro-plots

Surface cover	Total plant species	Dominated plant species	Height (cm)	Abundance	Coverage (%)	Evenness	Total coverage (%)	Slope gradient (%)
Grass	14	Common leymus	45	5	29	2	69	12
		Solidleaf bluegrass	46	5	28			
		Capillary wormwood	32	5	16			
		Bryophytes	N.A.	N.A.	81	N.A.		
SH-U	10	Seabuckthorn	87	1	41	2	73	13
		Capillary wormwood	34	3	19			
		Common leymus	51	3	11			
		Bryophytes	N.A.	N.A.	66	N.A.		
SH-M	14	Seabuckthorn	112	1	73	2	86	11
		Common leymus	47	3	32	2		
		Solidleaf bluegrass	39	2	6	2		
		Bryophytes	N.A.	60	N.A.	N.A.		
SH-L	12	Seabuckthorn	86	2	45	2	72	13
		Common leymus	52	3	27	2		
		Capillary wormwood	32	3	13	2		
		Bryophytes	N.A.	N.A.	56	N.A.		
BSCs	1	Bryophytes	N.A.	N.A.	67	N.A.	67	12
Bare soil	0	N.A.	N.A.	N.A.	N.A.	N.A.	0	12

Note: Field vegetation community investigation was conducted during rainfall simulation experiments in the growing season. The abundance of plants was classified by 5 levels; the evenness was defined as two classes (1 for uneven, and 2 for evenness). N.A. refers to no available data.

Fig. 1 The sketch-map regarding the specific micro-plot construction procedure

Fig. 2 Rainfall simulators and related experimental micro-plots

Note: "a" and "b" refer to rainfall simulator of LIRs and HIRs, respectively. The numbers①, ②, ③, ④, ⑤ and ⑥ represent SH-L, SH-M, SH-U, grass, BSCs and bare soil, respectively.

2.3 Rainfall simulation and data collection

In this study, a portable rainfall simulator, described by Luk et al.[37], was used with a SPRACO cone jet nozzle, mounted 4.57m above the soil surface (Fig. 2). Due to the field convenience and data reliability of this method, it has been used extensively in surface hydrology research by many scholars[14,15]. Specifically, the median volume of rain-drop size obtained by this simulator was 2.4mm, and the uniformity coefficient of rainfall reached 0.897 in most cases.

Windless days were selected for conducting the rainfall simulations in the field. One rainfall simulator was used to obtain lowintensity rainfalls (LIRs), while two rainfall simulators were used together for obtaining high-intensity rainfalls (HIRs) (Fig. 2(a) and (b)). During each experiment, surface runoff was collected at 3min intervals and the total runoff volume was measured during each simulated rainfall event. In order to save water, most of the simulated rainfall experiments were performed within 45min. In general, surface runoff ceased immediately after each rainfall simulation was finished, and this is mainly because of the initial dry soil condition and weak water-flow velocity. Basic rainfall variables and surface hydrological indicators such as rain depth, duration, antecedent soil moisture, total runoff discharge and time to runoff generation were all taken into consideration.

Among these hydrological variables, rainfall depth and duration was measured and recorded by a traditional SM_1 pluviometer and a SJ_1 auto-siphon udometer. Initial soil water content was measured by a portable TDR (Time Domain Reflectometer) instrument before and after each field experiment. On the basis of the above hydrological data, other selected variables, including rainfall intensity, runoff depth, mean/peak flow velocity, and runoff coefficient (RC) were all calculated and used for further analysis. Herein, the final consequence of runoff was represented by total runoff discharge and runoff depth. The other runoff variables refer to the detailed runoff dynamics and specific processes.

$$\mathrm{PSF}_{ij} = \frac{1}{i}\sum_{i=1}^{i}(\mathrm{PSF}_{ij}/\mathrm{MFV}_{ij}) \tag{1}$$

where i is the number of simulated rainfall events, and j is the number the underlying surface types. PSF_{ij}, PFV_{ij} and MFV_{ij} is the peak shape factor, peak flow velocity and mean flow velocity, respectively, when the i time simulated rainfall was conducted at j site.

$$\mathrm{RUD}_{ij} = \frac{1}{i}\sum_{i=1}^{i}(\mathrm{TRD}_{ij}/\mathrm{MPA}_j) \tag{2}$$

where RUD_{ij} and TRD_{ij} represents runoff depth and total runoff discharge, respectively, when the i time simulated rainfall was conducted at j site. MPA_j is the measured plot area of j site.

$$RC_{ij} = \frac{1}{i}\sum_{i=1}^{i}(RUD_{ij}/SRD_{ij}) \qquad (3)$$

where RC_{ij} and RD_{ij} represent runoff coefficient and simulated rainfall depth when the i time simulated rainfall was conducted at j site.

2.4 Statistical analysis

Descriptive statistical analysis was used for detecting the general features of simulated rainfall and related runoff characteristics. The mean value and standard deviation of each rainfall variable and runoff indicator under two major rain-intensities were quantified for further analysis. The above mentioned statistical analyses were conducted in the SPSS16.0 for Windows. In addition, histograms and line charts which depicted the features of hydrological responses were all drawn in the Excel 2007.

3 Results

3.1 Simulated rainfall and related runoff characteristics

In total, 126 rainfall simulations (24 times for grass and shrub plots, 16 times for biological crust plots and 14 times for bare land) were conducted during the growing season (from July to September) of 2011. Features of the simulated rainfall events and related runoff responses were captured (Table 2). Two major rainfall intensity events (LRIs and HIRs) were gathered by using one or two rainfall simulators together in the same plots (Fig. 2), respectively. In general, high-intensity rainfalls (HIRs) were recorded with a mean intensity value of 52mm/h, ranged from 42mm/h to 58mm/h, respectively. The recorded rainfall depth and duration on the other hand, ranges from 33.2 to 50.7mm and from 42.4min to 63.3min, respectively. For the low-intensity rainfall events (LIRs), the recorded rainfall depth and duration varied from 9.9mm to 57.6mm and from 19.7min to 122min, with a mean value of 26.5min and 57.4min, respectively. The calculated intensities of such rainfalls were captured with a mean value of 28mm/h, ranging from 20mm/h to 37mm/h. According to a comparison with local measured precipitation data from the field station, the two classified rainfall types can be defined as moderate-intensity events and high-intensity storms, both of which are prone to cause severe runoff and water loss in local sloping conditions in most cases.

Table 2 Two major simulated rainfalls and related characteristics of surface runoff response

Rainfall/runoff indicators	High-intensity rainfalls (HIRs)		Low-intensity rainfalls (LIRs)		HIRs/LIRs
	Mean	S.D.	Mean	S.D.	
Rain depth (mm)	41.3	3.738	26.5	8.605	1.56

Continued

Rainfall/runoff indicators	High-intensity rainfalls (HIRs)		Low-intensity rainfalls (LIRs)		HIRs/LIRs
	Mean	S.D.	Mean	S.D.	
Duration (min)	48.2	2.987	57.4	20.147	0.84
Intensity (mm/h)	51.5	4.023	28.3	4.747	1.82
ASM (%)	19.2	7.94	18.5	5.024	1.04
Time to runoff (s)	191.8	161.469	639.4	268.254	0.30
Total discharge (ml)	8134.1	7569.191	582.5	569.436	13.96
MFR (ml/min)	176.4	169.078	14.5	15.876	12.17
PFR (ml/min)	219.4	205.013	23.5	24.485	9.34
Runoff depth (mm)	4.9	5.156	0.3	0.351	16.33
RC (%)	12	12.195	1.3	1.459	9.23

Note: The abbreviations ASM, MFR, PFR and RC refer to antecedent soil moisture, mean flow rate, peak flow rate and runoff coefficient, respectively.

According to the results shown in Table 2, surface runoff varied highly within the two mentioned rainfall types. The mean value of time to runoff generation was 192s under the HIRs, while the related time value for runoff starting became 639s under the LIRs, which is 3.3 times more lag than that under HIRs. Meanwhile, other key runoff indicators such as total discharge, mean flow velocity, peak flow velocity, runoff depth and runoff coefficient were all far higher under HIRs than those under LIRs, with a mean value of 8.13L, 176ml/min, 219ml/min, 4.9mm and 12% in this study, respectively. Furthermore, according to the value of HIR/LIR shown in Table 2, the hydrological consequences between high-intensity and low-intensity rainfalls became more distinct. Specifically, an increase in less than 2 times the rainfall intensity has caused about 14 times the increase of total discharge; 12 times the increases of MFR and 9 times of PFR; 16 times of runoff depth; and more than 9 times the increase of RC.

3.2 Runoff dynamics under different plant species and soil surface cover

In general, our study found that runoff under different plant species and soil surface cover varied highly both under HIRs and LIRs (Fig. 3). Total discharge, mean flow velocity, peak flow velocity, runoff depth and runoff coefficient were all in the decreasing order of bare soil>BSCs>grass>shrub, both under HIRs and LIRs conditions. Meanwhile, the related values of runoff indicators under bare soil and BSCs were far higher than those under grass and shrub species, regardless of the rainfall intensities. Moreover, shrub plant species (seabuckthorn, *Hippophae rhamnoides* L.), rather than grass species, are more effective in reducing total discharge and specific water processes such as peak flow rate. The time-to-runoff generation, however, responded differently across various surface soil features and plant species. Namely, biological soil crust is mostly prone to generating runoff, followed by bare soil surface and grass, while the seabuckthorn shrub species was observed to play the best role in delaying the occurring time of runoff.

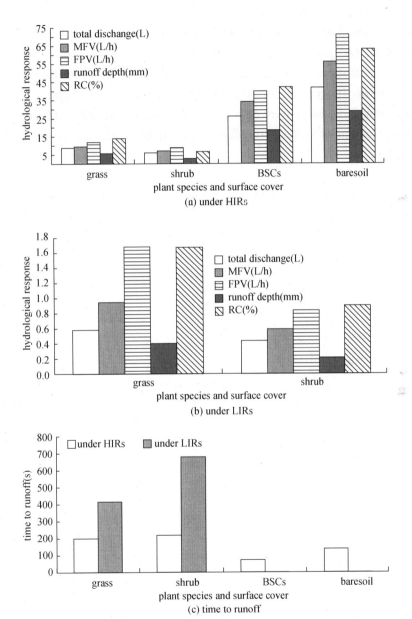

Fig. 3 Responses of runoff indicators to plant species and soil surface cover

Meanwhile, the specific processes of runoff rate and accumulative runoff discharge at each 3min time interval were also captured (Fig. 4). It can be found that distinct response of runoff processes appeared among different plant species and soil surface cover. The runoff rate increased sharply under bare soil conditions during the whole simulation time. For BSCs, although the changing trend of runoff rate was less than that of bare soil, also increased far more quickly than grass and shrub during the first several minutes. A 9-12min after runoff generation, however, the flow velocity decreased and stayed at a relatively stable level almost throughout the remaining phases. Compared with bare soil and BSCs, the response of runoff rate under shrub and grass species seemed to be less sensitive to rainfall. Only at the first 3-6min after runoff starting, a small increasing trend of runoff rate appeared, and then kept a very stable level. In general, for the four total surface cover types, runoff rate and accumulated runoff discharge were both in the same decreasing order of bare soil > BSCs > grass > shrub. No similar values of runoff rates among the four surface covers appeared at any time intervals.

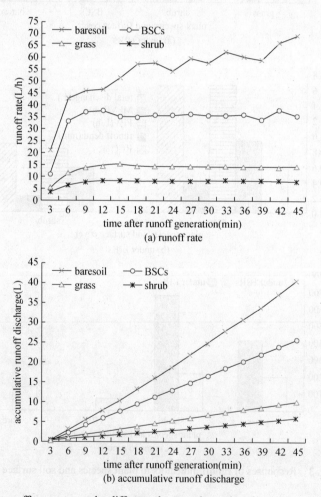

Fig. 4 Specific runoff processes under different plant species and surface cover at each time interval

3.3 Runoff responses to different vegetation distribution and slope position

Our study found that the specific position of shrub species in the plots played an important role in altering runoff consequences (Fig. 5). Under both HIRs and LIRs conditions, the total discharge, runoff depth, runoff coefficient, mean flow velocity, and peak flow velocity all captured the highest value when shrub was located in the upper area of the plots (SH-U). The runoff coefficient, which refers to the ratio of runoff depth to total rainfall depth,

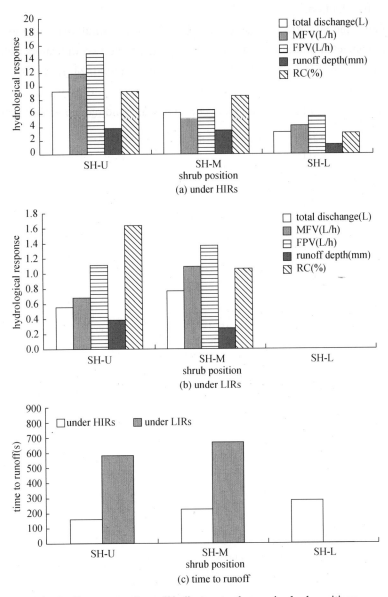

Fig. 5   Responses of runoff indicators to changes in shrub positions

also followed the same dynamic rule with other key runoff variables. Shrub in the middle area of the plot (SH-M) was much better at controlling runoff and reserve surface water than shrub in the upper areas (SH-U). The lowest values of runoff processes and volume were captured when shrub was located in the lower part of the plots (SH-L). Conversely, the time to runoff generation was found in the order of ST-L>ST-M>ST-U. No runoff occurred under ST-L under LTRs conditions, and thus there were no available data to show the time to runoff generation (Fig. 5). This consequence, however, re-confirmed the powerful buffering role of SH-L on runoff reduction.

Meanwhile, the specific process of runoff rate and accumulative volume at 3min time intervals were detected and analyzed (Fig. 6). In general, the two indicators both declined in the following order of SH-L<SH-M<SH-U at each time interval. Moreover, the runoff rate in SH-U was far higher than the other two plant positions at each time interval, which re-confirmed that SH-L was the best one to conserve water and reduce surface runoff, followed

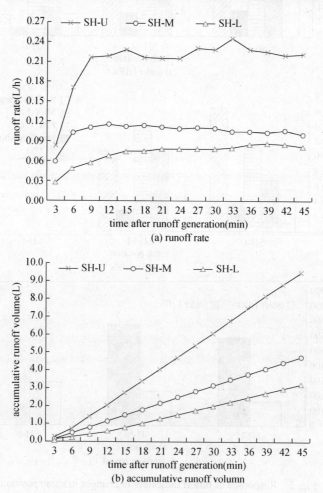

Fig. 6   Specific processes of runoff indicators to changes in shrub positions in the plots

by SH-M and SH-U. Specifically, runoff rate under SHU increased sharply from 0.083L/min to 0.217L/min at the first 9min after runoff start-up, and then kept at a relatively stable level with an average rate of 0.224L/min. At the same time, the runoff peak rate under SH-U was captured in the 33rd minute after runoff generation, with a value of 0.246L/min. The runoff rate at SH-M, however, sharply increased from 0.103L/min to 0.110L/min within the first 6min, and gently reached the 0.115L/min peak value, then declined a little bit with a mean value of 0.108L/min in the remaining time. Interestingly, the curve of the runoff rate under SH-L changed very slowly and gently throughout the whole process, and no rapid increasing trend was detected, especially in the first several minutes after runoff was generated. It started with a mean rate of 0.027L/min in the first 3min after runoff starting, and slowly reached the peak rate of 0.087L/min 42min later. This result showed that SH-L had a far stronger ability for reducing the sensitivity of runoff to rainfall than SH-M and SH-U.

4 Discussions

4.1 Effects of changes in rainfall intensity on surface runoff

In this study, the two types of rainfall intensities can be seen to represent moderate rainfalls and heavy rainstorms in the local region based on former research and historical records, which may induce severe overland flow and water erosion problems in most cases[12,13]. In general, our study confirmed that high-intensity rainfalls (HIRs), rather than low-intensity rainfalls (LIRs), are generally prone to quicker responses of runoff generation under similar conditions of underlying surface. For example, the mean value of time to runoff under HIRs was 191.8s, which only accounted for 30% of that under LIRs (Table 2). This average value of time to runoff under grass and shrub changed from 200.1 s to 413.8s and from 220.8s to 685.7s when underlying surface encountered HIRs and LIRs, respectively (Fig. 2(c)). Time to runoff under SH-U and SH-M was much higher than that of grass and shrub, but it still observed the decreasing order of LIRs and HIRs. Namely, runoff was more difficult to generate and be delayed under LIRs. No runoff loss was recorded in the SH-L covered plots under LIRs (Fig. 4(c)). Meanwhile, key indicators of water yield (e.g. runoff depth and total discharge), indicators of runoff-processes (mean flow and peak flow velocity) and indicators of runoff production ratio (RC) all confirmed that HIRs induced higher risk and consequences of peak flooding and surface water loss than those under LIRs. This indicates that rainstorms not only affect the specific dynamics of runoff, but also contribute to total water loss. More importantly, a threshold of rainfall inducing different runoff extents may exist, because the response between increases in rainfall variables and related runoff indicators was not parallel (Table 2). For example, an only 1.8 times an increase in rainfall intensity and 16% decrease in rainfall duration have caused a sharp increase in total discharge, runoff depth, mean flow velocity, peak flow velocity and runoff coefficient, with a value of 13.96, 16.33, 12.17, 9.34 and 9.23 times, respectively (Table 2).

This result may have important implications for the arid and semiarid ecosystems in the real world, especially when the region suffers accelerated rainfall variations in the context of climatic change [6]. According to our former research in the same region, for example, the destructive effects of rainfall extremes on surface runoff generation exceeded 8 times that of ordinary rainfall pulses in the sloping farmlands [13]. Increases in both rainfall intensity and total evapotranspiration may further aggregate the frangibility of the land surface, and increase the potential risks of runoff loss and the difficulties of vegetation rehabilitation [2,4,5]. This phenomenon has been monitored to be a fact and predicted to be continued in many regions of the world [7]. Small increases in the intensity and erosivity of rainfalls at large scales may induce severe peak flow and great loss of water resources in such regions, particularly in the poor surface covered and fragile ecosystems of the arid and semiarid regions [8,33]. Such situations, however, will certainly increase the potential risk of flooding, worsen the already stressed water conditions and eventually hamper ecological restoration processes.

4.2 Effects of plant species and morphological features on surface runoff

Our study indicates that the xerophytic seabuckthorn shrub and natural grass species act as more powerful tools in reducing water loss in sloping conditions than non-vegetated land surfaces, such as biological soil crusts (mainly dominated by moss) and bare soil conditions (Fig. 1 and Fig. 2). BSCs are much better than bare soil. Namely, shrub species in the plots can be considered as strong and effective landscape sink for retaining runoff and reducing surface water loss through overland flow, followed by grass species, then biological soil crusts. The risk of runoff generation under bare soil is highest. Several aspects may contribute to explaining the reasons for this phenomenon.

First, the difference of plant morphology between grass and shrub species is a crucial factor on raindrop interception and rainwater redistribution. Compared with grass and BSCs, shrub species have higher height, stronger woody stems, more developed branches and denser canopies [19] (Table 1). Our field investigation during the growing season showed that the height and coverage of a single one-year-old seabuckthorn can reach 1.1m and 73% respectively, which were far higher than those of grass species (Table 1). Local investigation in Dingxi also found that after 3-5 years of plant growth, the mean height, diameter at breast height (D.B.H.), crown width and ground diameter of seabuckthorn can reach 2.2m, 2.8cm, 1.8m and 4.0cm, respectively (Unpublished data). Due to this reason, total rainwater captured by canopy interception and stemflow may help to reduce raindrop energy and total runoff depth [38-40]. The Hortonian mechanism (rainfall excess infiltration) under such shrub species can be greatly weakened [41]. For example, Li et al. [42] found that stemflow under some shrubs (e.g. Reaumuria soongorica) in the Loess Plateau can reserve about 7.2% of the total gross precipitation, which becomes an important water source for soil moisture enhancement surrounded root systems and thus benefits plant survival. In turn, better plant growth can help to control runoff loss and improve soil conditions through root-network development and litter accumulation, which forms a virtuous cycle [19,43,44]. Studies from other dry-land regions also draw similar conclusions [2,25,39]. Secondary natural grass (e.g. Common leymus and Solidleaf bluegrass), on the other hand, is better than BSCs in controlling runoff and water loss (Fig. 1 and Fig. 2). Such kinds of perennial herbage are suitable for the semiarid region and can grow well even in severely dry conditions. In this study, the role of BSCs in reducing runoff is far weaker than shrub and grass, but better than bare soil. Although the conclusions regarding the effect of BSCs on infiltration and runoff was inconsistent and even conflicted[17], our result clearly supports the opinion that BSCs can be considered as biological carpets to protect soil and save water in situ.

Second, different features of plant physiology can markedly affect runoff consequence. According to field investigation, seabuckthorn was found to have a powerful ability of tillering propagation, which induced a fast growing and quick spreading. A strong root system of seabuckthorn not only can ensure its survival rate under multiple pressures such as a rigorous climate and poor soil conditions, but it can also help to improve the properties of the soil surface [20,40,44]. This is believed to be an important means for rainwater infiltration enhancement and flood reduction [45,46]. Third, surface roughness among plant species differs. The surface roughness coefficient, which is popularly used in the hydrodynamics field, can reflect the buffering function of different underlying surfaces on runoff, to some extent. Changes in surface roughness may have a great influence on overland flow and water transport capacity [47]. According to some studies, the coefficient of surface roughness decreased in the order of shrub, grass, BSCs and bare soil [25]. This changing trend is consistent with our findings (Fig. 2), confirming the best role of shrub in runoff control, followed by grass and BSCs. Last, different plant species may influence soil hydrological properties. For example, our former study in the same region also confirmed that seabuckthorn (shrub) and wheatgrass (natural grass) worked best for improving water holding capacity, litter depth accumulation and soil organic content, followed by alfalfa and Chinese pine, then spring wheat [13]. Meanwhile, seabuckthorn can increase 1.7-5.5times the soil organic matter and decrease soil buck density after 10 years of accumulation [48]. High SOM and low buck density, on the other hand, will increase soil aggregate stability and infiltration capacity, finally achieving the goal of water retention and runoff reduction [49].

4.3 Effect of surface cover on runoff generation and process

Our result indicated that the heterogeneity of effective surface cover may contribute to runoff differences. For instance, it is clear that bare soil has no effective protection compared to BSCs, grass and shrub. The effectiveness of BSCs cover is better than bare soil, but less than grass-and shrub-covered soil surface. In this study, it is interesting to ascertain the role of BSCs in surface hydrology. Being important components for sustaining ecological function in the arid and semiarid ecosystems [17], such microorganisms are widely distributed in the dry-land regions, including the Loess Plateau. Runoff velocity and water loss in bare soil were the highest, which is a strong source for runoff generation, and thus should be avoided in most cases. Other supplemental measures, such as mulching and land closure for natural vegetation succession are encouraged to be used in the field [47].

Meanwhile, field investigation indicated that the vertical protection of surface cover was very different across different plots (Table 1). For shrub covered plots, the underlying surface was protected not only by seabuckthorn, but also by grass and BSCs. Within such plots, the coverage of seabuckthorn was far higher than other plant species (Table 1). Grass covered plots were actually protected by the coupled role of grass species and BSCs communities. Although there was no shrub or grass protection in the BSCs covered plot, it was still far better than bare soil conditions. The total mean coverage of the plots also decreased in the order of shrub, grass, BSCs and bare soil, with a mean value of 77%, 69%, 67% and 0% (Table 1), which contributed to the changes of runoff generation.

Many field experiments from different regions of the world also found that increased surface cover is an important measure for controlling water loss, improving soil environments, and altering surface hydrological processes and consequences [12,13,19,50]. The amount of soil cover or its effectiveness is also significant in reducing the velocity of overland flow [11]. For example, a laboratory experiment conducted with rainfall simulation found that different mulches can protect surface soil effectively, increase rainwater infiltration and soil moisture content, decrease the threshold of runoff generation and delay the time to runoff at the same time [47]. According to this result, mulch covers of $2t/hm^2$ and $4t/hm^2$ caused reductions of 21% and 51% in the runoff peak, respectively [47]. In the sub-tropical region of Florida, clear-cutting of a forest markedly increased the peak flow and total discharge, and made the groundwater level increase at least 100cm [16]. Meanwhile, most of the paired catchment studies around United States and European countries have indicated that deforestation will greatly remove the surface cover and reduce total evapotranspiration, thus increased flooding risk and total water loss [51-54]. In the state of Colorado, 40% removal of forest coverage increased 50%-170% of total annual runoff, and increased the peak flooding in most years [40]. The predatory destruction of vegetation during the growing season induced a flood crest increase of 15%-60% [55]. In the humid Coweeta watershed of Southern America, plant removal increased 7%-30% of peak flow [56].

4.4 Effects of vegetation spatial position on runoff yield and process

Our study found that the specific position of shrub species in the experimental plots play a crucial role in affecting the process and consequence of runoff performance (Fig. 3 and Fig. 4).

Shrub in the lower part of the plots (SH-L) was the best to control runoff, followed by shrub in the middle (SH-M) and shrub in the upper positions (SH-U). Meanwhile, our study also confirmed SH-L can be considered as the strongest sink of rainwater-runoff within the mentioned land surfaces. The difference of runoff consequence under SH-M and SH-U, on the other hand, was not significant. This result indicated that the buffering role of shrub in altering surface hydrology is greatly dominated by its specific position and spatial distribution.

As stressed before, understanding the placement of plant species across the hillslope and detecting how the related runoff dynamics reflect and respond to rainfall pulses is of great value for runoff source-sink transformation in the arid and semiarid ecosystems[11]. In our study, the function of single shrub position in the experimental plots seems similar to the role of so-called "vegetative filter strips" (VFS), "plant belts" or "biological buffer strips" (BBS) in reducing losses of surface water and nutrient materials[3,23,34]. According to Borin et al.[31], for example, the buffer strips (BS) of biological materials can reduce the total runoff amount by 78% compared to no-BS, mainly derived from the decreased flow hydraulics resulting from the increased soil infiltration and surface roughness. In semiarid South-eastern Australia, tree belts can reduce runoff loss by 32%-68% in a one-in-tenyear rainstorm event and by 100% in a one-in-two-year storm event from a pastured hill-slope[34], and the researchers attributed the reason of runoff reduction to the role of tree roots performing in increasing infiltration in the root and surface soil zones. In Mediterranean rangeland, trees planted in the grassland slopes were detected to increase water storage beneath their canopy and reduce erosion as well as surface water runoff[57]. Liu et al.[58] also declared that protective measures worked best when located at the lowest position along slopes. According to our field investigation shown in Table 1, the surface coverage of SH-U, SH-M and SH-L was recorded as 73%, 88% and 72%, respectively. Meanwhile, the height of a single species of seabuckthorn in SH-U, SH-M and SH-L was 87cm, 112cm and 86cm respectively, with the related surface coverage being 41%, 73% and 45%, respectively (Table 1). This showed that, although the coverage of single shrub and total plant community in the SH-L plot was not higher (sometimes even lower) than SHU and SH-M, the related runoff loss under SH-L was still the lowest. Consequently, this result reconfirmed the importance of plant position buffers in reducing surface runoff generation and flooding risk in practice in dry-land ecosystems.

5 Conclusions and implications

In this study, micro-plot rainfall simulations were implemented to detect how runoff responds to various plant species, surface cover and vegetation distribution under high or low intensity rainfalls in a semiarid loess hilly area in China. Several major interesting findings were discovered. Firstly, runoff response was generally more sensitive to HIRs than LIRs under similar underlying surfaces. According to our simulations, a 1.8 times increase in rainfall intensity can induce many-folds (9-16 times, for example) of increases in peak flooding discharge and severe water loss. Changes in rainfall variables and the responses of key runoff indicators are not equal or coordinate. Secondly, the effects of different plant species and surface cover on runoff differed significantly. Seabuckthorn, being a suitable shrub species in the semiarid area, was proved to be more powerful for reducing flooding rates and retaining water loss, and delayed the time to runoff significantly, followed by secondary natural grass and biological soil crusts (BSCs). Bare soil is the worst, which can be attributed to the strong source of surface runoff generation and water loss, mainly due to no effective surface cover. Thirdly, the spatial position and distribution of plants play crucial roles in affecting runoff dynamics. Shrub at lower position (SH-L) has the most powerful buffering role in minimizing runoff loss, followed by SH-M, then SH-U.

Our findings have important implications for overland flow reduction, water conservation and ecological restoration in the arid and semiarid ecosystems. Three highlighted inspirations, which may benefit surface runoff control in practice, are stressed here. First, small increases in rainfall intensity may induce severe flooding disasters and great losses of water resources based on our results. Special attention, therefore, should be paid to the spatiotemporal pattern of natural stochastic rainfall (e.g. extreme rainfall events) in the real world, and valid measures should be taken to prevent its destructive effects, which may possibly bring nature to society, especially in the context of accelerated climatic change. Second, suitable plant species should be carefully selected and planted in terms of the sloping conditions for runoff reduction and ecological rehabilitation. Such plant selection measures must depend upon scientific field experimental studies. Shrub plantations were suggested as good choices in practice, while natural grasses and BSCs should be protected to develop well under the shrub canopies. Bare soil conditions, being strong runoff sources, should be avoided at full stretch through complicated management such as artificial plantation, surface mulching or land closure for natural succession. Third, since the specific positions of plants play effective roles in regulating water yield and hydrological processes, higher attention should be paid to the spatial distribution and landscape structure of vegetation on the slopes. The powerful buffer of shrub in the lower position of the hill-slope should be taken into consideration, particularly when an ecological and landscape design project is conducted in the loess hilly area and other fragile mountainous regions. Namely, the importance of vegetation pattern design should be emphasized and implemented scientifically in ecological restoration practices.

Acknowledgements

The Dingxi Institute of Soil and Water Conservation in Gansu province and several anonymous farmers are acknowledged for their warm-hearted assistance in plots construction, rainfall simulation and field cooperation. This research was supported by the National Natural Science Foundation of China (41390462, 41371123).

References

[1] Chen L D, Huang Z L, Gong J, et al. The effect of land cover/ vegetation on soil water dynamic in the hilly area of the loess plateau, China. Catena , 2007, 70 (2): 200-208.

[2] Wang L, Wei S P, Horton R, et al. Effects of vegetation and slope aspect on water budget in the hill and gully region of the Loess Plateau of China. Catena, 2011, 87: 90-100.

[3] Maetens W, Poesen J, Vanmaercke M. How effective are soil conservation techniques in reducing plot runoff and soil loss in Europe and the Mediterranean? Earth-Science Reviews, 2012, 115: 21-36.

[4] Richard A K. Global warming is changing the world. Science, 2007, 316: 188-190.

[5] Putten W H V D, Bardgett R D, Bever J D, et al. Plant-soil feedbacks: the past, the present and future challenges. Journal of Ecology, 2013, 101: 265- 276.

[6] Weltzin J F, Loik M E, Schwinning S, et al. Assessing the response of terrestrial ecosystems to potential changes in precipitation. Bioscience, 2003, 53 (10): 941-952.

[7] Sun G, McNulty S G, Moore J, et al. Potential Impacts of Climate Change on Rainfall Erosivity and Water Availability in China in the Next 100 Years. Beijing: The 12th international soil conservation conference, 2002.

[8] Zhao F, Deng H, Zhao X. Rainfall regime in three Gorges area in China and the control factors. International Journal of Climatology, 2009, 30: 1396-1406.

[9] Yang L, Wei W, Chen L D, et al. Response of deep soil moisture to land use and afforestation in the semi-arid Loess Plateau, China. Journal of Hydrology, 2012, 475: 111-122.

[10] Sinoga J, Diaz A, Bueno E, et al. The role of soil surface conditions in regulating runoff and erosion processes on a metamorphic hillslope (Southern Spain). Catena, 2010, 80: 131-139.

[11] Imeson A C, Prinsen H A M. Vegetation patterns as biological indicators for identifying runoff and sediment source and sink areas for semiarid landscapes in Spain. Agriculture Ecosystems and Environment, 2004, 104: 333-342.

[12] Wei W, Chen L D, Fu B J, et al. The effect of land use and rainfall regimes on runoff and erosion in the loess hilly area, China. Journal of Hydrology, 2007, 335 (3-4): 247-258.

[13] Wei W, Chen L D, Fu B J, et al. Responses of water erosion to rainfall extremes and vegetation types in a loess semiarid hilly area, NW China. Hydrological Processes, 2009, 23: 1780-1791.

[14] Shi Z H, Yan F L, Li L, et al. Interrill erosion from disturbed and undisturbed samples in relation to topsoil aggregate stability in red soils from subtropical China. Catena, 2010, 81: 240-248.

[15] Shi Z H, Fang N F, Wu F Z, et al. Soil erosion processes and sediment sorting associated with transport mechanisms on steep slopes. Journal of Hydrology, 2012, 454-455: 123-130.

[16] Sun G, Riekerk H, Comerford N B. Groundwater table rise after forest harvesting on cypress-pine flatwoods in Florida. Wetlands, 2000, 20 (1): 101-112.

[17] Chamizo S, Cantón Y, Rodríguez-Caballero E, et al. Runoff at contrasting scales in a semiarid ecosystem: a complex balance between biological soil crust features and rainfall characteristics. Journal of Hydrology, 2012, 452-453: 130-138.

[18] Cantón Y, Solé-Benet A, de Vente J, et al. A review of runoff generation and soil erosion across scales in semiarid South-eastern Spain. Journal of Arid Environments, 2011,75: 1254-1261.

[19] Xu X L, Ma K M, Fu B J, et al. Influence of three plant species with different morphologies on water runoff and soil loss in a dry-warm river valley, SW China. Forest Ecology and Management, 2008, 256: 656-663.

[20] de Baets S, Poesen J, Knapen A, et al.Root characteristics of representative Mediterranean plant species and their erosion reducing potential during concentrated runoff. Plant and Soil, 2007, 294: 169-183.

[21] Xu X L, Ma K M, Fu B J, et al. Soil and water erosion under different plant species in a semiarid river valley, SW China: the effects of plant morphology. Ecological Research, 2009, 24: 37-46.

[22] Bochet E, Rubio J L, Poesen J. Relative efficiency of three representative matorral species in reducing water erosion at the microscale in a semiarid climate (Valenicia, Spain). Geomorphology, 1998, 23: 139-150.

[23] Ma L, Pan C Z, Teng Y G, et al. The performance of grass filter strips in controlling high-concentration suspended sediment from overland flow under rainfall/non-rainfall conditions. Earth Surface Processes and Landforms, 2013, 38: 1523-1534.

[24] Mayor Á G, Bautista S, Bellot J. Scale-dependent variation in runoff and sediment yield in a semiarid mediterranean catchment. Journal of Hydrology, 2011, 397 (1-2): 128-135.

[25] Xiao P Q, Yao W Y, Römkens M J M. Effects of grass and shrub cover on the critical unit stream power in overland flow. International Journal of Sediment Research, 2011, 26: 387-394.

[26] Shi Z H, Yue B J, Wang L, et al. Effects of mulch cover rate on interrill erosion processes and the size selectivity of eroded sediment on steep slopes. Soilence Society of America Journal, 2013, 77: 257-267.

[27] Huang J, Wu P T, Zhao X N. Effects of rainfall intensity, underlying surface and slope gradient on soil infiltration under simulated rainfall experiments. Catena, 2013, 104: 93-102.

[28] Cerdà A. Effect of climate on surface flow along a climatological gradient in Israel: a field rainfall simulation approach. Journal of Arid Environments, 1998, 38: 145-159.

[29] Abudi I, Carmi G, Berliner P. Rainfall simulator for field runoff studies. Journal of Hydrology, 2012, 454-455: 76-81.

[30] Pérez-Latorre F, de Castro L, Delgado A. A comparison of two variable intensity rainfall simulators for runoff studies. Soil and Tillage Research, 2010,107: 11-16.

[31] Borin M, Vianello M, Morari F, et al. Effectiveness of buffer strips in removing pollutants in runoff from a cultivated field in North-East Italy. Agriculture Ecosystems and Environment, 105, 2005: 101-114.

[32] Li Z, Liu W B, Zhang X C, et al. Impacts of land use change and climate variability on hydrology in an agricultural catchment on the Loess Plateau of China. Journal of Hydrology, 2009, 377: 35-42.

[33] Zhang X C, Liu W Z. Simulating potential response of hydrology, soil erosion, and crop productivity to climate change in Changwu tableland region on the Loess Plateau of China. Agricultural and Forest

Meteorology, 2005, 131: 127-142.

[34] Ellis T W, Leguédois S, Hairsine P B, et al. Capture of overland flow by a tree belt on a pastured hillslope in South-eastern Australia. Australian Journal of Soil Research, 2006, 44: 117-125.

[35] FAO-UNESCO, Soil Map of the World (1 : 5000 000). Paris: Food and Agriculture Organization of the United Nations, UNESCO, 1974.

[36] Rodríguez-Caballero E, Cantón Y, Chamizo S, et al. Effects of biological soil crusts on surface roughness and implications for runoff and erosion. Geomorphology, 2012, 145-146: 81-89.

[37] Luk S, Abrahams A D, Parsons A J. A simple rainfall simulator and trickle system for hydro-geomorphical experiments. Physical Geography, 1986, 7: 344-356.

[38] Meng Q H, Fu B J, Tang X, et al. Effects of land use on phosphorous loss in the hilly area of the Loess Plateau, China. Environmental Monitoring and Assessment, 2008, 139: 195-204.

[39] Návar J. Stemflow variation in Mexico's northeastern forest communities: its contribution to soil moisture content and aquifer recharge. Journal of Hydrology, 2011, 408: 35- 42.

[40] Yang F S, Cao M M, Li H E, et al. Simulation of sediment retention effects of the single seabuckthorn flexible dam in the Pisha Sandstone area. Ecological Engineering, 2013, 52: 228-237.

[41] Wei X H, Sun G. Watershed Ecosystem Processes and Management. Higher Education Press, 2009: 1-419.

[42] Li X Y, Liu L Y, Gao S Y, et al. Stemflow in three shrubs and its effect on soil water enhancement in semiarid loess region of China. Agricultural and Forest Meteorology, 2008, 148: 1501-1507.

[43] Descheemaeker K, Muys B, Nyssen J, et al.Litter production and organic matter accumulation in exclosures of the Tigray highlands, Ethiopia. Forest Ecology and Management, 2006, 233 (1), 21-35.

[44] Gyssels G, Poesen J, Bochet E, et al. Impact of plant roots on the resistance of soils to erosion by water: a review.Progress in Physical Geography, 2005, 29 (2), 189-217.

[45] Jost G, Schume H, Hager H, et al. A hillslope scale comparison of tree species influence on soil moisture dynamics and runoff processes during intense rainfall. Journal of Hydrology, 2012, 420-421: 112-124.

[46] Tesfuhuney W A, Van Rensburg L D, Walker S.In-field runoff as affected by runoff strip length and mulch cover. Soil and Tillage Research, 2013, 131: 47-54.

[47] Montenegro A, Abrantes J, de Lima J, et al.Impact of mulching on soil and water dynamics under intermittent simulated rainfall. Catena, 2013, 109: 139-149.

[48] Li H E, Li G J, Zhang K, et al.Research on the effect of seabuckthorn flexible dam on the improvement of soil organic matter. Bull. Soil Water Conservation, 2007, 12 (6): 1-4.

[49] Xue Z J, Man C, An S S. Soil nitrogen distributions for different land uses and landscape positions in a small watershed on Loess Plateau, China. Ecological Engineering, 2013, 60: 204-213.

[50] Durán Zuazo V H.Impact of vegetation cover on runoff and soil erosion at hillslope scale in Lanjaron, Spain. Environmentalist, 2004, 24: 39-48.

[51] Scott D F, Lesch W. Streamflow responses to afforestation with Eucalyptus grandis and Pinus patula and to felling in the Mokobulaan experimental catchments, South Africa. Journal of Hydrology, 1997, 199: 360-377.

[52] Zhang L, Dawes W R, Walker G R. Response of mean annual evapotranspiration to vegetation changes at catchment scale. Water Resources Research, 2001, 37: 701-708.

[53] Robinson M, Cognard-Plancq A L, Cosandey C, et al. Studies of the impact of forests on peak flows and baseflows: a European perspective. Forest Ecology and Management, 2003, 186: 85-97.

[54] Brown A E, Zhang L, McMahon T A, et al. A review of paired catchment studies for determining changes in water yield resulting from alterations in vegetation. Journal of Hydrology, 2005, 310: 28-61.

[55] Hornbeck J W, Adams M B, Corbett E S. et al. Long-term impacts of forest treatments on water yield: a summary for northeastern USA. Journal of Hydrology, 1993, 150 (2-4): 323-344.

[56] Bosch J M, Hewlett J D. A review of catchment experiments to determine the effect of vegetation changes on water yield and evapotranspiration. Journal of Hydrology, 1982, 55: 23-33.

[57] Joffre R, Rambal S. How tree cover influences the water balance of Mediterranean Rangelands. Ecology, 1993, 74: 570-582.

[58] Liu S L, Dong Y H, Li D, et al. Effects of different terrace protection measures in a sloping land consolidation project targeting soil erosion at the slope scale. Ecological Engineering, 2013, 53: 46-53.

Effects of rainfall change on water erosion processes in terrestrial ecosystems: a review

Wei Wei   Chen Liding   Fu Bojie

Abstract

Water erosion is the most destructive erosion type worldwide, causing serious land degradation and environmental deterioration. Against a background of climate change and accelerated human activities, changes in natural rainfall regimes have taken place and will be expected to become more pronounced in future decades. Long-term shifts may challenge the existing cultivation systems worldwide and eventually alter the spatiotemporal patterns of land use and topography. Meanwhile, specific features of soil crusting/sealing, plant litter and its decomposition, and antecedent soil moisture content (ASMC) will accompany rainfall variability. All these changes will increase pressures on soil erosion and hydrological processes, making accurate erosion prediction and control more difficult. An improved knowledge and understanding of this issue, therefore, is essential for dealing with the forthcoming challenges regarding soil and water conservation practices. In this paper, the characteristics of changes in natural rainfall, its role on terrestrial ecosystems, the challenges, and its effect on surface water erosion dynamics are elaborated and discussed. The major priorities for future research are also highlighted, and it is hoped that this will promote a better understanding of water erosion processes and related hydrological issues.

Keywords: climate change; process; rainfall; terrestrial ecosystem; water erosion

1 Introduction

Soil erosion, a geophysical process involving the detachment, transport and deposition of soil materials, is one of the most serious ecoproblems in many parts of the world, and has put great pressures on the earth surface and its environment[1-3]. Some scholars even describe erosion and related land degradation as 'earth cancer', which has persistently plagued ancient cultures and modern civilization[4]. Statistics show that over 85% of global land degradation is associated with erosion, most of which occurred after the second world war, causing approximately a 17% reduction in crop productivity and environmental damage[5,6]. Generally, there are two major types of erosion: water erosion and wind erosion[7-9]. Water erosion, however, is more serious than wind erosion on account of the geographical extent and severity of impacts globally[10]. Studies report that erosion by water affects a land area of $1.1 \times 10^9 hm^2$ annually, accounting for more than 55% of world erosion in total[4,11]. Other studies estimate that accelerated erosion by water and wind is responsible for one-half and one-quarter of all kinds of soil degradation, respectively[12]. The entire eroded area in the Chinese Loess Plateau, for example, is calculated as $4.54 \times 10^5 km^2$, of which $3.37 \times 10^5 km^2$ (74.23% in total) is affected by water erosion[13].

New evidence shows that water erosion is a major contributor to atmospheric CO_2 release and is partly responsible for the accelerated greenhouse effect[14]. The prevention of water erosion thus is of paramount importance in the management and conservation of natural resources. Fortunately, special attention has been paid to this issue during the past several decades, and erosive processes have been studied intensively from microscales (e.g., raindrop, soil crusting and sealing) to mesoscales (i.e., field and slope) and even larger scales such as watershed, regional or even global levels[15].

In natural environments, many factors play important roles in affecting the dynamics of soil erosion and surface water loss, of which soil conditions, land use/vegetation cover, topography and rainfall characteristics are expected to be the most important[16]. Among these factors, however, the most cited is rainfall[7-9,17-19]. Rainfall is the initial and essential driving force for natural runoff generation and erosion variation. Rainfall variables such as amount, intensity and erosivity all affect water flow and sediment budgets significantly[18,20]. Responses of net erosion rates to environment-climate changes show various trends across different timescales[1]. Studying the role of changes in natural rainfall is thus important for water erosion analysis and for its control.

Although much information regarding the influence of rainfall change on water erosion dynamics is recorded, former studies have been largely limited to the role of simulated rainfall, single rainfall events and its regime heterogeneity on hydrological dynamics[18,21,22]. How and why natural rainfall change can affect the terrestrial ecosystem and related erosional processes over different spatial scales and the long term remains uncertain. However, this is essential for evaluating earth surface change comprehensively (e. g., for water resource cycles, ecosystem health and land-use management), especially with the context of accelerated global change [3,23].

2 Rainfall change: a challenging issue facing the 21st century

Changes in rainfall parameters and regimes over time and space have increasingly become recognized as a component in climate change [3,24]. According to the consensus of atmospheric scientists and hydrologists, current climate change is occurring both in terms of air temperature and precipitation [8,25,26]. Recently, some studies declare that precipitation, rather than rising temperature or elevated CO_2 concentration, is likely to have a more significant impact on the physiochemical processes of earth surface systems [20,24]. This conclusion has been supported by other scholars. For example, based on climatic scenario analysis, Wang and Xia [27] found that the sensitivity of runoff to rainfall change is far higher than that of runoff to temperature variation. Meanwhile, earth surface features such as land use/cover and topography have also changed significantly under intensive agriculture techniques, which alter local water-heat-energy balances and eventually the regional or even global rainfall regimes [28,29] (Fig.1).

So far, changes in rainfall have taken place and have been monitored across different spatiotemporal scales worldwide [24,25]. The patterns of precipitation, with global redistribution in rainfall amounts and increases in the frequency of heavy rainfall events, as well as increases in the number and duration of droughts have all been shown to have changed, and will 0continue to change [29]. Rainfall in NW China was found to be marked by a tendency of increase~decrease~increase~decrease from the 1960s and 1970s to the 1980s and 1990s, respectively [30]. During the twentieth century, total precipitation in the USA increased, of which 53% was in the form of extreme rainfall events [31]. Precipitation coming from daily rainfall in excess of 50mm also increased significantly [9]. Another study revealed that rainfall in Urumchi (the capital of Xinjiang Uygur Autonomous Region, China) was lowest during the 1960s, then increased in the 1970s and 1980s, but has become more uncertain and changeable since the 1990s [32]. Moreover, this dynamic trend will continue and even accelerate, both in the number of wet days and the percentage of precipitation in intense convective storms as opposed to longer-duration, less intense storms [8,25,33].

Reports estimate that mean temperatures near the earth surface are likely to increase 1.4-5.8 ℃ in the following 10 decades and beyond [28]. This warming is alarming for accelerating global atmospheric and hydrologic circulations, which may possibly change existing rainfall patterns [31]. For example, global warming may bring more precipitation to high latitudes in both winter and summer and less precipitation to low latitudes [24]. Similar results have also been found by other researchers. Oki and Kanae [34] pointed out that global

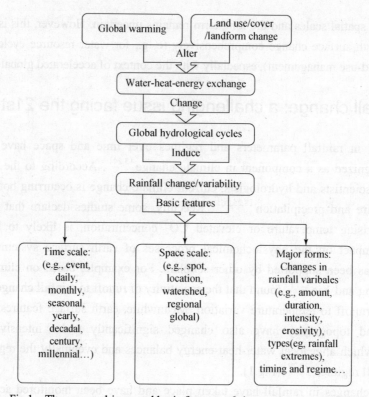

Fig.1 The causes, drivers and basic features of changes in natural rainfall

hydrological cycles will be enhanced, and finally increase natural rainfall events with high intensities. Rainfall variables (e. g., depth, duration, intensity and erosivity), mean or extreme values, and their patterns and distributions all will change markedly over time and space in the following years [25].

Shifts in rainfall over widespread spatial scales and varied time series (i. e., several, tens, hundreds of decades and more) have influenced many physiochemical, biological and ecological processes in terrestrial ecosystems [24,35], such as potential changes in the distributions of plant species, surface vegetation cover and type, litter properties, soil sealing/crusting and overground hydrological processes [36-38]. Globally, changes in the patterns of rainfall and temperature have played key roles in reshaping and driving geographical phenomena and surface processes [23] (e. g., hydrologic, desertification and erosion processes). Increased rainstorms between prolonged dry periods, as well as reduced water resources for irrigation, may promote the activities and spread of pests and diseases on crops and livestock, as well as soil erosion and desertification [39].

However, studying the effect of rainfall fluctuation and evolution on these surface processes is challenging. The major reasons are threefold. First, changes in precipitation are not uniform around the world, both from short to long timespans, and from small to large spatial scales. Rainfall variables and distributions vary greatly over time and space, which is especially true in the key arid and semi-arid fragmentized environments [18,20,24]. The features of ecosystems and earth surfaces in such areas are much more unstable and fragile than those in humid regions with rich vegetation, making the responses of surface processes more sensitive and changeable to local rainfall variability [40]. This sensitivity will certainly reduce the predictability of water erosion processes, and thus increase the difficulties of runoff-erosion prevention and control. Second, scale effects regarding water erosion dynamics exist over different spatial and temporal scales. Factors and drivers which influence, or even determine, the erosional consequences at various scales differ from one another [41]. Results from field and slope scales thus cannot be simply applied to the watershed or much larger scales [42]. Although many efforts have been made during the past several decades, methods and theories of scale transformation remain problematic. Third, responses of terrestrial ecosystems to changes in natural rainfall are always accompanied by timelags [24]. This kind of timelag mainly embodies slow and complex responses of vegetation succession, new adaptive crop cultivation system re-establishment, surface falling/plant litter and soil physiochemical features, and long-term rainfall dynamics [41]. This further increases the difficulties in rainfall-ecosystem evaluation.

3 Rainfall change as a determinant to water erosion variation

Generally, raindrop splash detachment and surface overland flow are the two basic drivers for soil peel-back, flood generation, water loss and sediment mobilization [17,43]. The two drivers, however, are both rooted in rainfall energy, its intensity and volume. For instance, rainfall aggressiveness is a typical characteristic in the arid and semi-arid Mediterranean zones, which causes severe water erosion problems [40]. Rainfall regimes with strong intensities and low frequencies, rather than those events with weak intensities and high frequencies, were shown to be the highest inducers of severe runoff and soil erosion in the loess hilly area of China [18]. Hence, natural rainfall is considered as the principal climatic agent on surface water flow and eroding processes in natural environments [44,45]

The specifi c evolution of rain parameters such as depth, duration, intensity, erosivity and spatiotemporal distribution (inter/intraannual or event variability) plays a key role in inducing various water erosion rates [19,24,45,46]. Ramos and Martínez-Casasnovas [47] discovered that rainfall distribution (time and concentration), together with preceding soil moisture conditions, effectively controls the surface hydrological response. The intensity of rainfall was found to be a major parameter controlling runoff response in fragile and poor landcover areas, while total rainfall amount is more closely correlated with runoff in coarsertextured and highly permeable soils with a denser plant cover [18,40]. Kosmas et al. [48] found that the depth/volume of surface runoff produced from a single rain event is relatively minor if annual rainfall does not exceed 280mm, whereas, when rainfall amount exceeds 700mm, higher runoff coefficients (the proportion of rainfall converting into surface runoff) will appear. Therefore, changes in key rainfall variables will cause variation in the degree of water erosion. In order to avoid under/overestimating erosional changes, studies thus suggest that shifts in rainfall must be reflected as a combination of storm intensity and the number of rainy days[33]. Moreover, the effects of rainfall characteristics on hydrological responses has distinct spatiotemporal variability. As a consequence, the result varies highly from place to place and from period to period.

4 Predicting the effects of rainfall change on water erosion

This question of how changes in precipitation increase runoff risks and aggregate erosion rates has been addressed by different scholars in many areas of the world (Table 1). For example, using climate change models, studies estimate that a 7% increase in precipitation will lead to a 26% increase in erosion in the United Kingdom[8]. Based on the Erosion Productivity Impact Calculator (EPIC) model, Lee et al.[49] calculated that a 20% increase in precipitation would probably lead to a 37% increase in soil erosion and 40% in surface runoff. Runoff in South Africa will possibly increase by 20%-40% if the local precipitation increases by 10%[9]. Rainfall erosivity in the Chinese Loess Plateau is predicted to increase 8%-35% during the twenty-first century, which is expected to further degrade local conditions and make water loss and soil erosion more serious[50].

Recently, new studies have claimed that the intensities, amounts and seasonality variations of event precipitation will continue to increase across the globe, and may accelerate surface runoff generation and erosion, mainly due to the ongoing climate change and its stimulus to the atmospheric hydrological cycles[12].

Table 1 Major findings and research regarding rainfall change and water erosion dynamics around the world

Covered geographical areas	Major conclusions/findings	Methodology	Source
Midwest USA	10%-310% increase in runoff and 33%-274% increase in erosion due to increased rainfall and reduced land coverage	Water Erosion Prediction Project (WEPP) model	[31]
Meuse basin, Europe	3% increase in rain erosivity inducing 333% increase in water erosion	WATEM/SEDEM model	[51]
South Korea	20% increase in storm depths and occurrence causing 54%-60% and 27%-62% increase in runoff and soil loss, respectively	Climate generator (CLIGEN); Water Erosion Prediction Project (WEPP) model	[52]
Saxony, Germany	22%-66% increase in erosion due to increased intensity and extreme events	ECHAM4-OPYC3 and EROSION2D model	[53]
Brazil	22%-33% increase in mean annual sediment yield caused by 2% increase in annual rainfall	Hadley Center climate model (HadCM2)	[54]
Different locations in USA	Each 1% change in rainfall may cause 2% and 1.7% changes in runoff and erosion, respectively	CLIGEN model and regression equations	[33]

Continued

Covered geographical areas	Major conclusions/findings	Methodology	Source
Global scale	7% increase in rainfall during the twenty-first century	GCMs (general circulation models)	[55]
South Downs, UK	7% increase in precipitation causing 26% increase in water erosion	Erosion Productivity Impact Calculator (EPIC) model	[56]
Changwu tableland, Loess hilly area, China	23%-37% increase in annual rainfall, 29%-79% increase in runoff and 2%-81% increase in soil erosion	HadCM3, WEPP and stochastic weather generator (CLIGEN)	[50]
South Africa	A 10% increase in rainfall may lead to a 20%-40% increase in runoff	CERES-Maize and ACRU models	[57]
Loess Plateau, China	4%-18% increase in rainfall with runoff increasing from 6% to 112% and erosion increasing from −10% to +167%	GCM	[3]
Greece	The length and frequency of flood are predicted to increase twofold and threefold, respectively	Goddard Institute for Space Studies climate change model	[58]
Dingxi, Gansu province, northwestern China	Runoff and erosion rates under rainfall extremes were 2.68 and 53.15 times the mean ordinary rates, respectively	Statistics on long-term consecutive field data in situ	[59]
Global scale	About 40% erosion potential due to increased precipitation	GIS-based RUSLE model	[60]

The major role of natural rainfall on water erosion dynamics can be approached from the perspective of both direct and indirect effects (Fig.2). Several pivotal aspects regarding this issue are discussed below.

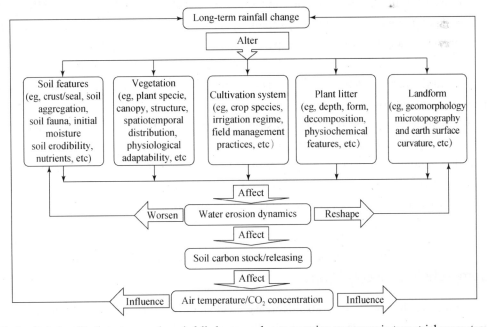

Fig.2 Relationship between secular rainfall change and water erosion processes in terrestrial ecosystems

4.1 The enhancement of rainfall extremes

The enhancement of rainfall extremes is one of the most important aspects of rainfall change in the past and also in future predictions [26,59]. Extreme events with heavy rain intensities play crucial roles in increasing water erosion and degrading surface water quality and environments [61]. For example, thousands of landslides were triggered on unstable and poorly covered slopes due to enhanced rainfall extremes in northern Europe [62]. Soil erosion, landslides, mass movements and debris flows are known to be more severe when largescale heavy rainstorms occur [63]. The annual water erosion and nutrient loss under extremely heavy rainfall exceeds 50% of the total in many environments [61]. Other studies also report that the dominant seasonal concentration of water erosion is limited to a few major storms [21].

4.2 Long-term dynamics of rainfall regimes influence species' natural selection, and human cultivation systems

Studies indicate that changes in rainfall patterns and inputs play a significant role in shaping the productivities, compositions, structures, patterns and services of terrestrial plant communities [64]. For example, precipitation, rather than temperature, is considered as the key natural factor in determining the dynamics of NDVI in northern China [65]. Rainfall distribution, coupled with temperature, is the major factor determining the morphological and physiological characteristics of plant species, as well as vegetation patterns [64]. Alterations of rainfall regimes and temperature, therefore, are likely to challenge the existing spatiotemporal patterns of plant species, cultivated crop systems and their management. Changing rainfall or temperature alters local suitability for specific crops and the need for irrigation and fertilization [41], and farmers have to adapt their management practices to the new climate [31,35,39,66]. These transformations, in turn, will lead to possible changes in ground cover.

Changes in vegetation (canopy and structure, etc) will alter rainfall interception rates as well as secondary rainfall redistribution under the vegetation canopy. For example, statistics show that rainfall interception ranges from 134.0mm to 626.7mm, with a mean rate of 19.85%±7.16% among different forest ecosystems [67]. However, the sparse vegetation with less than 10% plant cover is claimed to have an insignifi-cant effect on runoff and soil loss control [68].

4.3 Rain erosivity and the reshaping of terrestrial landforms and geomorphological conditions

Landform change [69] may also decrease local slope stability, reduce rainfall infiltration rates, increase soil erodibility, and finally induce higher water loss and soil erosion problems [43,62]. For example, after hundreds of years of rainfall erosion coupled with unwise human disturbance, the Chinese Loess Plateau, which was once flat and vegetation-rich, become a notoriously poor vegetation-covered area full of deep gullies and ravines [2,7,70].

4.4 Dynamics in the parameters and patterns of natural rainfall and surface sealing and crusting

Many studies assert that the timing and kinetic energy of rainfall is significant for soil surface seals and crusts in natural conditions [24,38]. Meanwhile, studies also confirm that the first few millimetres of the soil layer largely control the soil's response to the eroding forces of water and sediment in situ [17,38]. Therefore, the rates and risks of water erosion variations are partly dependent upon the characteristics of surface sealing and crusting.

4.5 Changes in rainfall and fluctuations in plant litter and soil organic matter accumulation

This effect is mainly due to the decomposition of plant litter which is also greatly restricted by the near earth-surface temperature and soil moisture content in situ [29,37]. For example, responses of litter decomposition and surface living organisms to wet and dry years caused by rainfall variability are different, especially in the arid and semi-arid environments [71]. Meanwhile, rainfall is an important limitation and determining factor for different plants in certain climatic zones, which influences the physical features (e. g., figure and size) and chemical components (e. g., grease ratio in leaves) of plant fall and litter [24]. Surface fall and litter, however, plays a significant role in absorbing rain water and reducing soil water loss [36]. Litter not only directly protects the surface soil from splash erosion, weakens the kinetic energy of raindrops, and slows the thresholds of runoff velocities, but it also conserves surface rainwater due to its strong moisture holding capacity [72]. Zhu et al. [73] found that litter could intercept about 6%-13% of the total rainfall amount. Other studies show that litter contributes to humus formation, increased surface roughness, and decreased thresholds of soil erosion and water loss generation [74].

4.6 Changes in rainfall and antecedent soil moisture content (ASMC)

The dynamics of timing, frequency, distribution and other rainfall variables all play crucial roles in the fluctuations of antecedent soil water content [75]. The latter, however, has been found to have a positive correlation with runoff thresholds and water erosion rates [17,76]. Unfortunately, due to a lack of observational data, high spatiotemporal heterogeneity and changeable features of soil water, the interactions between rainfall variability and ASMC, as well as their coupling role in water erosion dynamics remains unclear [22,48,77], which poses an important challenge for future research.

5 Concluding remarks: facing this issue

Soil erosion is a widespread form of land and environmental deterioration, responsible for enormous losses in soil fertility and degradation of water resources. Changes in natural rainfall, however, contribute greatly to the dynamics of water erosion, and are expected to become more significant against a background of increased global climate and environmental change. This will increase the difficulties of erosion prediction and control. In future years, therefore, the following issues regarding rainfall change and water erosion processes should be prioritized and receive more attention as important environmental and economic challenges.

1) More attention should be paid to changes in rainfall arising from anthropogenic gas emissions and other global processes. As noted above, changes in rainfall variables and regimes may have greater effects on terrestrial ecosystems and surface hydrological processes than other climatic factors such as air temperature and CO_2. Little research, however, has focused on how anticipated changes in rainfall might affect water erosion processes. Finding a more effective way to study the potential impacts of rainfall variation on the Earth's surface and related biophysical processes still remains a big challenge.

2) Greater efforts should be made in the field of scaling effects and scale transformations regarding changes in rainfall and related erosive dynamics. To solve this problem, data measurements and collection from different regions and different scales should be enhanced. Meanwhile, integrated mathematical and geohydrological models with accurate parameter verifications, which are used for upscaling, should be given more attention, with the goal of providing results for extrapolation to predict future rainfall scenarios.

3) More attention should be paid to the role of rainfall variability in the dynamics of plant litter, soil features such as surface crusts and seals, and in antecedent soil moisture content (ASMC) across different scales and landscape positions. All of these factors play crucial roles in determining the specific dynamics of water erosion processes.

4) Because water erosion dynamics in natural environments are complex, distinguishing the effect of precipitation from the effects of other factors that vary independently during observational intervals is a crucial problem. To deal with this, long-term observations and cross site comparisons regarding population and community changes, considered in conjunction with secular records of historical precipitation data, should be studied.

Acknowledgements

This research was supported by the National Natural Science Foundation of China (40801041; 40621061), the National Basic Research Program of China (2009CB421104) and the National Advanced Project of the Eleventh Five-year Plan of China (2006BAC01A06).

References

[1] Diodato N. Modeling net erosion responses to enviroclimatic changes recorded upon multisecular timescales. Geomorphology, 2006, 80: 164-177.

[2] Chen L D, Wei W, Fu B J, et al. Soil and water conservation on the loess plateau in China: review and prospective. Progress in Physical Geography, 2007, 31: 389-403.

[3] Zhang X P, Zhang L, Zhao J, et al. Responses of streamflow to changes in climate and land use/cover in the Loess Plateau, China. Water Resources Research, 2008, 44 (7) :2183-2188.

[4] Rattan L. The urgency of conserving soil and water to address 21st century issues including global warming. Journal of Soil and Water Conservation, 2008, 63: 140A-141A.

[5] Busnelli J, del V Neder L, Sayago J M. Temporal dynamics of soil erosion and rainfall erosivity as geoindicators of land degradation in northwestern Argentina. Quaternary International, 2006, 158:147-161.

[6] Vaezi A R, Sadeghi S H R, Bahrami H A, et al. Modeling the USLE K-factor for calcareous soils in northern Iran. Geomorphology, 2008, 97: 414-423.

[7] Wei W, Peng H, Li D Z. Current status of eco-environment and countermeasures in the loess hilly area. Journal of Northwest Forestry University, 2004, 19: 179-182.

[8] Nearing M A, Jetten V, Baffaut C, et al. Modeling response of soil erosion and runoff to changes in precipitation and cover. Catena, 2005, 61: 131-154.

[9] Nearing M R O, Nearing M A, Viing R C, et al. Climate change impacts on soil erosion in Midwest United States with changes in crop management. Catena, 2005, 61: 165-184.

[10] Lü Y H, Fu B J, Chen L D, et al. Nutrient transport associated with water erosion: progress and prospect. Progress in Physical Geography, 2007, 31(6): 607-620.

[11] Bridges E M, Oldeman L R. Global assessment of human-induced land degradation. Arid Soil Research and Rehabilitation, 1999, 13: 319-325.

[12] Berhe A A, Harte J, Harden J W, et al. The significance of the erosion-induced terrestrial carbon sink. Bioscience, 2007, 57:337-346.

[13] Yellow River Conservancy Commission (YRCC). Yellow River Near Future Management and Development Plan. Beijing: The State Council of China, 2002.

[14] Polyakov V O, Lal R. Soil organic matter and CO_2 emission as affected by water erosion on field runoff plots. Geoderma, 2008, 143: 216-222.

[15] Peeters I, Oost K V, Govers G, et al. The compatibility of erosion data at different scales. Earth and Planetary Science Letters, 2008, 265: 138-152.

[16] Raclot D, Albergel J. Runoff and water erosion modeling using WEPP on a Mediterranean cultivated catchment. Physics and Chemistry of the Earth, 2006, 31: 1038-1047.

[17] Wei W, Chen L D, Fu B J, et al. Mechanism of soil and water loss under rainfall and earth surface characteristics in a semiarid loess hilly area. Acta Ecologica Sinica, 2006, 26: 3847-3853.

[18] Wei W, Chen L D, Fu B J, et al. The effect of land uses and rainfall regimes on runoff and soil erosion in the semi-arid loess hilly area, China. Journal of Hydrology, 2007, 335: 247-258.

[19] Nyssen J, Poesen J, Moeyersons J, et al. Dynamics of soil erosion rates and controlling factors in the Northern Ethiopian Highlands—towards a sediment budget. Earth Surface Processes and Landforms, 2008, 33: 695-711.

[20] Cantón Y, Domingo F, Solé-Benet A, et al. Hydrological and erosion response of a badlands system in semiarid SE Spain. Journal of Hydrology, 2001, 252: 65-84.

[21] Mannaerts C M, Gabriels D. A probabilistic approach for predicting rainfall soil erosion losses in semiarid areas. Catena, 2000, 40: 403-420.

[22] Merz R, Blöschl G, Parajka J. Spatiotemporal variability of event runoff coefficient. Journal of Hydrology, 2006, 331: 591-604.

[23] Verschuren D, Laird K R, Cumming B F. Rainfall and drought in equatorial east Africa during the past 1100 years. Nature, 2000, 403: 410-413.

[24] Weltzin J F, Loik M E, Schwinning S, et al. Assessing the response of terrestrial ecosystems to potential changes in precipitation. Bioscience, 2003, 53: 941-952.

[25] Richard A K. Global warming is changing the world. Science, 2007, 316: 188-190.

[26] IPCC. 2007. Freshwater Resources and their Management//Parry M L, Canziani O F, Palutikof J P, et al. Climate Change 2007: Impacts, Adaptation and Vulnerability: Contribution of Working Group II to the Fourth Assessment Report of the Intergovernmental Panel on Climate Change. Cambridge: Cambridge University Press, 2007: 174-210.

[27] Wang M L, Xia J. The effect of land use change and climatic Oscillation on water circulation in the East River Watershed. Pearl River, 2004, 2: 4-7.

[28] Foley J A, Ruth D, Asner G P, et al. Global consequences of land use. Science, 2005, 309: 570-74.

[29] Rustad L E. The response of terrestrial ecosystems to global climate change: towards an integrated approach. Science of the Total Environment, 2008, 404: 222-235.

[30] Wei Z G, Dong W J, Hui X Y. Evolution of trend and international oscillatory variabilities of precipitation over NW China. Acta Meteorologica Sinica, 2000, 58: 234-243.

[31] O'Neal M R, Nearing M A, Vining R C, et al. Climate change impacts on soil erosion in Midwest United States with changes in crop management. Catena, 2005, 61: 165-184.

[32] Liu H Y, Wang X M, Xiao S J, et al. Study on the change of precipitation in Urumqisince recent 40 years. Arid Zone Research, 2007, 24 (6): 785-789.

[33] Pruski F F, Nearing M A. Climate-induced changes in erosion during the 21st century for eight USA locations. Water Resources Research, 2002, 38: 1-11.

[34] Oki T, Kanae S. Global hydrological cycles and world water resources. Science, 2006, 313: 1068-1072.

[35] Salinger M J, Stigter C J, Das H P. Agrometeorological adaptation strategies to increasing climate variability and climate change. Agricultural and Forest Meteorology, 2000, 103: 167-184.

[36] Eagleson P S. Ecological optimality in waterlimited natural soil-vegetation systems, 1. Theory and hypothesis. Water Resources Research, 1982, 18: 325-340.

[37] Berg B, McLaugherty C. Plant Litter: Decomposition, Humus Formation, Carbon Sequestration. Berlin: Springer, 2003.

[38] Singer M J, Shainberg I. Mineral soil surface crusts and wind and water erosion. Earth Surface Processes and Landforms, 2004, 29: 1065-1075.

[39] Ittersum M K, Howden S M, Asseng S. Sensitivity of productivity and deep drainage of wheat cropping systems in a Mediterranean environment to changes in CO_2, temperature and precipitation. Agriculture, Ecosystems and Environment, 2003, 97: 255-273.

[40] Martínez-Mena M, Albaladejo J, Castillo V M. Factors influencing surface runoff generation in a Mediterranean semi-arid environment: Chicamo watershed, SE Spain. Hydrological Processes, 1998, 12: 741-54.

[41] Fuhrer J. Agroecosystem responses to combinations of elevated CO_2, ozone, and global climate change. Agricultural, Ecosystems and Environment, 2003, 97: 1-20.

[42] Chen L D, Lü Y H, Fu B J, et al. A framework on landscape pattern analysis and scale change by using pattern recognition approach. Acta Ecologica Sinica, 2006, 26: 663-670.

[43] Sanchis M P S, Torri D, Borselli L, et al. Climate effects on soil erodibility. Earth Surface Processes and Landforms, 2008, 33: 1082-1097.

[44] Soil and Water Conservation Society SWCS. Conservation Implications of Climate Change: Soil Erosion and Runoff from Cropland. Ankeny: SWCS, 2003.

[45] Piccarreta M, Capolongo D, Boenzi F, et al. Implications of decadal changes in precipitation and land use policy to soil erosion in Basilicata, Italy. Catena, 2006, 65: 138-151.

[46] Zhang G H. Response of rainfall erosivity to climate change in the Yellow River Basin. Journal of Mountain Science, 2005, 23: 420-424.

[47] Ramos M C, Martínez-Casasnovas J A. Impact of land leveling on soil moisture and runoff variability in vineyards under different rainfall distributions in a Mediterranean climate and its influence on crop productivity. Journal of Hydrology, 2006, 321: 131-146.

[48] Kosmas C, Gerontidis St, Marathianou M. The effect of land use change on soils and vegetation over various lithological formations on Lesvos (Greece). Catena, 2000, 40: 51-68.

[49] Lee J L, Phillips D L, Dodson R F. Sensitivity of the US Corn Belt to climate change and elevated CO_2: II Soil erosion and organic carbon. Agricultural Systems, 1996, 52: 503-521.

[50] Zhang X C, Liu W Z. Simulating potential response of hydrology, soil erosion, and crop productivity to climate change in Changwu tableland region on the Loess Plateau of China. Agricultural and Forest Meteorology, 2005, 131: 127-142.

[51] Ward P J, van Balen R T, Verstraeten G, et al. The impact of land use and climate change on late Holocene and future suspended sediment yield of the Meuse catchment. Geomorphology, 2009, 103: 389-400.

[52] Kim M K, Flanagan D C, Frankenberger J R, et al. Impact of precipitation changes on runoff and soil erosion in Korea using CLIGEN and WEPP. Journal of Soil and Water Conservation, 2009, 64: 154-62.

[53] Michael A, Schmidt J, Enke W, et al. Impact of expected increase in precipitation intensities on soil

loss-results of comparative model simulations. Catena, 2005, 61: 155-164.

[54] Favis-Mortlock D T, Guerra A J T. The implications of general circulation model estimates of rainfall for future erosion: a case study from Brazil. Catena, 1999, 37: 329-354.

[55] Houghton J T, Ding Y, Griggs D J, et al. Climate Change 2001: The Scientific Basis Contribution of Working Group 1 to the Third Assessment Report of the Intergovernmental Panel on Climate Change. Cambridge: Cambridge University Press, 2001.

[56] Favis-Mortlock D T, Boardman J. Nonlinear responses of soil erosion to climate change: a modeling study on the UK South Downs. Catena, 1995, 25: 365-387.

[57] Schulze R. Transcending scales of space and time in impact studies of climate and climate change on agrohydrological responses. Agriculture, Ecosystems and and Environment, 2000, 82: 185-212.

[58] Panagoulia D, Dimou G. Sensitivity of flood events to global climate change. Journal of Hydrology, 1997, 191: 208-222.

[59] Wei W, Chen L D, Fu B J, et al. Responses of water erosion to rainfall extremes and vegetation types in a loess semiarid hilly area, NW China. Hydrological Processes, 2009, 23: 1780-1791.

[60] Yang D W, Kanae S, Oki T, et al. Global potential soil erosion with reference to land use and climate change. Hydrological Processes, 2003, 17: 2913-2928.

[61] Gao C, Zhu J, Hosen Y, et al. Effects of extreme rainfall on the export of nutrients from agricultural land. Acta Geographica Sinica, 2005, 60: 991-997.

[62] Brierley G, Stankoviansky M. Editorial: geomorphic responses to land use change. Catena, 2003, 51: 173-179.

[63] Cheng J D, Lin L L, Lu H S. Influences of forests on water flows from headwater watersheds in Taiwan. Forest Ecology and Management, 2002, 165: 11-28.

[64] Bates J D, Svejcar T, Miller R F, et al. The effects of precipitation timing on sagebrush steppe vegetation. Journal of Arid Environments, 2006, 64: 670-697.

[65] Li Z, Yan F L, Fan X T. The variability of NDVI over Northwest China and its relation to temperature and precipitation. Journal of Remote Sensing, 2005, 9 (3): 308-313.

[66] Southworth J, Randolph J C, Habeck M, et al. Consequences of future climate change and changing climate variability on maize yields in the mid western United States. Agriculture, Ecosystems and Environment, 2000, 82: 139-158.

[67] Shi P L, Li W H. Influence of forest cover change on hydrological process and watershed runoff. Journal of Natural Resources, 2001, 16(5): 481-487.

[68] Juan P. The role of vegetation patterns in structuring runoff and sediment fluxes in drylands. Earth Surface Processes and Landforms, 2005, 30: 133-147.

[69] Goudie A S. Global warming and fluvial geomorphology. Geomorphology, 2006, 79: 384-394.

[70] Fu B J. Soil erosion and its control in the Loess Plateau of China. Soil Use and Management, 1989, 5: 76-82.

[71] Yahdjian L, Sala O E. Do litter decomposition and nitrogen mineralization show the same trend in the response to dry and wet years in the Patagonian steppe? Journal of Arid Environments, 2008, 72:

687-695.

[72] Hou X, Bai G, Cao Q. Study on benefits of soil and water conservation of forest and its mechanism in loess hilly region. Research of Soil and Water Conservation, 1996, 3: 98-103.

[73] Zhu J, Liu J, Zhu Q, et al. Hydroecological functions of forest litter layers. Journal of Beijing Forestry University, 2002, 24: 30-34.

[74] Boer M, Puigdefábregas J. Effects of spatially structured vegetation patterns on hill-slope erosion in a semiarid Mediterranean environment: a simulation study. Earth Surface Processes and Landforms, 2005, 30: 149-167.

[75] Koster R D, Dirmeyer P A, Guo Z C, et al. Regions of strong coupling between soil moisture and precipitation. Science, 2004, 305: 1138-1140.

[76] Luk S. Effect of antecedent soil moisture content on rainwash erosion. Catena, 1985, 12: 129-139.

[77] Rudolph A, Helming K, Diestel H. Effect of antecedent soil water content and rainfall regime on microrelief changes. Soil Technology, 1997, 10: 69-81.

The effect of land uses and rainfall regimes on runoff and soil erosion in the semi-arid loess hilly area, China[*]

Wei Wei   Chen Liding   Fu Bojie   Huang Zhilin

Wu Dongping   Gui Lida

Abstract

The main purpose of this article is to analyze runoff and soil loss in relation to land use and rainfall regimes in a loess hilly area of China. Based on 14 years of field measurements and K-means clustering, 131 rainfall events were classified into three rainfall regimes. Rainfall Regime II is an aggregation of rainfall events with such features as high intensity, short duration and high frequency. Regime I is the aggregation of rainfall events of medium intensity, medium duration and less frequent occurrence. Regime III is the aggregation of events of low intensity and long duration and infrequent occurrence. The following results were found. ①Mean runoff coefficient and erosion modulus among the five land use types are: cropland > pastureland > woodland > grassland > shrubland. ②The sensitivity of runoff and erosion to the rainfall regimes differ. Rainfall Regime II causes the greatest proportion of runoff and soil loss, followed by Regime I and Regime III. ③The processes of runoff and soil loss, however, are complicated and uncertain with the interaction of rainfall and land use. This is mainly due to the different stages of vegetation succession. Based on these results, it was suggested that more attention should be paid to Rainfall Regime II since it had the most erosive effect. Shrubland is the first choice to control soil erosion when land use conversion is implemented, whereas pastureland (alfalfa) is not.

---

[*] 原载于: Journal of Hydrology, 2007, 335: 247-258.

Large-scale plantation of alfalfa therefore, should be avoided. Grassland and woodland can be used as important supplements to shrubland.

Keywords: Runoff and soil erosion; Rainfall regime; Clustering; Land use; Loess hilly area

1 Introduction

Soil erosion, defined as the detachment and displacement of soil particles from the surface to another location [1,2], continues to be a primary cause of soil degradation throughout the world [3], and has become an issue of significant and severe societal and environmental concern [4,5]. Land use/cover, as one of the most important factors, influences the occurrence and the intensity of runoff and sediment yield [6-8]. Non-uniform variations in land use/vegetation coverage proved to be closely related to hydrological responses over catchments [9]. By properly adjusting of land use/land cover patterns, soil properties can be greatly improved, consequently reducing soil erosion to the allowed threshold [10,11], and the improved soil physical properties can also positively affect the establishment of vegetation [12]. On the other hand, improper land use and/or cover patterns can cause severe water, soil and nutrient losses, and further land degradation [13,14].

Runoff and erosion processes, however, are strongly affected by many other factors besides land use/land cover. Among these factors, the one most mentioned is rainfall. Rainfall can cause soil erosion and runoff when it reaches the ground [15-17]. Also, the spatiotemporal heterogeneity and uneven characteristics of rainfall play a key role in soil erosion [18-21]. Morin et al. [22] found that complex interactions exist between the spatiotemporal distributions of rainfall systems and watershed hydrological responses. Local storm patterns are important in determining the shape of the runoff hydrograph [23]. Runoff and sediment generation in different land use types may thus vary greatly with various rainfall types. Addressing the response of runoff/erosion to different land use/land cover types and different rainfall types is therefore important for land use structure adjustment and vegetation restoration.

Rainfall classification, however, is an important problem, which needs to be solved. Most studies focused on the response of the runoff/erosion process to single rainfall pattern and different vegetation types [24-26]. Undoubtedly, controlling soil erosion requires much more detailed and accurate data in the real world [4]. Many studies, however, are based on rainfall simulations, and thus the conclusions are often not applicable to the real world. For example, some authors have suggested that the nozzles of rainfall simulators produce low kinetic energies relative to natural rainfall [27]. Madden et al. [28] also found that the kinetic energy and erosivity of rainfall produced by simulators could be lower than that of natural rainfall. This insufficiency of energy plays an important role in infiltration capacity, preventing surface crusting and sediment detachment [29]. Other results also show that rainfall simulators are unable to reproduce natural rainfall conditions [30-32]. Accordingly, finding real rainfall-runoff-sediment patterns based on measurements are important for soil erosion control.

Soil loss and runoff studies at plot scales have been confirmed to be of crucial importance [33]. Reliable and consistent erosion measurements and extensive field data have played primary roles in soil erosion analysis and prediction on larger scales [34,35]. In addition, soil properties are always affected by land uses/vegetation evolution over long time scales (e.g., months-centuries), which then further influence runoff and soil erosion [36,37]. For example, the accumulation of litter under plants contributes to increased surface roughness, higher infiltration rates, and decreased runoff generation thresholds [38]. Bochet et al. [39] also found that topsoil modification and erosion processes are mainly due to the differential influences of species morphology (i.e., aboveground structure) and components (i.e., litter cover and organic matter). Moreover, the impact of plant roots on soil resistance to erosion by water is also significant [40-43]. In general however, these kinds of long-term consecutive studies in arid and semi-arid areas are relatively scarce.

In this study, based on 14 years of field measurements in plots in a semi-arid loess hilly area, 131 rainfall events that produced runoff were recorded. On the basis of rainfall depth, duration and maximum 30-min intensity, all the events were classified into three categories. They were then used to analyze the effects of varying land uses and rainfall regimes on runoff and soil erosion. The specific objectives were: ① to analyze the effects of land use/land cover on soil and water loss, ② to determine the response of runoff and soil erosion to different rainfall regimes, and ③ to study the role of different land use types on soil erosion control under different rainfall regimes.

2 Material and methodology

2.1 Study area

Our experiments were all conducted in a small catchment, Anjiapo Catchment, Dingxi, Gansu province, China (35°35′N, 104°39′E) in the middle reaches of the Yellow River. This region is dominated by a temperate terrestrial climate with warm-humid summers and colddry winters. The average annual precipitation is about 427mm, of which more than 80% falls from May to September. The potential annual transpiration, however, can reach 1510mm. The mean monthly temperature ranges from 7.6℃ to 27.5℃, with an average annual temperature of 6.3℃. Approximately 141 days annually are frost-free.

Local soil develops from wind-accumulated loess parent material, which is about 40-60m on depth. The dominant soil in the region belongs to calcic Cambisol [44] with clay of 33.12%-42.17%, organic matter of 3.7-13.4g/kg, and soil density is from 1.09 to 1.36g/cm^3 within 2m depth [45]. This type of soil has week resistant to erosion [3]. Vegetation in the study area is poor due to lack of water. The dominant plant species are: Chinese pine (*Pinus tabuliformis* Carr.), seabuckthorn (*Hippophae rhamnoides* L.), pea shrub (*Caragana kansuensis* Pojark.), Chinese arborvitae (*Platycladus orientalis* L.), apricot (*Prunus armeniaca* L.), alfalfa (*Medicago sativa* L.), sainfoin (*Onobrychis vichfolia* Scop.), and bunge needlegrass (*Stipa bungeana* Trin.). The main crop species are potatoes (*Solanum tuberosum* L.), spring wheat (*Triticum aestivum* L. cv Leguan), maize (*Zea mays* L.) and flax (*Linum usitattissimum* L.).

2.2 Experimental design

Fifteen experimental plots were planted on northfacing hill-slopes, on which rain-fed crops (wheat, potatoes, beans and millet) were grown before the plots were established. To reduce the effects of position, all plots were established on the same slope. They were oriented parallel to the slope and adjacent to each other.

The following five land cover types with three replications (10°, 15° and 20°) in the experimental plots were investigated. ① Cropland (*Triticum aestivum* L. cv Leguan): Field management was similar to that used by local farmers, and the seeds of spring wheat were sown in April and harvested manually in early August. ② Pastureland (*Medicago sativa* L.): The seed was drilled or broadcasted in April and harvested in late July. Alfalfa was replanted annually from 1993 to 1999 due to low yield. ③ Shrubland (*Hippophea rhamnoides* L.): Saplings of sea buckthorn were planted in 1.0m by 1.0m spacing in March, 1986. The litter remained on the plots during the experiment. ④ Woodland (*Pinus tabulaeformis* Carr.): Saplings of Chinese pine were planted in 3.0m rows and 1.5m columns in March, 1986. ⑤ Grassland (*Stipa bungeana* Trin.): This natural species was left to grow without human disturbance.

Plots used for shrubland and woodland were 10m×10m, while plots used for slope cropland, pastureland and grassland were 10m×5m. In each plot, cement ridges (30cm above ground) were constructed at the borders to isolate plot runoff and sediment. A discharge ditch was created at the top of each plot to control runoff and sediments from the upper slope. At the base of each plot, a marked H-flume and two volumetric tanks were built at the outlet of each plot for surface runoff and sediments collection.

Precipitation during the rainy season was measured by a SM_1 pluviometer and a SJ_1 auto-siphon udometer. The depth, duration and intensity of each rainfall event were recorded. The runoff and erosion of each rainfall event were monitored. In total, 131 rainfall events with runoff (1986-1999) were recorded.

The possible initial spatial variations of runoff/erosion effects caused by slope aspect, slope position, elevation, and the physiochemical features of local soil were not considered, since all the plots were situated in a similar environment.

2.3 Clustering methodology

Clustering approach is a fundamental and important tool in statistical analysis. In the past, statistical clustering techniques have been widely used in such diverse scientific fields as psychology, zoology, biology, botany, sociology, meteorology, physiognomy, etc.[24,46]. It aims to group objects based on their similarities. There are two methods of clustering, the hierarchical clustering method, and the non-hierarchical method or K-means clustering. In the former, the number of clusters is obtained by automatic statistical analysis[47]. Two types of hierarchical cluster methods are included, i.e., R type and Q type[48,49]. R type clustering is used for variables classification, and Q type clustering for different cases.

In this study, the non-hierarchical clustering method, i.e., the K-means clustering was used to classify the rainfall events. This method is suitable for a large number of cases[47], and a cluster number is required before classification. To determine the number of clusters in a data set, numerous criteria were proposed[50]. In our study, attempts were made until the most suitable clusters appeared. Normally, the classification must meet the ANOVA criterion of significant level ($P < 0.05$).

2.4 Statistical analysis

Fourteen years of consecutive data (1986-1999) were used in this study. In order to analyze the surface runoff and sediment loss, two indices including runoff coefficient (ratio of rainfall excess to rainfall) and erosion modulus were used

$$C = (SR/P) \times \% \qquad (1)$$

where C, SR and P denote runoff coefficient, surface runoff and precipitation, respectively.

$$E_m = (SL/P_A)/n \qquad (2)$$

where E_m, SL, P_A and $1/n$ refer to erosion modulus, sediment loss, area of experimental plot and number of years, respectively.

All results were calculated with SPSS13.0 for windows.

3 Results

3.1 Rainfall regimes

Using K-means clustering, the 131 rainfall events were divided into three groups based upon three rainfall eigenvalues, including rainfall depth, duration and maximum 30-min intensity (Table 1).

Table 1 Statistical features of different rainfall regimes

Rainfall regime	Eigenvalue	Mean	Standard deviation	Variation coefficient	Sum	Frequency (times)
I	P (mm)	26.72	9.68	0.36	962	36
	D (min)	932	209	0.22	33 563	
	I_{30} (mm/min)	0.14	0.12	0.86	—	
II	P (mm)	14.09	7.87	0.56	1 253	89
	D (min)	216	166	0.77	19 190	
	I_{30} (mm/mn)	0.26	0.17	0.65	—	
III	P (mm)	31.75	7.44	0.23	191	6
	D (min)	1737	296	0.17	10 420	
	I_{30} (mm/min)	0.11	0.07	0.64	—	

Note: P, D, I_{30} separately represents precipitation depth, duration and maximum 30-min intensity.

In general, Rainfall Regime III has the highest values of mean rainfall depth and duration, followed by Rainfall Regime I and Rainfall Regime II. Mean maximum 30-min intensity, however, decreases in the order of Rainfall Regime II, Rainfall Regime I and Rainfall Regime III. Average rainfall eigenvalues represent the general characteristics of rainfall events. We thus conclude that Rainfall Regime II is the group of rainfall events with strong intensity, frequent occurrence of rainfall and very short duration, while Rainfall Regime III consists of rainfall events with low intensity, long duration and infrequent occurrence. Rainfall Regime I, however, is composed of rainfall events, which have moderate rainfall eigenvalues, i.e., higher intensity and shorter duration than Rainfall Regime III, but lower intensity and longer duration than Rainfall Regime II.

During the period of measurement, Rainfall Regime II occurred 89 times with a total of 1253mm from May 1986 to September 1999. Rainfall Regime I occurred 36 times with a total of 962mm. Rainfall Regime III, however, was observed only 6 times.

The distributions of three rainfall regimes in different years varied (Fig. 1). Fig. 1(a) shows the total depths of the three rainfall regimes in different years. In most years, the rainfall regime sequence was Rainfall Regime II > Rainfall Regime I > Rainfall Regime III, except for 1986, 1993 and 1999, when Rainfall Regime I had higher values than Rainfall Regime II. This was because the frequency of Rainfall Regime I were the highest [Fig. 1(b)]. Fig. 1(c) shows that the mean maximum 30-min intensity of Rainfall Regime II was the highest, followed by Rainfall Regimes I and III.

Generally, the regime and distribution of local rainfall are highly varied in different years. Rainfall Regime III, for example, most obviously shows the highest inter-annual variations. All six occurrences appeared in 1989, 1990, 1993, 1995, 1996, and 1999, respectively. No such rainfall events appeared in the other years.

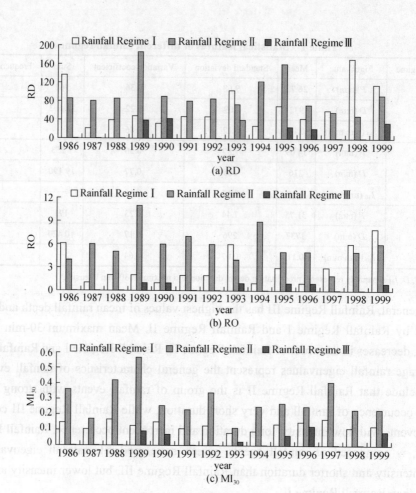

Fig.1 Characteristics of three rainfall regimes in different years

Note: RD, rainfall depth (mm); RO, rainfall occurrences (times); MI_{30}, maximum 30-min rainfall intensity (mm/min). Meanwhile, in some years no data are shown for Rainfall Regime III, and it is because this kind of rainfall did not occur during these years.

3.2 Soil and water loss in different land use types

Mean runoff and soil losses differ among different land use types based on 14 years of measurements (Table 2). Cropland has the highest surface runoff and soil erosion, far more than those in the other four land use types. Second is pastureland, followed by woodland and grassland, then shrubland. It was found that both mean runoff coefficients and mean erosion moduli showed similar changing trend among the five land use types (Table 2).

Table 2 Average 14-year runoff and soil erosion indexes in the study area

Indexes	Cropland	Pastureland	Shrubland	Woodland	Grassland
Runoff coefficient (%)	8.40	7.16	2.61	5.46	3.91
Erosion modulus [t /(km²·a)]	8599	3392	131	760	534

In Table 3, the criterion for soil erosion classification, given by the Ministry of Water Resources of China (SL190-96) based on the mean annual erosion modulus, was used. It was found that cropland has a very high erosion modulus [$8599t/(km^2 \cdot a)$], higher than the lower boundary of highly intense erosion [$8000t/(km^2 \cdot a)$]. Pastureland, however, has a medium soil erosion level, and the other three land uses have a feeble level. Shrubland is best for soil and water control, with a $131t/(km^2 \cdot a)$ erosion modulus, which can even be neglected in most cases.

Table 3 Criterion of soil erosion severity classification on the loess Plateau

Level	Feeble	Gentle	Middle	Intense	Highly intense	Severe
Mean erosion modulus [$t/(km^2 \cdot a)$]	<1000	1000-2500	2500-5000	5000-8000	8000-15,000	>15,000

The role of mean annual runoff and erosion effects among different land use types in different years were captured by line-drawing method under SPSS13.0 (see Fig. 2). Fig. 2 shows the mean annual runoff coefficients and erosion moduli from 1986 to 1999. Runoff coefficients and erosion moduli differ in different years. For instance, the lowest and highest runoff coefficients of cropland were 1.84 in 1991 and 8.79 in 1986, respectively. The other land use types had the same characteristics in surface runoff and erosion moduli. However, the lower and peak value of surface runoff coefficient and erosion moduli in different land use types were not synchronous. In most years, runoff and soil loss ranked in order of: cropland > pastureland > woodland > grassland > shrubland. Pastureland (alfalfa), however, had the highest runoff coefficient in 1994 and 1995, and the highest erosion modulus in 1994. Woodland had

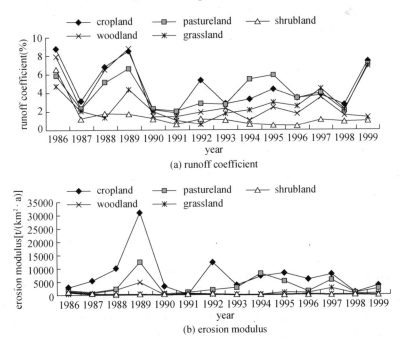

(a) runoff coefficient

(b) erosion modulus

Fig.2 Mean soil and water losses under different land uses over the period of 1986-1999.

a higher runoff coefficient and a lower erosion modulus than cropland in 1989, and grassland had the highest runoff coefficient in 1997.

3.3 Soil and water loss of different rainfall regimes

The characteristics of runoff and soil loss under these three rainfall regimes are indicated in Fig. 3. We found that the values of the runoff coefficients and erosion moduli among different land use types were as follows: Rainfall Regime II > Rainfall Regime I > Rainfall Regime III. Rainfall Regime II created the most runoff and soil erosion, far more than that in Rainfall Regime I. Rainfall Regime III, however, created little runoff and erosion.

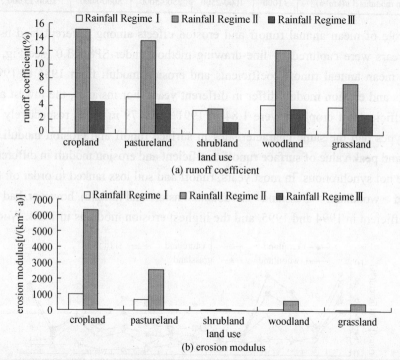

Fig. 3 Features of runoff and erosion of five land uses under different rainfall regimes.

Furthermore, the effects of land use on runoff and erosion in different years were also captured (Fig. 4 and Fig. 5). From 1986 to 1999, different land use types had different responses to rainfall regimes. We found that runoff coefficients and erosion moduli fluctuated greatly over time in Rainfall Regime II, followed by Rainfall Regime I and Rainfall Regime III.

From 1986 to 1999, different land use types indicated various changing trends. Cropland and pastureland showed similar fluctuations among different years. Shrubland showed clearly decreasing trends in Rainfall Regime II, but not in Rainfall Regime I and Rainfall Regime III. Woodland had a sensitive response to Rainfall Regime II. Grassland, however, had a complex dynamic than did the other land use types. Its runoff coefficient varied irregularly under Rainfall Regimes I and II. Nevertheless, its erosion moduli under Rainfall Regime II were higher for several years, and then decreased sharply like those of shrubland and woodland. No decreasing trends were found in the other two rainfall types.

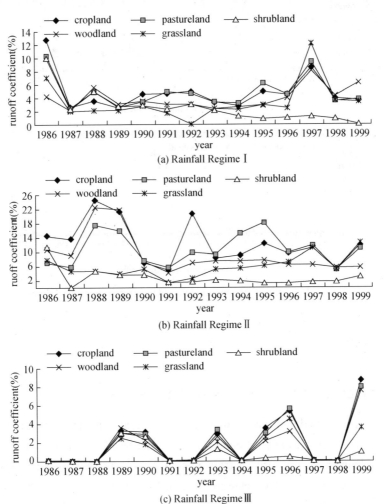

Fig. 4　Inter-annual runoff coefficients under different rainfall regimes

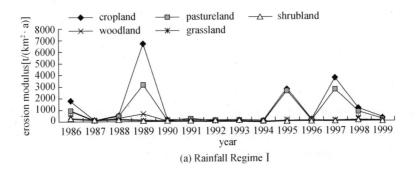

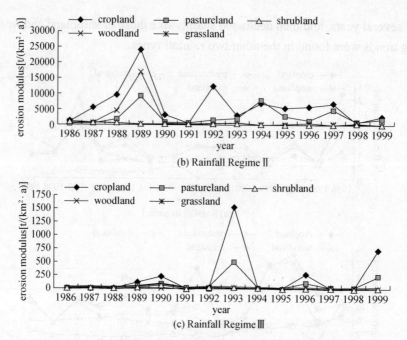

Fig. 5 Inter-annual erosion moduli under different rainfall regimes

4 Discussion

4.1 Effects of land uses on runoff and soil loss

In this study we found that runoff and soil loss varied among land use types (Table 2). This was explained in various ways by different scholars. First of all, vegetation canopy was thought to play a key role in protecting surfaces from erosion[6,7,37,51]. For example, Hou et al.[52] found that when the coverage rate of vegetation increased from 10%, to 28%, 56% and 60%, soil erosion decreased from 1523t/km^2 to 527t/km^2, 218t/km^2 and 107t/km^2, respectively. Zhou et al.[53] also found that soil erosion was negatively linearly correlated with vegetation coverage ($r = 0.99$) in the loess hilly area. Other studies also draw the similar conclusions[42].

Secondly, it was recognised that litter production and organic matter accumulation could reduce soil-water loss[36,38]. Litter not only directly protects the surface soil from splash erosion, weakens the kinetic energy of raindrop, and slows runoff velocities, but also conserves surface rainwater due to its strong moisture holding capacity[52]. Zhu et al.[54] found that litter depth differed among land use types, and could intercept 6%-13% of total rainfall amount. Meanwhile, studies show that litter contributes to humus formation, increased surface roughness, and decreased runoff generation thresholds[55].

Thirdly, plant roots can decrease soil and water loss by increasing soil anti-scouribility, soil anti-shear strength and enhancing penetrability [40,43,56]. Meanwhile, roots may form a dense network (e.g., grass roots in topsoil) that physically binds soil particles, and the soil-root matrix has proven to be stronger than the soil or roots separately [42]. A mechanical barrier to soil and water movement is thus created [57,58]. Yun et al.[59] reports that root systems can loosen soil, enhance its porosity, and thus reduce runoff and erosion.

Lastly, human activities play an important role in runoff and soil erosion [60,61]. In our study, because of its high density (Table 4) and prickliness, shrubland is seldom disturbed by human activity. Cropland however, is strongly affected by human disturbance (e.g., sown in April and harvested at the end of July), which disturbs the soil layer and reduces land coverage.

Table 4 Mean annual vegetation coverage rate (%) of land uses in different years

Year	Cropland	Pastureland	Shrubland	Woodland	Grassland
1986	50	90	40	15	80
1987	57	85	60	24	85
1988	52	75	95	37	100
1989	62	45	98	39	100
1990	68	40	99	45	100
1991	53	50	99	55	98
1992	43	40	99	65	95
1993	36	40	99	70	93
1994	30	35	99	74	93
1995	30	38	99	78	90
1996	35	40	99	80	87
1997	46	40	99	82	85
1998	70	35	99	84	85
1999	67	50	99	86	80

Measurement: The herbaceous coverage was estimated in the maximum growth period by 5 quadrates of 1m×1m within each plot; the coverage of seabuckthorn and Chinese pine was calculated by means of the measured values of canopy width. The mean value of three replications was used to represent coverage rate of each land use.

Alfalfa (*Medicago sativa* L.) is widely cultivated since it has a high yield, wide adaptability and high drought tolerance [62,63]. As the major plant species of pastureland, however, alfalfa has a poor effect on soil and water conservation (Table 2). There are two reasons for this. Firstly, alfalfa is often harvested by local farmers for animal use. It results in the destruction of alfalfa, making soil sensitive to erosion. Secondly, alfalfa absorbs water from deep soil layers creating dry soil layers[64,65], which leads to plant degradation[66,67]. This is unhelpful to soil and water conservation. From this point of view, large-scale alfalfa plantation is thus not recommended for semi-arid areas.

4.2 Response of runoff and soil loss to different rainfall regimes

According to our study, Rainfall Regime II is the most frequent rainfall events. It has a destructive effect on the soil surface and tends to induce higher runoff and erosion. Other researchers have reported this as well [68,69]. Rainfall Regime III constitutes only a small portion of total rainfall events and has slightly erosive effects on soil. Hence, it cannot produce severe erosion and water loss. This means that the response of runoff and erosion to land uses is most sensitive to Rainfall Regime II, then Rainfall Regime I, followed by Rainfall Regime III.

Under Rainfall Regime II, the runoff and soil losses of land use types, such as shrubland and woodland decreased as plants grew. In the first 3-4 years after plantation, runoff and erosion were very serious (even higher than cropland and pastureland in some years), then the runoff coefficients and erosion moduli decreased sharply, and finally reached a relatively stable low level after 1990 (Fig. 4 and Fig. 5). This phenomenon is similar to that observed in several former studies [52,70,71].

We found that the decreasing trend under Rainfall Regime II is closely related to plant succession. According to local investigation, vegetation coverage of shrubland and woodland gradually increased with plant growth (Table 4), while runoff and erosion decreased. Other authors have also drawn similar conclusions. For example, based on consecutive experiments in a hilly loess area, Hou et al. [52] found that plant succession could gradually increase coverage rate, accumulate litter layer, and improve soil properties, which notably reduced runoff and soil erosion with plant evolution. Zou et al. [41] indicated that a significant correlation ($r = 0.998$) between root amounts and fixed soil amount existed, and root amounts increased quickly with plant growth. The measured runoff and erosion reduction, therefore, actually result from the combined effect of both above-ground (canopy, stems and leaves, and litter layers) and below-ground (roots) biomass [42,58].

The decreasing trend shown under Rainfall Regime II, however, did not exist in Rainfall Regimes I and III. This was mainly due to lower rainfall intensity in these two rainfall regimes, whereas the mean precipitation depth and duration of single rainfall events under these two regimes are higher than those under Rainfall Regime II (Table 2). From this standpoint, rainfall intensity is the most important indicator in predicting or indicating degrees of soil erosion, although other eigenvalues such as rainfall depth and duration are indispensable to determine rainfall types. Other studies also confirmed that rainfall intensity played a vital role in runoff and sediment generation [23,25,72]. Furthermore, rainfall frequency is also very important (Table 2). Low frequency combined with weak intensity can result in less soil loss, especially under Rainfall Regime III.

4.3 Process of soil and water loss under the interaction of land use and rainfall regime

Our study indicates that comprehensive inter-relationships exist between rainfall, land use/vegetation cover and soil and water loss (Fig. 4 and Fig. 5). Due to the influence of land cover types and vegetation development dynamics, the relationship between rainfall and the process of overland flow and soil loss becomes uncertain. On the one hand, rainfall is a destructive force to the land surface making the soil prone to splash erosion [16,72-74]. On the other hand, land use/vegetation cover has positive effects on the surface [75], which plays a significant role in the accumulation of fine particles and protecting soil surface from erosion [76,77].

Generally, our results indicate two cases. In the first one, soil and water loss is dominated by rainfall itself, and land use/vegetation plays a less important role, i.e., the runoff and soil loss are controlled by rainfall erosivity rather than by erosion-resistance of land use and vegetation cover [37]. According to our study, the responses of runoff and erosion in this case are sensitive to the rainfall regimes, especially to Rainfall Regime II (Fig. 4 and Fig. 5). Due to poor surface coverage and other factors such as human disturbance, cropland, pastureland, the initial sapling stages of shrubland and woodland both belong to the first case.

In the second case, soil and water loss are dominated by vegetation/land use, and the effect of the rainfall regime is slight, then the runoff and soil erosion are insensitive to rainfall and its regimes. The coverage rate of shrubland (*Hippophae rhamnoides* L.) reached 99% after 1990 (Table 4), which decreased runoff and erosion significantly, and induced less sensitivity to rainfall. Xu [37] also indicated that vegetation cover can protect surface soil from erosion. Meanwhile, other studies have also found that vegetation succession can improve soil physiochemical properties and hence lower erodibility [71,78-80]. After several years' growth (after 1990, in Fig. 4 and Fig. 5) therefore, the runoff and soil erosion of shrubland (*Hippophea rhamnoides* L.) and woodland (*Pinus tabulaeformis* Carr.) decreased and kept at a lower level. This situation resembles the second case.

5 Conclusion and suggestion

In this study, three rainfall regimes were classified using K-means clustering based on rainfall depth, intensity and duration. Rainfall Regime II is the dominant aggregation of rainfall events, which have such features as high intensity, short duration and high frequency. Rainfall Regime I is the aggregation of rainfall events of medium intensity, duration and frequency. Rainfall Regime III is the aggregation of rainfall events of weak intensity, long duration and low frequency.

Results showed that the runoff and soil losses of different land uses varied greatly under different rainfall regimes. Generally, runoff and sediment loss under Rainfall Regime II were greatest, followed by Rainfall Regimes I and III. Furthermore, the runoff coefficient and erosion modulus of shrubland were lowest, followed in increasing order by grassland and woodland. Pastureland had an unsatisfactory erosion control effect, slightly weaker than cropland and far greater than the other three land use types.

Under Rainfall Regime II, shrubland and woodland showed clearly that runoff and sediment yields decrease quickly with plant growth. Runoff and erosion were very severe during the first 3-4 years after plantation, then decreased and stabilized at a low level. This indicates that the growth periods of plant species in each land use play pivotal roles in resisting surface runoff and soil erosion.

The findings in this study have important implications for surface runoff and soil loss control in the semi-arid loess hilly area. Firstly, different practical countermeasures should be laid out according to rainfall types. More attention should be paid to the seasonal distribution of the most erosive rainfall type in further studies, and effective measures need to be taken to control its destructive effects. Secondly, in order to control soil erosion, the most suitable land use types can be selected based on scientific observation. Scrubland (e.g., seabuckthorn) should be recommended as the first key plant species in land use adjustment and vegetation restoration. Grassland and woodland, however, can be used as important supplement to shrubland. Meanwhile, results indicate large-scale plantation of alfalfa should be avoided. More studies should focus on the relationship between alfalfa's water exhaustion and its erosion control ability. Thirdly, more attention must be paid to the plants' succession stages. Due to severe runoff and soil loss at the initial sapling stages after planting (or seeding), other measures such as straw mulch, plastic film or prevention of human disturbance should be taken into account.

Acknowledgements

The authors thank the Dingxi Institute of Soil and Water Conservation in Gansu province for providing experimental plots and pure-hearted field assistance. Sincere thanks are also expressed to Gansu Research Institute of Forestry. Ms. Victoria Wilhoite, at the University of South Florida School of Library and Information Science, is acknowledged for her valuable comments and English improvement. The authors express their appreciation to the reviewers by whose constructive remarks this paper has been improved. This research was supported by National Natural Science Foundation of China (40321101; 90502007) and the National Advanced Project of the Tenth Five-year Plan of China (2001BA606A-03).

References

[1] Govers G, Everaert W, Poesen J, et al. A long flume study of the dynamic factors affecting the resistance of a loamy soil to concentrated flow erosion. Earth Surface Processes and Landforms, 1990, 11:

515-524.

[2] Lal R. Erosion Encyclopedia of Soil Science. New York: Marcel Dekker, 2002: 395-398.

[3] Fu B, Gulinck H. Land evaluation in an area of severe erosion: the Loess Plateau of China. Land Degradation and Development, 1994, 5: 33-40.

[4] Elsen E, Hessel R, Liu B. Discharge and sediment measurements at the outlet of a watershed on the Loess plateau of China. Catena, 2003, 54: 147-160.

[5] Singha R, Tiwarib K N, Malb B C. Hydrological studies for small watershed in India using the ANSWERS model. Journal of Hydrology, 2006, 318: 184-199.

[6] Hovius N. Controls on sediment supply by large rivers, relative role of Eustasy, climate, and tectonism in continental rocks. Society of Sedimentary Geology, 1998, 59: 3-16. (Special Publication).

[7] Karvonen T, Koivusalo H, Jauhiainen M. A hydrological model for predicting runoff from different land use areas. Journal of Hydrology, 1999, 217: 253-265.

[8] Chen L, Wang J, Fu B, et al. Land use change in a small catchment of northern Loess Plateau, China. Agriculture, Ecosystem and Environment, 2001, 86: 163-172.

[9] Siriwardena L, Finlayson B L, McMahon T A. The impact of land use change on catchment hydrology in large catchment: the Comet River, Central Queensland, Australia. Journal of Hydrology, 2006, 326: 199-214.

[10] Fu B. Soil erosion and its control in the Loess Plateau of China. Soil Use and Management, 1989, 5: 76-82.

[11] Chen L, Messing I, Zhang S. Land use evaluation and scenario analysis towards sustainable planning on the Loess Plateau in China—case study in a small catchment. Catena, 2003, 54: 303-316.

[12] Kosmas C, Gerontidis St, Marathianou M. The effect of land use change on soils and vegetation over various lithological formations on Lesvos (Greece). Catena, 2000, 40: 51-68.

[13] Luk S H, Chen H, Cai Q, et al. Spatial and temporal variations in the strength of loess soils, Lishi, China. Geoderma, 1989, 45: 303-317.

[14] Costa M H, Botta A, Cardille J A. Effects of large scale changes in land cover on the discharge of the Tocantins River, Southeastern Amazonia. Journal of Hydrology, 2003, 283 (4): 206-217.

[15] Sharma P P, Gupta S C, Foster G R. Predicting soil detachment by raindrops. Soil Science Society of America Journal, 1993, 57: 674-680.

[16] Dijk A I J M, Bruijnzeel L A, Rosewell C J. Rainfall intensity-kinetic energy relationships: a critical literature appraisal. Journal of Hydrology, 2002, 261: 1-23.

[17] Kinnell P I A. Raindrop-impact-induced erosion processes and prediction: a review. Hydrological Processes, 2005, 19: 2815-2844.

[18] Li F, Cook S, Geballe G T, et al. Rainwater harvesting agriculture: an integrated system for water management on rainfed land in China's semiarid areas. AMBIO, 2000, 29 (8): 477-483.

[19] Nearing M A. Potential changes in rainfall erosivity in the US with climate change during 21st century. Journal of Soil and Water Conservation, 2001, 56 (3): 229-232.

[20] Bürger G. Selected precipitation scenarios across Europe. Journal of Hydrology, 2002, 262: 99-110.

[21] Endale D M, Fisher D S, Steiner J L. Hydrology of a zero-order Southern Piedmont watershed through 45 years of changing agricultural land use. Part 1. Monthly and seasonal rainfall-runoff relationship. Journal of Hydrology, 2006, 316: 1-12.

[22] Morin E, Goodrich D C, Maddox R A. Spatial patterns in thunderstorm rainfall events and their coupling with watershed hydrological response. Advances in Water Resources, 2006, 29 (6): 843-860.

[23] de Lima J L M P, Singh VP. The influence of the pattern of moving rainstorms on overland flow. Advances in Water Resources, 2002, 25: 817-828.

[24] Yeh H-Y, Wensel L C, Turnblom E C. An objective approach for classifying precipitation patterns to study climatic effects on tree growth. Forestry Ecology and Management, 2000,139: 41-50.

[25] de Lima J L M P, Singh V P, Isabel M. The influence of storm movement on water erosion: storm direction and velocity effects. Catena, 2003, 52: 39-56.

[26] Kirkby M J, Bracken L J, Shannon J. The influence of rainfall distribution and morphological factors on runoff delivery from dryland catchments in SE Spain. Catena, 2005, 62: 136-156.

[27] Luk S H, Abrahams A D, Parsons A J. A simple rainfall simulator and trickle system for hydro-geomorphological experiments. Physical Geography, 1986, 7 (4): 344-356.

[28] Madden L V, Wilson L L, Ntahimpera N. Calibration and evaluation of an electronic sensor for rainfall kinetic energy. Phytopathology, 1998, 88 (9): 950-959.

[29] Mathys N, Klotz S, Esteves M. Runoff and erosion in the Black Marls of the French Alps: observations and measurements at the plot scale. Catena, 2005, 63: 261-281.

[30] Aizen V, Aizen E, Glazirin G. Loaiciga simulation of daily runoff in Central Asian alpine watersheds. Journal of Hydrology, 2000, 238: 15-34.

[31] Mazi K, Koussis A D, Restrepo P J.Erratum to "A groundwater-based, objective-heuristic parameter optimization method for a precipitation-run-off model and its application to a semi-arid basin" [Journal of Hydrology 290 (2004) 243-258].Journal of Hydrology, 2004, 299: 160-161.

[32] Nearing M A, Jetten V, Baffaut C. Modeling response of soil erosion and runoff to changes in precipitation and cover. Catena, 2005, 61: 131-154.

[33] Licznar P, Nearing M A. Artificial neural networks of soil erosion and runoff prediction at the plot scale. Catena, 2003, 51: 89-114.

[34] Zhang X C, Nearing M A, Risse L M, et al. Evaluation of WEPP runoff and soil loss predictions using natural runoff plot data. Transaction of the ASAE, 1996, 39 (3): 855-863.

[35] Nearing M A, Govers G, Norton D L. Variability in soil erosion data from replicated plots. Soil science Society of America Journal, 1999, 63 (6): 1829-1835.

[36] Eagleson P S. Ecological optimality in water-limited natural soil-vegetation systems, 1. Theory and hypothesis. Water Resources Research, 1982, 18: 325-340.

[37] Xu J. Precipitation-vegetation coupling and its influence on erosion on the Loess Plateau, China. Catena, 2005, 64, 103-116.

[38] Boer M, Puigdefa'bregas J. Effects of spatially structured vegetation patterns on hill-slope erosion in a

semiarid Mediterranean environment: a simulation study. Earth Surface Processes and Landforms, 2005, 30: 149-167.

[39] Bochet E, Rubio J L, Poesen J. Modified topsoil islands within patchy Mediterranean vegetation in SE Spain. Catena, 1999, 38: 23-44.

[40] Hou X. Study on the benefits of plants to reduce sediment in the loess rolling gullied region of north Shaanxi. Bulletin of Soil and Water Conservation,1990,10 (2): 33-40.

[41] Zou D, Yu T, Zhou Q, et al. Mechanism of soil and water conservation and its benefit of Eulaliopsis binata. Research of Agricultural Modernization, 2000, 21 (4): 210-213.

[42] Gyssels G, Poesen J, Bochet E, et al. Impact of plant roots on the resistance of soils to erosion by water: a review. Progress in Physical Geography, 2005, 29 (2): 189-217.

[43] Mao R, Meng G, Zhou Y. Mechanism of plant roots on soil erosion control. Research of Soil and Water Conservation, 2006,13 (2): 241-243.

[44] FAO-UNESCO. Soil map of the world (1 : 5000 000). Paris: Food and agriculture organization of the United Nations, UNESCO, 1974.

[45] Huang Y, Chen L, Fu B, et al. The wheat yield and water-use efficiency in the Loess Plateau: straw mulch and irrigation effects. Agricultural Water Management, 2005, 72: 209-222.

[46] Anderberg M R. Cluster Analysis for Applications. New York: Academic Press Inc., 1973.

[47] Hong N. Products and Servicing Solution Teaching Book for SPSS of Windows Statistical. Beijing: Tsinghua University Press, Beijing Communication University Press, 2003:300-311.

[48] Horváth Sz. Spatial and temporal patterns of soil moisture variations in a sub-catchment of River Tisza. Physics and Chemistry of the Earth, 2002, 27: 1051-1062.

[49] Yu J, He X. Data Statistical Analysis and Application of SPSS Software. Beijing: People's Post Press, 2003: 253-276.

[50] Perruchet C. Constrained agglomerative hierarchical classification. Pattern Recognition, 1983, 16 (2): 213-217.

[51] Pizarro R, Araya S, Jordán C, et al. The effects of changes in vegetative cover on river flows in the Purapel river basin of central Chile. Journal of Hydrology, 2006, 327: 249-257.

[52] Hou X, Bai G, Cao Q. Study on benefits of soil and water conservation of forest and its mechanism in loess hilly region. Research of Soil and Water Conservation, 1996, 3 (2): 98-103.

[53] Zhou Z, Shangguan Z, Zhao D. Modeling vegetation coverage and soil erosion in the Loess Plateau Area of China. Ecological Modeling, 2006, 198: 263-268.

[54] Zhu J, Liu J, Zhu Q, et al. Hydro-ecological functions of forest litter layers. Journal of Beijing Forestry University, 2002, 24: 30-34.

[55] Descheemaeker K, Muys B, Nyssen J, et al. Litter production and organic matter accumulation in exclosures of the Tigray highlands, Ethiopia. Forest Ecology and Management, 2006, 233 (1): 21-35.

[56] Famiglietti J S, Rudnicki J W, Rodell M. Variability in surface moisture content along a hillslope transect: Rattlesnake Hill, Texas. Journal of Hydrology, 1998, 210: 259-281.

[57] Gyssels G, Poesen J, Nachtergaele J, et al. The impact of sowing density of small grains on rill and

ephemeral gully erosion in concentrated flow zones. Soil and Tillage Research, 2002, 64: 189-201.

[58] De Baets S, Poesen J, Gyssels G, et al. Effects of grass roots on the erodibility of topsoils during concentrated flow. Geomorphology, 2006, 76: 54-67.

[59] Yun X, Zhang X, Li J, et al. Effects of vegetation cover and precipitation on the process of sediment produced by erosion in a small watershed of loess region. Acta Ecologica Sinica, 2006, 26 (1): 1-8.

[60] Poesen J W A, Verstraeten G, Soenens R, et al. Soil losses due to harvesting of chicory roots and sugar beet: an underrated geomorphic process? Catena, 2001, 43: 35-47.

[61] Ruysschaert G, Poesen J, Verstraeten G. Inter-annual variation of soil losses due to sugar beet harvesting in West Europe. Agriculture, Ecosystems and Environment, 2005, 107: 317-329.

[62] Li Y. Productivity dynamic of alfalfa and its effects on water eco-environment. Journal of Pedology, 2002, 39 (3): 404-411.

[63] Yu J, Li F, Wang X, et al. Soil water and alfalfa yields as affected by altering ridges and furrows in rainfall harvest in a semiarid environment. Field Crop Research, 2006, 97:167-175.

[64] Hou Q, Han R, Han S. Primary studies on the soil dry layer issue of artificial woodland and grassland on the Chinese Loess Plateau. Soil and Water Conservation in China, 1999, 5: 11-14.

[65] Wang Z, Liu B, Lu B. A study on water restoration of dry soil layers in the semi-arid area of Loess Plateau. Acta Ecologica Sinica, 2003, 23 (9): 1944-1950.

[66] Grimes D W, Wiley P L, Sheesley W R. Alfalfa yield and plant water relations with variable irrigation. Crop Science, 1992, 32: 1381-1387.

[67] Jia Y, Li F, Wang X, et al. Soil water and alfalfa yields as affected by alternating ridges and furrows in rainfall harvest in a semiarid environment. Field Crops Research, 2006, 97: 167-175.

[68] Zhu X, Ren M. The shaping process and countermeasures of the Chinese Loess Plateau. Soil Conservation in China, 1992, 2: 2-10.

[69] Zheng F, He X, Gao X, et al. Effects of erosion patterns on nutrient loss following deforestation on the Loess Plateau of China. Agriculture, Ecosystems and Environment, 2005, 108: 85-97.

[70] Hou X, Bai G, Cao Q. Contrastive experiments on the soil penetrability and anti-scourability of *Robinia pseudoacacia* L., *Hippophae rhamnoides* L., *Caragana korshinskii*. Journal of Soil Water Conservation 1995, 9 (3): 90-95.

[71] Li Y, Shao M. Changes of soil physical properties under long-term natural vegetation restoration in the Loess Plateau of China. Journal of Arid Environment, 2006, 64: 77-96.

[72] Jiao J, Wang W, Hao X. Precipitation and erosion features of rainstorms in different patterns on the Chinese Loess Plateau. Journal of Arid Land Resources and Environment, 1999, 13 (1): 34-41.

[73] Jackson I J. Relationship between rainfall parameters and interception by tropical forest. Journal of Hydrology, 1975, 24: 215-238.

[74] Salles C, Poesen J. Rain properties controlling soil splash detachment. Hydrological Processes, 2000, 14: 271-282.

[75] Braud I, Vich A I J, Zuluaga J. Vegetation influence on runoff and sediment yield in the Andes region: observation and modeling. Journal of Hydrology, 2001, 254: 124-144.

[76] Bradshaw A. Restoration of mined lands—using natural processes. Ecological Engineering, 1997, 8: 255-269.

[77] Bradshaw A. The use of natural processes in reclamation—advantages and difficulties. Landscape and Urban Planning, 2000, 51:89-100.

[78] Douglas I. Man, vegetation and sediment yield of river. Nature, 1967, 215: 925-928.

[79] Coppus R, Imeson A C, Sevink J. Identification, distribution and characteristics of erosion sensitive areas in three different Central Andean ecosystems. Catena, 2003, 51: 315-328.

[80] Kosmas C, Danalatos N, Cammeraat L H. The effect of land use on runoff and soil erosion rates under Mediterranean conditions. Catena, 1997, 29: 45-59.

Responses of water erosion to rainfall extremes and vegetation types in a loess semiarid hilly area, NW China*

Wei Wei   Chen Liding   Fu Bojie   Lü Yihe   Gong Jie

Abstract

Rainfall extremes (RE) become more variable and stochastic in the context of climate change, increasing uncertainties and risks of water erosion in the real world. Vegetation also plays a key role in soil erosion dynamics. Responses of water erosion to RE and vegetation, however, remain unclear. In this article, on the basis of the data measured on 15 plots (area: 10m × 10m and 10m × 5m) and the definition of World Meteorological Organization (WMO) on rainfall extremes, 158 natural rainfall events from 1986 to 2005 were analysed, and rain depth and maximal 30-min intensity (MI_{30}) were used to define RE. Then, water erosion process under RE and five vegetation types (spring wheat, alfalfa, sea buckthorn, Chinese pine, and wheatgrass) were studied in a key loess semiarid hilly area, NW China. The following findings were made: ① The minimal thresholds of depth and MI_{30} for defining RE were determined as 40.11mm and 0.55 mm/min, respectively. Among the studied rainfall events, there were four events with both the variables exceeding the thresholds (REI), five events with depths exceeding 40.11mm (REII), and four events with MI_{30} exceeding 0.55 mm/min (REIII). Therefore, not only extreme rainstorm, but also events with lower intensities and long durations were considered as RE. Moreover, RE occurred mostly in July and August, with a probability of 46 and 31%, respectively. ② Extreme events, especially REI, in general caused severer soil-water loss. Mean extreme runoff and erosion rates were 2.68

---

* 原载于：Hydrological Processes, 2009, 23: 1780-1791.

and 53.15 times of mean ordinary rates, respectively. The effect of each event on water erosion, however, becomes uncertain as a result of the variations of RE and vegetation. ③ The buffering capacities of vegetation on RE were generally in the order of sea buckthorn > wheatgrass > Chinese pine > alfalfa > spring wheat. In particular, sea buckthorn reduced runoff and erosion effectively after 3-4 years of plantation. Therefore, to fight against water erosion shrubs like sea buckthorn are strongly recommended as pioneer species in such areas. On the contrary, steep cultivation (spring wheat on slopes), however, should be avoided, because of its high sensitivities to RE.

Keywords: rainfall extremes; vegetation; runoff; soil erosion; loess semiarid hilly area

1 Introduction

Water erosion in the Chinese loess hilly area is still very severe, challenging the sustainability of social and environmental security in the locality [1-4]. Special meteorological status and intensified human activity, coupled with harsh natural conditions and fragile ecosystems, are considered as the major trigger and driver of water erosion in this semiarid environment [5-7].

Among these factors, precipitation extremes, in particular, play important roles in inducing severer water erosion in the semiarid ecosystems than in the temperate areas. For example, studies show that the temporal distribution of heavy rainstorms is especially necessary for assessing runoff amount and erosion rate [8,9]. The severity, frequency, extent, and impact of soil erosion processes are likely to be altered by the high degree of fluctuation and variation of rainfall depth, intensity and frequencies [10,11]. The number, timing and magnitude of storms also greatly impact hydrological processes [12]. Soil erosion, landslides, mass movements, and debris flows have been monitored and found to be more severe when large-scale heavy rainstorms occur [13]. The annual nutrient loss due to water erosion, under extremely heavy rainfall, exceeds 50% of the total amount [14]. Mannaerts and Gabriels [15] also found that the dominant seasonal concentration of water erosion was limited to a few major storms only.

More notably, the episodes of rainfall extremes (RE) in nature have been monitored to be more capricious, stochastic, and unpredictable during the past several decades [12]. The intensity, frequency, and regimen of extreme precipitation events, which have already increased globally, are predicted to increase further [16,17]. This variability will probably increase the complexity and uncertainty of water erosion process [10,18]. For example, studies found that changeable precipitation is the main factor in runoff and sediment alteration [19-21]. Recently, a new prediction declares that rain erosivity, which is the major driving force of water erosion, will increase 8%-35% on the Chinese Loess Plateau in the next 10 decades [18]. Another study also found high levels of seasonal and annual runoff fluctuations in this region during the past several decades, mainly owing to the varied precipitation patterns [7]. High degree of variations and occurrences of RE coupled with fragile earth surface can cause severer consequences of soil erosion and land degradation.

The type, cover, structure, and age of vegetation, on the other hand, are another key factor, besides RE, influencing water erosion processes [22-24]. Studies found that various hydrological responses occur under different vegetation communities [1,25-27]. Vegetation plays a key role in intercepting rainfall, increasing water infiltration, fixing soil by roots, providing mechanical protection by reducing raindrop energy and 'splash' effects, and trapping sediment [28-30]. Water erosion will decrease, if adapted vegetation restoration is applied. In particular, proper plant selection and management (mulching and land closure) can increase surface cover and root energy, improve soil property, and therefore reduce soil erosion [13,31-33]. Vegetation destruction (steep cultivation and bare zones caused by deforestation) or improper species use, on the contrary, may deteriorate soil features and induce severe soil loss as well as land degradation on larger scales [34-36]. For example, sloping farmlands converted from forests and grasslands in the loess hilly area have suffered from severe erosion, and water and nutrient loss for a long time [37,38]. Decrease or even disappearance of vegetative cover plays negative role in modifying the processes of rainfall interception and overland flow, which affect the dynamics of surface water cycle and deposition of sediment yields [2,39,40].

However, responses of water erosion to RE remain unclear, and this is especially true under different vegetation types. Studying RE and their erosion responses among vegetation types is thus significant for understanding the comprehensive mechanisms and dynamics of hydrological and erosion processes. Furthermore, consecutive measurements from plot scales for long periods have been confirmed to be of crucial importance for reliable water erosion analysis and prediction on larger scales [41,42]. In this study, 158 natural rainfall events producing runoff were used for RE selection. These data sets were based on field measurement and collection extending from 1986 to 2005 (data in 2000 and 2001 missing). The main purpose of this article is to highlight and discuss the responses of water erosion processes to RE and vegetation types in a key loess semiarid hilly area, NW China. Three specific sub-objectives are expected to be achieved: ① to study the role of RE on water erosion processes; ② to analyse the buffering capacities of different vegetation types on extreme rates of runoff and erosion; ③ to discuss the combined effect of RE and vegetation types on water erosion processes.

2 Materials and Methods

2.1 Study area

The study area and experimental plots are located at the field station of Dingxi Soil and Water Conservation Institute (1895m a.s.l., 35°33′N, 104°35′E) in Anjiagou catchment, a typical loess hilly area in Gansu, NW China. This region is dominated by a semiarid continental temperate climate, with two distinct seasons, cold-dry winters and warm-wet summers [7]. The mean annual precipitation has been about 426.6mm over the past five decades, over 80% of which has fallen in the rainy season from May to September [34], with an annual variability of 26%. The maximum and minimum precipitations have been 721.8mm, in 1967 and 245.7 mm, in 1982 [43]. However, the potential annual evaporation can reach 1510mm. The mean monthly temperature ranges from −7.6℃ to 27.5℃, with an average annual temperature of 6.3℃. Annually, about 141 days are frost-free.

Local soil is developed from the wind-accumulated loess parent material, which is about 40-60m deep. The dominant soil is calcic Cambisol [44] with clay content of 33.12%-42.17%, organic matter of 3.7-13.4g/kg, and soil buck density of 1.09-1.36 g/cm^3 within 2-m depth. Meanwhile, the topography in the study area is fragmentized and complex (Table 1), with a mean slope of 14.3° and a gully density of 3.14km/km^2. This type of soil and landform is easy to suffer from serious water erosion [45,46].

Table 1 Topography and gradient features of the study area

Index	Slope gradient						Landform			
	<5°	5°-10°	10°-15°	15°-20°	20°-25°	>25°	Hilltop	Hill slope	Terrace	Gully slope
Ratio(%)	11.7	14.1	38.9	15.1	9.3	11.2	0.9	74.6	10.8	13.7

After hundreds of years of cultivation and other strong human disturbances, most of the natural vegetation in this area has been converted into farmland, causing loss of ground cover protection. The majority of the remainder consists of natural secondary grass species [wheatgrass (*Agropyron cristatum* L. Gaertn), needle-grass (*Stipa capillata* Linn.), etc.] and planted vegetation types such as sea buckthorn (*Hippophae rhamnoides* L.), pea tree (*Caragana Korshinskii* Kom), poplar (*Populus* spp.), Chinese arborvitae (*Platycladus orientalis* L.), Chinese pine (*Pinus tabuliformis* Carr.), and apricot (*Prunus armeniaca* L.). The main crop species are spring wheat (*Triticum aestivum* L. cv Leguan), alfalfa (*Medicago sativa* L.), potatoes (*Solanum tuberosum* L.), and maize (*Zea mays* L.). Because of severe drought and poor soil conditions *in situ*, most of the artificial plants can not grow well and the productivity remains relatively low.

2.2 Experimental manipulation

In 1986, 15 plots were collocated on a north-facing hill-slope, about 100m from the meteorological station of Dingxi Soil and Water Conservation Institute. Before constructing the plots, former cultivated rain-fed crops, such as wheat, potatoes, beans, and millet in this area were cleared off. To reduce the effects of position and topography, all the runoff plots were established on the same slope. They were oriented parallel to the slope and adjacent to one another.

The following five vegetation types with three replications have been investigated on the 15 experimental plots (slope degree: 10°, 15° and 20°), which were installed in the same year (1986): ① Spring wheat (*T. aestivum* L. cv Leguan): field management was similar to that used by local farmers, and the seeds of spring wheat were sown in April each year and harvested manually in late July or early August. Because of bad climate and severe drought, spring wheat remained low in productivity during the periods of measurement. ② Alfalfa (*M. sativa* L.): the seeds were sown by drilling or broadcasting in April, 1986 and harvested in late July of the same year. Alfalfa was replanted annually from 1993 to 1999, because of low productivity. ③ Sea buckthorn (*H. rhamnoides* L.): saplings were planted at 1.0m by 1.0m spacing in March, 1986. Litter remained on the plots without clearance. A few artificial disturbances, due to high density and prickly branches, existed in this plant community. ④ Chinese pine (*P. tabulaeformis* Carr.): Chinese pine saplings were planted in 3Ð0-m rows and 1Ð5-m columns in March 1986, without artificial pruning and irrigation. ⑤ Wheatgrass (*A. cristatum* L. Gaertn): This natural species was left to grow without any human disturbance since 1986, when the cropland was abandoned, and wheatgrass has now become the dominant species in the patches. It is well known as a good grass species growing in cold arid regions. Meanwhile, the mean annual coverage rate of each vegetation type from 1986 to 2005 was measured (Fig.1).

Plots used for sea buckthorn and Chinese pine were 10m×10m, while plots used for spring wheat, alfalfa, and wheatgrass were 10m×5m. On each plot, cement ridges (about 30cm above ground) were constructed at the borders to isolate plot runoff and sediment yield. A discharge ditch was created at the top of each plot to control runoff and sediment yield from the upper slope. At the base of each plot, a marked H-flume and two volumetric tanks were built at the outlet of each plot for surface runoff and sediment collection.

2.3 Data measurement

Daily rainfall during the rainy seasons was measured and recorded by a SM_1 pluviometer and a SJ_1 auto-siphon udometer in the study area. A rainfall event would be separated into two parts and defined as two independent events, if an interval exceeding 6h existed during the event. According to the basic information provided by the data logger, several

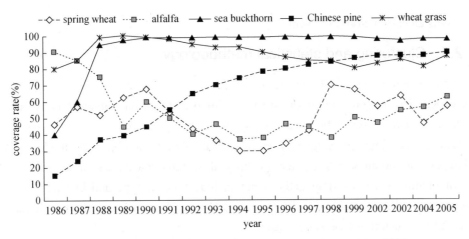

Fig. 1 Average annual coverage of vegetation types from 1986 to 2005

Note: Measurement: The coverage of alfalfa and wheatgrass was estimated in the maximal growth period by 5 quadrates of 1m×1m within each plot; the coverage of sea buckthorn and pine was measured by means of values of canopy width; the coverage of spring wheat was estimated by the row spacing. The mean coverage rate of three replications was used for each vegetation type.

key rainfall variables including depth, duration, average intensity, and maximum 30-min intensity (MI_{30}) were all calculated and recorded. Runoff volume from each plot in each rainfall event was recorded simultaneously. Sediments from the 15 plots were first sampled using 250-ml bottles, and then deposited, separated from the water, and dried at 105 ℃ in a forced air oven to a constant weight. The monitored data were then used to calculate the related hydrological indices representing the degrees of soil erosion and water loss.

2.4 Definition of RE

In this study, RE are defined on the basis of the criterion of the World Meteorological Organization (WMO), i.e. when the difference between the specific rainfall variables and corresponding mean values of 30-year measurements exceeds the double variance of each variable, the rainfall events are defined as RE [47]. Moreover, depth and MI_{30} are used to determine RE in this article. The corresponding calculated equation is as follows:

$$\chi - \varpi > 2\theta \tag{1}$$

where χ represents rainfall variable such as depth or intensity, ϖ is the mean value of this variable in 30 years, and θ is the variance of the related variable.

On the basis of the definition mentioned above, two specific sub-criteria for determining RE were carried out as follows: ① Rainfall depth exceeds its mean annual value (calculated as 18.87mm in this study), and MI_{30} exceeds the summation of its double variance and annual value. ② If MI_{30} can not reach the WMO criterion, rainfall depth must exceed the corresponding criterion. Rainfall events agreeing with one or both of the two criteria will be regarded as an extreme rainfall event.

2.5 Clustering and statistical methodology

In this study, hierarchical clustering method was used to analyse and classify different styles of RE. This method is a fundamental and important tool for statistical analysis, widely used in such diverse scientific fields as psychology, zoology, biology, hydrology, botany, sociology, meteorology, physiognomy, etc.[48]. It aims to group objects on the basis of their similarities, and the number of clusters is obtained by automatic statistical analysis[49]. Two types of hierarchical clustering methods are included, i.e. R type and Q type[48]. R type clustering is used for variable classification, and Q type clustering for different cases. In this article, R type clustering procedure is used.

Meanwhile, to open out more specific features of the defined RE, the descriptive statistical method was used for analyzing three key rainfall variables (i.e. rain depth, duration, and MI_{30}) shown in Fig. 2 and Table 2. Furthermore, in order to calculate and analyse the relationships between RE, runoff, erosion, and vegetation types clearly, two indicators including REr (r = runoff) and REe (e = erosion), which respectively represent the ratio of extreme runoff coefficient to the mean runoff coefficient and the ratio of extreme erosion modulus to mean erosion modulus, are established in this study and can be determined by the following two mathematical equations:

$$REr = MERC/MRC = (ESR/EPD)/(SR/PD) \qquad (2)$$

where MERC, MRC, ESR, EPD, SR and PD refer to mean extreme runoff coefficient (%), mean runoff coefficient (%), extreme surface runoff, extreme precipitation depth, surface runoff and precipitation depth, respectively.

$$REe = MEEM/MEM = (ESL/PA)/(SL/PA) \qquad (3)$$

where MEEM, MEM, ESL, SL and PA refer to mean extreme erosion modulus (t/km^2), mean erosion modulus (t/km^2), soil loss under extreme rainfall events, soil loss under ordinary rainfall events, and area of experimental plot, respectively.

All the calculations were processed under SPSS 13.0 for windows.

3 Results

3.1 Defined RE

According to the consecutive data sets and WMO criterion, the thresholds of depth and MI_{30} of RE in the study area are firstly determined, i.e. 40.11mm and 0.55mm/min, respectively. Meanwhile, the mean depth for the ordinary events is calculated as 18.87mm. To be an extreme event, each of the two variables should separately exceed the threshold of 40.11mm or 0.55mm/min. According to this criterion, 13 extreme events were finally screened out from a total of 158 rainfall events from 1986 to 2005 (2000 and 2001 are excluded). All the remaining 145 rainfall events belonged to ordinary events, and are not to be involved in this article.

From the results shown in Fig. 2, we can see that extreme events appeared almost every year during the past two decades. More importantly, this kind of events experienced clear variation among different months. During this period, RE occurred six times in July with a probability of 46%, four times in August with a probability of 31%, and three times in May with 24% probability. No events of RE however, were captured in June and September during the past 20 years. From this point of view, the most important time for paying special attention is July, followed by August, and May.

Meanwhile, it can be seen that the selected rainfall variables experienced high fluctuations among different extreme events during the years from 1986 to 2005 (Table 2). The rainfall depths of extreme events ranged from 22.00 to 66.60 mm with a mean value of 39.92mm, and this value is far higher than the mean depth (18.87mm) of ordinary rainfall events. The duration varied from 22 to 1435min with a high coefficient of variation of 0.78. The maximal intensity in 30 min, however, experienced a highest value of 0.95mm/min and a lowest value of 0.11mm/min. These quantitative results indicate that not only extremely heavy storms, but also some events with lower intensities and long durations are here considered as RE.

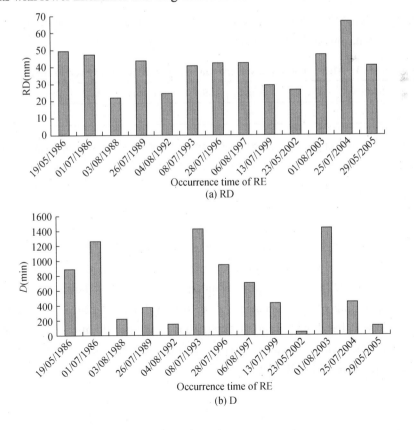

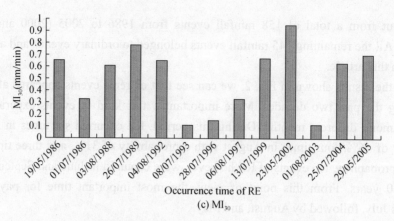

(c) MI$_{30}$

Fig. 2 Characteristics of rainfall variables of extreme events

Note: RD, rainfall depth; D, duration; MI$_{30}$, maximal intensity in 30 min.

Table 2 Descriptive Statistics of rainfall variables

Variable	Range	Minimum	Maximum	Sum	Mean	SD	CV
RD(mm)	44.60	22.00	66.60	519.00	39.92	12.22	0.31
D(min)	1407.0	22.0	1435.0	8354.0	642.6	501.9	0.78
MI$_{30}$(mm/min)	0.84	0.11	0.95	—	0.51	0.27	0.53

RD, D, MI$_{30}$, SD and CV separately refer to rainfall depth, duration and maximal intensity in 30 min, standard deviation and coefficient of variation.

Table 3 Mean runoff and soil erosion under extreme events and ordinary events

Index	MERC(%)	MRC(%)	RErM	MEEM [t/(km² · a)]	MEM [t/(km² · a)]	REe
Value	14.74	5.51	2.68	1.43×10⁵	2683	53.15

Furthermore, the captured 13 extreme events have been classified into three categories on the basis of their similarities under the procedure of hierarchical clustering method. This is used for analyzing the common features of different types of RE and then discussing their erosive role (Fig. 3).

The first category (REI) has such features as depth and MI$_{30}$, all above the relevant threshold. This kind of events holds a total percentage of 30.77%. The second category (REII) is the aggregation of RE with depth exceeding the related threshold (40.11mm), but the value of MI$_{30}$ does not reach the corresponding standard (0.55mm/min). Herein, five extreme events were captured, holding a total percentage of 38.46%. Although the MI$_{30}$ of this kind of events is relatively low, it has higher rainfall depths due to longer durations. The third category (REIII) of RE has such features as MI$_{30}$ exceeding the threshold of 0.55mm/min and the depths in the ranges of 18.87-40.11mm.

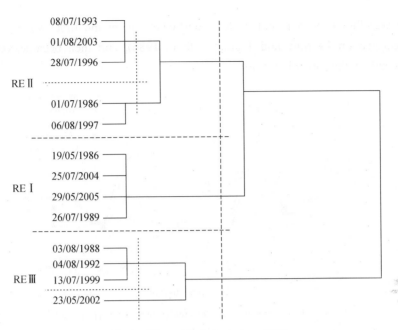

Fig.3 Hierarchical clustering of RE

This result also revealed that secondary classifications existed in the last two categories (the lightest broken line shown in Fig.3). Two sub-levels appeared in RE II and RE III. In RE II, for example, two extreme rainfall events belonged to the same sub-category, and the other three extreme events pertain to the other sub-category. The main reason for this can be attributed to the difference of MI_{30}. For category III, however, the MI_{30} once reached 0.95mm/min, which was the highest maximal intensity of the monitored years, its depth, however, was only 26.5mm. This event is obviously known from other events in this category.

3.2 Runoff under RE and vegetation types

It is found that the ratios of extreme runoff coefficients to mean runoff coefficients (REr) under different vegetation types and RE are relatively complex and differ from each other (Fig.4). Firstly, in most cases, the runoff coefficients of extreme rainfall events were far higher than those of ordinary rainfall events. For example, the highest value of REr was 9.50 on 26 July 1989 (Fig.4), which appeared for Chinese pine. Secondly, the results show that the ratios of runoff coefficients of different vegetation types under the same extreme event differed greatly. For example, the values of REr of spring wheat and alfalfa on 8 July 1993 were respectively monitored as 4.08 and 4.76, whereas the related REr values of sea buckthorn, Chinese pine, and wheatgrass were only 0.50, 0.61, and 0.59. Thirdly, surface runoff for the same vegetation type has different consequences regarding water loss under different extreme rainfall features. For example, the value of REr on 28 July 1996 under sea buckthorn was only 0.38, whereas the value on 1 July 1986 reached 7.13 (Fig.4). Lastly, sea buckthorn was found to have high runoff coefficients under the first two extreme rainfall events (occurred on 19 May and 1 July, both in 1986), and then became very low and relatively stable in the following years.

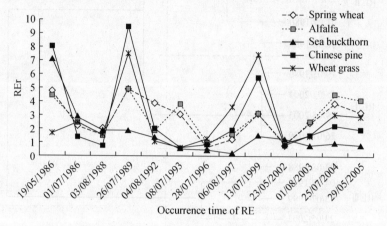

Fig.4 The degrees of water loss under vegetation types and RE

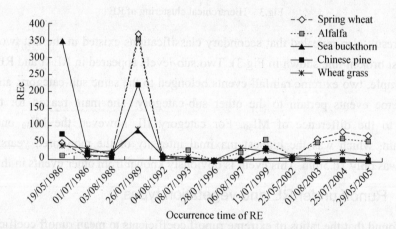

Fig.5 The degrees of soil loss under vegetation types and RE

3.3 Erosion under RE and vegetation types

From Fig.6, we can see that the degree and dynamics of soil loss rate caused by water erosion under the interaction of vegetation types and RE differ from each other. Generally, the ratio of extreme erosion modulus to the mean erosion modulus (REe) is very high under the impact of most of the events. The highest values of REe in the five vegetation types are in the order of spring wheat (368.53) > alfalfa (353.60) > sea buckthorn (345.18) > Chinese pine (222.46) > wheatgrass (90.45). The peak value under each vegetation type however, did not always appear in the same rainfall event. For example, the highest erosion modulus under sea buckthorn was observed on 19 May 1986, whereas all the related highest moduli under other four vegetation types appeared on 26 July 1989. Meanwhile, results also showed that the degree of extreme erosion was different among different vegetation types, even under the same extreme rainfall event.

4 Discussion

4.1 Responses of water erosion process to RE

According to the results, soil erosion and water loss are severer under RE than under ordinary rainfall events in most cases (Fig.4 and Fig.5). Generally, mean extreme runoff coeffiecient (MERC) and mean extreme erosion modulus (MEEM) are respectively 14.74% and $1.43 \times 10^5 t/km^2$, which are 2.68 times and 53.15 times of those under ordinary rainfall events (Table 3). The results therefore, re-confirmed the destructive role of RE on water erosion processes reported in other studies[14,50]. Meanwhile, as elaborated previously (Fig.2), the key periods for extreme events to occur are July, August and May, with percentages of 46.15, 30.77 and 23.08, respectively. Therefore, in order to minimize the destructive role of these kinds of events on surface soil and water, much attention should be paid to the most vital and sensitive periods when soil erosion potentials are highest, i.e. July and August. Furthermore, in order to control severe soil erosion caused by RE more successfully, other countermeasures such as straw mulch, plastic film mulch, and land closure should be taken into account. Such measures are highly valuable in improving soil character and reducing severe water erosion [32,33,51].

In this study, extreme rainfall events are grouped into three major categories based on the thresholds of depth and MI_{30} (Fig.3). Results indicate that the sensitivity of runoff and erosion to different extreme events differs. In general, surface water loss and soil erosion, are more sensitive to the first category of extreme events (REI) than those of other two categories (RE II and RE III), whose depth and MI_{30} both exceed the thresholds of 40.11mm and 0.55mm/min (Fig.2 and Fig.3). For example, the values of REr and REe under rainfall events which occurred on 26 July 1989 are higher than those under the majority of other extreme events, although some variances may appear among different vegetation types. Therefore, from the point of view of rainfall features, RE with high intensity and large amount play a vital role in causing severe soil erosion and water loss in most cases. REII, which is the aggregation of extreme rainfall events with long duration but low intensity, also causes high sensitivity and variation regarding water erosion rates. This important finding indicates that extreme runoff and erosion are not always caused by heavy rainstorms with strong intensities only. Events with long durations but relatively low intensities also play important roles in inducing severe water erosion. This result highlights the characteristics of RE in the key loess semiarid hilly area, mainly because previous studies only considered rainstorms with extremely strong intensities as extreme events [14,50,52]. The effect of RE III on runoff and soil erosion, moreover, is also significant, although it has shorter duration and lower depth than the other two RE types. The frequency, intensity and duration of extreme events, which are more consequential to water erosion than the mean values, are stressed in this study. Moreover, according to our results, the specific amount and degree of runoff and erosion under these three extreme rainfall categories are complicated, and it is mainly because the factors which influence water erosion rates and risks are too complex to predict in the real world [1,6,25,53].

4.2 Responses of extreme water erosion to vegetation types

In this study, the extreme rates of runoff and soil loss in the five vegetation types mentioned have been investigated (Table 4). Results show that the types and ages of vegetation play a key role in controlling extreme hydraulic destruction, and this conclusion is very similar to those in other studies [29,30]. Herein, to understand their specific responses to RE, the five xeric vegetations and their effects on water erosion are separately addressed in detail.

Table 4 Extreme runoff and erosion features in each vegetation type

Index	Spring wheat	Alfalfa	Sea buckthorn	Chinese pine	Wheatgrass
MERC(%)	23.02	19.98	4.31	15.83	10.56
MRC(%)	8.40	7.16	2.61	5.46	3.91
MERC/MRC	2.74	2.79	1.65	2.90	2.70
MEEM [t/(km^2·a)]	4.87×10^5	1.87×10^5	1.86×10^5	2.32×10^4	1.39×10^4
MEM [t/(km^2·a)]	8612	3321	135	600	427
MEEM/MEM	56.55	56.31	13.78	38.67	32.55

The first one is Chinese pine (*P. tabuliformis* Carr.). This plant is a traditional pine species in China. Its ability for holding soil water and preventing surface runoff, however, remains controversial. For example, the ratio of MERC and MRC (REr) under Chinese pine is the highest according to our research (Table 4). This implies that it was more sensitive to extreme runoff than the other four vegetation types. Two major reasons are possibly responsible for this phenomenon. First, it could be the long term effects of growing Chinese pine on the physicochemical characteristics of the surface soil. Huang et al.[51] found that the surface soil under Chinese pine gradually hardened with time, which decreased water infiltration and increased surface runoff. Second, more grease was observed in the litter of Chinese pine than in other plant litters, which eventually decreased the water holding capacity of the litter and created more surface runoff [54].

Alfalfa (*M. sativa* L.), an excellent fodder for livestock, is planted in large scale in the loess area of China. In this study however, alfalfa was found very sensitive to RE, sometimes even more than spring wheat (Table 3). There may be three major reasons for this. First, it was reported that the survival and growth of alfalfa are at the cost of soil water exhaustion [36], which deteriorates the soil status in situ and thus increases soil erosion, especially under extreme rainfall conditions [34]. For example, a study in a similar semiarid area found that 1100L of water are required to produce 1kg biomass of alfalfa [35]. Second, alfalfa always grows from July to late August. This time happens to be the sensitive season for soil and water loss, mostly because of high rainstorm frequencies and extreme events. Because of its high quality as forage for animal use, local farmers reap alfalfa, disturb the soil, and finally seriously decrease or even uncover the surface layer. In this situation, the risk of runoff and soil erosion dominated by heavy and frequent downpours sharply increases. Thirdly, we found that alfalfa grows well in flat areas (terrace, dam-land, etc.) where soil is fertile and humid, while it grows poorly in sloping areas because of water and nutrient shortages (Unpublished material). Because of these limitations, large-scale plantation of alfalfa with the major purpose of soil and water conservation should be carefully reexamined [36], and detailed measures and approaches such as land selection for alfalfa growth should also be of concern. Sea buckthorn (*H. rhamnoides* L.), famous for its high drought resistance and as an excellent drug raw material, is the most powerful species at fighting against extreme hydrological events among the five types of vegetation (Table 4). The possible reasons are as follows. First, sea buckthorn grows well and has a higher canopy density and coverage, especially after 3-4 years succession (Fig.1, Table 4). Effective vegetative coverage however, is proved to have a markedly negative correlation with erosion rates [24,31]. Second, soil under sea buckthorn is more improved than that under the other vegetation types. After about 20 years of succession, the soil buck density of sea buckthorn becomes the lowest of the five vegetation types, at only 1Ð17g/cm^3 (Table 4). Its water holding capacity and organic matter content in the soil surface are also higher than those of others. Higher water holding capacity and organic matter content, however, have been found to play key roles in increasing water infiltration and surface roughness [55], and they, therefore, increase the runoff threshold and decrease the potential erosion risk. Third, the litter under sea buckthorn was the highest found in a local study in September, 2005 (Table 5). Plant litter plays an effective role in weakening raindrop energy, improving soil physiochemical condition, increasing surface roughness and slowing runoff velocities [31,56]. Meanwhile, thick layer of plant litter may improve soil organic matter (SOM), and SOM may improve water infiltration rate by as much as 150%, thus decreasing runoff amount and velocity [35]. Fourth, the effective root network may help to increase slope stability and anti-splash erosion capability [13].

Spring wheat (*T. aestivum* L. cv Leguan) was found to play the poorest role in reducing water erosion rates amongst the five vegetation types (Table 4; Fig.4 and Fig.5). The reasons could be attributed to several aspects such as poor density and coverage, bad soil properties, and strong human disturbance [32,51]. For example, low density and poor coverage of spring wheat play ineffectual role in intercepting rainfall and reducing raindrop energy [57]. Meanwhile, as one of the major crop species in this area, the effect of spring wheat on runoff and soil erosion is different from that of the other vegetation types. For instance, the roots of spring wheat are proved to be less powerful in protecting and fixing soil than those of grasses, shrubs or arbors [37]. Furthermore, our study indicates that spring wheat is harvested in late July or early August each year, leaving the soil to be bare until the next year. This will certainly increase the risks and rates of water erosion. All these factors contribute a lot to the different roles of spring wheat and other four vegetation types in reducing water erosion rates.

Wheatgrass (*A. cristatum* L. Gaertn), an excellent native grass in fighting against drought and cold weather, always grows well on abandoned farmlands in the loess plateau. Studies indicate that wheatgrass has a powerful ability on controlling extreme water erosion rates, weaker than that of sea buckthorn but higher than those of the other three vegetation types (Table 3). Two major factors possibly account for this. One, soil conditions such as soil buck density (SBD), SOM, and water holding capacity (WHC) are all improved after nearly 20 years of succession (Table 5). However, lower SBD, and higher SOM and WHC are proved to reduce runoff and soil loss effectively, even under extreme rainfall events [33]. Second, the canopy coverage of wheatgrass however, was lower than that of sea buckthorn but higher than those of the other three vegetation types, according to field investigation (Fig.1). Other studies also drew similar conclusions. For example, because of poor rainfall interception and redistribution, grass communities like wheatgrass were found to play less effective roles in controlling soil erosion in the steep slopes than dense shrubs [26,27].

4.3 Responses of water erosion to the interaction of vegetation types and RE

Results indicate that processes of surface water flow and sediment transportation become complicated and dubious under the co-functions of RE and vegetation types (Fig.4 and Fig.5). The following hypothesis could be proposed to explain our results. First, the features and time-series of RE play key roles in the specific dynamics of water erosion (Fig.2 and Table 2). Rainfall variables, which influence the regimes and dynamics of runoff and erosion, are considerably complicated and instantaneous [9-13]. Second, the effects of vegetation types and growing conditions on soil erosion are also indicated. In general, our study highly emphasizes the combined roles of hydrological and erosional processes under vegetation types and RE. On the one hand, vegetation conditions such as coverage and density change highly with different vegetation ages (Fig.1), which eventually alter the regimes of surface water loss and soil erosion over space and time [22,58]. This has also been confirmed by other studies. For example, Li et al. [59] found that the improvement of vegetation coverage rate at landslide reached 89.69% after over 6 years of natural succession. On the other hand, the amounts/depths, intensities and distributions of extreme events also experienced high temporal variations (Table 2 and Fig.2). The coupled role of these two factors on water erosion thus became non-linear and difficult to determine in many cases. Although the mean values of runoff and erosion under extreme events are generally higher than those under common events, our study also indicates that there is no exact mathematical relationship between a specific extreme event and the consequent extreme hydrological response. Both previous and recent studies have achieved similar results. For example, many studies have confirmed that the largest rainfall events do not necessarily produce the maximum soil erosion and related extreme fluvial discharge [60-62]. Because of this, the relation between erosion process and extreme rainfall in the semiarid Mediterranean region is still hard to assess and uncover [62,63].

Meanwhile, our results also indicate that the stages of vegetation development play a key role in controlling and altering water erosion processes, resulting in the hydrological response becoming increasingly more insensitive to RE. For example, sea buckthorn was very sensitive to RE at initial stages after plantation, exhibiting the highest REr for extreme events that occurred on 19 May 1986, 1 July 1986 and 3 August 1988, and also REe on 19 May 1986 (Fig.4 and Fig.5). After three to four successive years, however, water erosion in sea buckthorn became less sensitive to RE, and thus little soil and water loss was created. Therefore, to fight against severe water erosion caused by RE in larger scales, more special attention should be paid not only to the spatial re-adjustment of vegetation types, but also to the different stages of growth of the vegetation.

5 Conclusions

In this study, on the basis of field measured hydrological data and the definition of WMO on extreme events, two rainfall variables including depth and MI_{30} were used to determine and classify RE. The following results were obtained: First, the thresholds of depth and MI_{30} for RE are separately calculated as 40.11mm and 0.55mm/min, with the mean depth of ordinary events being 18.87mm. Results indicate that not only extremely heavy rainstorms, but also some rainfall events with lower intensities and long durations should be considered as RE. Secondly, extreme events cause severer hydrological destruction in most cases, especially for the events whose depth and MI_{30} all above the thresholds (REI). The specific erosion consequence, however, becomes complex and uncertain due to the differences of rainfall variables, timing, frequency, vegetation types, etc. Thirdly, as a result of the influence of vegetation types and their ages, responses of water erosion to different extreme events differ. The buffering capacity of the five vegetation types on RE was generally in the order of sea buckthorn >wheatgrass > Chinese pine > alfalfa > spring wheat. Perennial plants such as sea buckthorn were monitored to be sensitive to RE in the first 3-4 years after plantation. However, with the vegetative succession going on, the role of sea buckthorn in reducing water erosion rates enhanced significantly.

These findings have important implications for controlling water erosion caused by RE in the semiarid environments. For example, although RE may generally cause severe water erosion and soil loss, the highest erosion rate are not necessarily caused by the severest extreme event. Therefore, to fight against water erosion more effectively special attention should be paid to the characteristics of RE and vegetation types. Other measures such as mulching and land closure are encouraged for protecting the surface soil from severe erosion. Meanwhile, land use readjustment and vegetation restoration are strongly needed. The shrub, sea buckthorn was intensely recommended to be pioneer species because of its high quality in fighting extreme hydrological events, especially after 3-4 years of plantation. Slope cultivation (spring wheat planted on deep slopes), however, should be avoided in most possible cases, because of its higher sensitivity to extreme rainfall events.

Acknowledgements

We thank the Dingxi Institute of Soil and Water Conservation in Gansu province, for their assistance on field work. Sincere thanks and appreciations are also transferred to Ms Vicky and the two anonymous reviewers, for their valuable and constructive comments. This research was supported by the National Basic Research Program of China (2009CB421104), Knowledge Innovation Project of the Chinese Academy of Sciences (kzcx2-yw-421), the National Natural Science Foundation of China (40801041), and the National Advanced Project of the Eleventh Five-year Plan of China (2006BAC10B05).

References

[1] Fu B J. Soil erosion and its control in the Loess Plateau of China. Soil Use and Management, 1989, 5: 76-82.

[2] Kang S Z, Zhang L, Song X Y, et al. Runoff and sediment loss responses to rainfall and land use in two agricultural catchments on the Loess Plateau of China. Hydrological Processes, 2001, 15(6): 977-988.

[3] Wei J, Zhou J, Tian J L, et al. Decoupling soil erosion and human activities on the Chinese Loess Plateau in the 20th century. Catena, 2006, 68: 10-15.

[4] Chen H, Guo S L, Xu C Y, et al. Historical trends of hydro-climatic variables and runoff response to climate variability and their relevance in water resource management in the Hanjiang basin. Journal of Hydrology, 2007, 344: 171-184.

[5] Shi H, Shao M A. Soil and water loss from the Loess Plateau in China. Journal of Arid Environments, 2000, 45: 9-20.

[6] Wang L, Shao M A, Wang Q J, et al. Historical changes in the environment of the Chinese Loess Plateau. Environmental Science and Policy, 2006, 9: 675-684.

[7] Xu Z X, Li J Y, Liu C M. Long-term trend analysis for major climate variables in the Yellow River basin.

Hydrological processes, 2007, 21: 1935-1948.

[8] Klik A, Truman C C. What is a typical rainstorm? Ghent, Belgium: International Symposium "25 Years of Assessment of Erosion", 2003.

[9] Apaydin H, Erpul G, Bayramin I, et al. Evaluation of indices for characterizing the distribution and concentration of precipitation: a case for the region of Southeastern Anatolia Project, Turkey. Journal of Hydrology, 2006, 328: 726-732.

[10] Gregory P, Ingram J, Campbell B, et al. Managed production systems// Walker B, Steffen W, Canadell J, et al. The Terrestrial Biosphere and Global Change. Implications for Natural and Managed Ecosystems. Synthesis Volume. International Geosphere—Biosphere Program Book Series 4. Cambridge: Cambridge University Press, 1999: 229-270.

[11] Sivakumar M V K. Interactions between climate and desertification. Agricultural and Forest Meteorology, 2007, 142: 143-155.

[12] Weltzin J F, Loik M E, Schwinning S, et al. Assessing the response of terrestrial ecosystems to potential changes in precipitation. Bioscience, 2003, 53: 941-952.

[13] Cheng J D, Lin L L, Lu H S. Influences of forests on water flows from headwater watersheds in Taiwan. Forest Ecology and Management, 2002, 165: 11-28.

[14] Gao C, Zhu J, Zhu J, et al. Effects of extreme rainfall on the export of nutrients from agricultural land. Acta Geographica Sinica, 2005, 60: 991-997.

[15] Mannaerts C M, Gabriels D. A probabilistic approach for predicting rainfall soil erosion losses in semiarid areas. Catena, 2000, 40: 403-420.

[16] Easterling D R, Meehl G A, Parmesan C, et al. Climate extremes: observations, modeling, and impacts. Science, 2000, 289: 2068-2074.

[17] IPCC (Intergovernmental Panel on Climate Change) Working Group I. Climate Change 2001: The Scientific Basis. Cambridge: Cambridge University Press, 2001.

[18] Zhang X C, Liu W Z. Simulating potential response of hydrology, soil erosion, and crop productivity to climate change in Changwu tableland region on the Loess Plateau of China. Agricultural and Forest Meteorology, 2005, 131: 127-142.

[19] Guo S L, Wang J X, Xiong L H, et al. A macro-scale and semidistributed monthly water balance model to predict climate change impacts in China. Journal of Hydrology, 2002, 268: 1-15.

[20] Jake F W, Michael E L, Susanne S, et al. Assessing the response of terrestrial ecosystems to potential changes in precipitation. Bioscience, 2003, 53(10): 941-952.

[21] Chen L D, Wei W, Fu B J, et al. Soil and water conservation on the Loess Plateau in China: review and perspective. Progress in Physical Geography, 2007, 31: 389-403.

[22] Cerdà A. The influence of geomorphological position and vegetation cover on the erosional and hydrological processes on a Mediterranean hillslope. Hydrological Processes, 1998, 12: 661-671.

[23] Zhou Z C, Shangguan Z P, Zhao D. Modeling vegetation coverage and soil erosion in the Loess Plateau area of China. Ecological modelling, 2006, 198: 263-268.

[24] Pizarro R, Araya S, Jord'an C, et al. The effects of changes in vegetative cover on river flows in the

Purapel river basin of central Chile. Journal of Hydrology, 2006, 327: 249-257.

[25] Chen L D, Messing I, Zhang S R, et al. Land use evaluation and scenario analysis towards sustainable planning on the Loess Plateau in China. Catena, 2003, 54: 303-316.

[26] Zhang X X, Yu X X, Wu S H, et al. Effects of forest vegetation on runoff and sediment production on sloping lands of loess area. Chinese Journal of Applied Ecology, 2005, 16: 1613-1617.

[27] Zhao H B, Liu G B, Cao Q Y, et al. Influence of different land use types on soil erosion and nutrient care effect in loess hilly region. Journal of Soil and Water Conservation, 2006, 20: 20-24.

[28] Gyssels G, Poesen J. The importance of plant root characteristics in controlling concentrated flow erosion rates. Earth Surface Processes and Landforms, 2003, 28: 371-384.

[29] Rey F, Ballais J L, Marre A, et al. Rôle de la végétation dans la protection contre l'érosion hydrique de surface [Role of vegetation in protection against surface hydric erosion]. Comptes Rendus Geosciences, 2004, 336: 991-998.

[30] Bochet E, Poesen J, Rubio J L. Runoff and soil loss under individual plants of a semiarid Mediterranean shrubland: influence of plant morphology and rainfall intensity. Earth Surface Processes and Landforms, 2006, 31: 536-549.

[31] Hou X L, Bai G, Cao Q. Study on benefits of soil and water conservation of forest and its mechanism in loess hilly region. Research of Soil and Water Conservation, 1996, 3(2): 98-103.

[32] Li F M, Wang J, Xu J Z, et al. Productivity and soil response to plastic film mulching durations for spring wheat on entisols in the semiarid Loess Plateau of China. Soil and Tillage Research, 2004, 78: 9-20.

[33] Adekalu K Q, Okunada D A, Osunbitan J A. Compaction and mulching effect on soil loss and runoff from two southwestern Nigeria agricultural soils. Geoderma, 2006, 137: 226-230.

[34] Wang Z Y, Wang G Q, Li C Z, et al. Primary study and implication of vegetation-erosion dynamics. Science in China (Series D), 2003, 33(10): 1013-1023.

[35] Pimentel D, Berger B, Filiberto D, et al. Water resources: agricultural and environmental issues. BioScience, 2004, 54: 909-918.

[36] Li J, Chen B, Li X F, et al. Effects of deep soil desiccation on alfalfa grasslands in different rainfall areas of the Loess Plateau of China. Acta Ecologica Sinica, 2007, 27: 75-89.

[37] Zheng F L, Merrill S D, Huang C H, et al. Runoff, Soil erosion and erodibility of conservation reserve program land under crop and hay production. Soil Science Society of America Journal, 2004, 68: 1332-1341.

[38] Zheng F L. Effect of vegetation changes on soil erosion on the Loess Plateau. Pedosphere, 2006, 16: 420-427.

[39] Pan C Z, Shangguan Z P. Runoff hydraulic characteristics and sediment generation in sloped grassplots under simulated rainfall conditions. Journal of Hydrology, 2006, 331: 178-185.

[40] Rulli M C, Rosso R. Hydrologic response of upland catchments to wildfires. Advances in Water Resources, 2007, 30: 2072-2086.

[41] Zhang X C, Nearing M A, Risse L M, et al. Evaluation of WEPP runoff and soil loss predictions using

natural runoff plot data. Transaction of the ASAE, 1996, 39: 855-863.

[42] Licznar P, Nearing M A. Artificial neural networks of soil erosion and runoff prediction at the plot scale. Catena, 2003, 51: 89-114.

[43] Huang Y L, Chen L D, Fu B J, et al. The wheat yields and water-use efficiency in the Loess Plateau: straw mulch and irrigation effects. Agricultural Water Management, 2005, 72: 209-222.

[44] FAO-UNESCO. Soil Map of the World (1 : 5000 000). Paris: Food and Agriculture Organization of the United Nations, UNESCO, 1974.

[45] Fu B J, Gulinck H. Land evaluation in an area of severe erosion: the Loess Plateau of China. Land Degradation and Development, 1994, 5: 33-40.

[46] Wei W, Chen L D, Fu B J, et al. Mechanism of soil and water loss under rainfall and earth surface characteristics in a semiarid loess hilly area. Acta Ecologica Sinica, 2006, 26: 3847-3853.

[47] CCCIN (China Climate Change Information Network). Climate change: observation facts and future scenarios. http://www.ccchina.gov. cn/cn/NewsInfo.asp?NewsId=3979. [2008-4-20].

[48] Yu J, He X. Data Statistical Analysis and Application of SPSS Software. Beijing: People's Post Press, 2003: 253-276.

[49] Hong N. Products and Servicing Solution Teaching Book for SPSS of Windows Statistical. Beijing: Tsinghua University Press, Beijing Communication University Press, 2003: 300-311.

[50] Wang M B, Li H J, Chai B F. Effect of extreme rainfall on the soil water cycle of forest land. Journal of Soil Erosion and Water Conservation, 1996, 2(3): 83-86.

[51] Huang Z L, Fu B J, Chen L D. Differentiation of soil erosion by different slope, land use pattern and variation of precipitation in loess hilly region. Science of Soil and Water Conservation, 2005, 3(4): 11-18.

[52] Wang W Z. Study on the relationship between rainfall character and soil loss: criterion of erosive rainfall. Bulletin of Soil and Water Conservation, 1984, 4(2): 58-62.

[53] He X B, Tian J L, Tang K L, et al. Bio-climatic imprints on a Holocene loess palaeosol from China. Journal of Asian Earth Sciences, 2004, 22: 455-464.

[54] Liu G Q, Wang H, Qin D Y, et al. Hydrological and ecological functions of litter layers for main forest-types in Qinling Mts. of the Yellow River. Journal of Natural Resources, 2002, 17: 55-61.

[55] Bissonnais Y L, Cerdan O, Lecomte V, et al. Variability of soil surface characteristics influencing runoff and inter-rill erosion. Catena, 2005, 62: 111-124.

[56] Neave M, Rayburg S. A field investigation into the effects of progressive rainfall-induced soil seal and crust development on runoff and erosion rates: the impact of surface cover. Geomorphology, 2007, 87: 378-390.

[57] Robichaud P R, Lillybridge T R, Wagenbrenner J W. Effects of postfire seeding and fertilizing on hillslope erosion in north-central Washington, USA. Catena, 2006, 67: 56-67.

[58] Xu J X. Coupling relationship between precipitation and vegetation and the implications in erosion on the Loess Plateau, China. Acta Geographica Sinica, 2006, 61: 57-65.

[59] Li W T, Lin C Y, Chou W C. Assessment of vegetation recovery and soil erosion at landslides caused by

a catastrophic earthquake: a case study in Central Taiwan. Ecological Modelling, 2006, 28: 79-89.

[60] Romero M A, Cammeraat L H, Vacca A, et al. Soil erosion at three experimental sites in the Mediterranean. Earth Surface Processes and Landforms, 1999, 24: 1243-1256.

[61] Nunes J P, Vieira G N, Seixas J, et al. Evaluating the MEFIDIS model for runoff, soil erosion prediction during rainfall events. Catena, 2005, 61: 210-228.

[62] José C G, José L P, de Luis M. A review of daily soil erosion in Western Mediterranean area. Catena, 2007, 71: 193-199.

[63] Michael A, Schmidt J, Enke W, et al. Impact of expected increase in precipitation intensities on soil loss-results of comparative model simulations. Catena, 2005, 61: 155-164.